International Perspectives on Aging

Series Editors: Jason L. Powell, Sheying Chen

For further volumes:
http://www.springer.com/series/8818

Sheying Chen · Jason L. Powell
Editors

Aging in China

Implications to Social Policy of a Changing Economic State

 Springer

Editors
Sheying Chen
Pace University
One Pace Plaza 18th Floor
New York, NY 10038, USA
sheyingchen@yahoo.com

Jason L. Powell
University of Lancashire
Livesey House LH 311
PR1 2HE Preston, Lancashire, UK
jpowell1@uclan.ac.uk

ISBN 978-1-4419-8350-3   e-ISBN 978-1-4419-8351-0
DOI 10.1007/978-1-4419-8351-0
Springer New York Dordrecht Heidelberg London

Library of Congress Control Number: 2011941702

Printed on acid-free paper

Springer is part of Springer Science+Business Media (www.springer.com)

This book is dedicated to
Our parents and grandparents

Preface

China is fast on its way to become the most powerful economic force in the world. Simultaneously, it has the fastest increase in the aging population in the world. Between 2010 and 2040, the portion of people 65 and older will rise from around 7% to between 25% and 30% of the population (Kim & Lee, 2007).

China has four unique characteristics that distinguish it from other countries in Asia. First, the proportion of aging population is growing faster than Japan, the country previously recognized as having the fastest rate, and much faster than nations in Western Europe for example. Second, an early arrival of an aging population before modernization has fully taken place, with social policy implications. It is certain that China will face a severely aged population before it has sufficient time and resources to establish an adequate social security and service system for older people. Third, there will be fluctuations in the total dependency ratio. The Chinese government estimates are that the country will reach a higher "dependent burden" earlier in the twenty-first century than was previously forecast (Powell & Cook, 2010). Fourth, the strong influence of the government fertility policy and its implementation on the aging process means that fewer children are being born. However, with more elderly people, a conflict arises between the objectives to limit population increase and yet maintain a balanced age structure (Peng & Guo, 2001).

The intersection of these fourfold factors means that the increased aging population is giving rise to serious concerns among Chinese social policy makers. There is a chronic lack of good resource materials that attempt to explain social policy in its relationship to examining the problems and possibilities of human aging grounded in an analysis of the political economy of social policy in China and impact on rural and urban spaces.

Such analysis of China will be covered by conceptual, theoretical, and empirical approaches in this volume. The book will also discuss substantive topics of pensions, family care, housing, health, and mental health. The book brings together an array of active researchers to provide discussions of critical implications of aging social policy and the economic impact in China. It also fits the core requirements of the book series by flagging *Aging in China* as a key component of international aging.

This book is aimed at second and third year advanced undergraduates in social policy, social gerontology, sociology, psychology, health studies, nursing, social work, family studies, urban studies, rural community development, Asian and international studies, and public and business administration. It will also appeal to postgraduates in these areas, including interdisciplinary master's level courses; to professionals in health and social care and public services; and to academics in the social and human sciences.

The book's cross-disciplinary appeal is one of its major strengths given the expertise of the contributors. Such discussions of social policy and effects on older people in China will allow debates around the impact of health and social welfare on contemporary social life to enter new realms: realms that students and practitioners can utilize to reflect on their own experiences in challenging assumptions about aging and relationship to health and social welfare, while learning from the experiences of other cultures. This book is timely given the recent proliferation of academic degrees focused specifically at the interplay of socio-economic policies and aging in China, but more generally in light of the current high profile of comparative studies of a wide range of subjects. It is our hope that the efforts reflected in this book will make a significant contribution to the research literature and teaching resources.

New York, NY, USA   Sheying Chen
Preston, Lancashire, UK   Jason L. Powell

References

Kim, S., & Lee, J-W. (2007). Demographic changes, saving and current account in East Asia. *Asian Economic Papers, 6*(2), 22–53.
Peng, X., & Guo, Z. (Eds.). (2001). *The changing population of China*. Oxford: Blackwell.
Powell, J., & Cook, I. (Eds.). (2010). *Aging in Asia*. New York: Nova Science Publishers.

Contents

Contributors

Cheung Ming Alfred CHAN Lingnan University, Tuen Mun, Hong Kong

Cecilia Lai Wan CHAN Centre on Behavioral Health, The University of Hong Kong, Pokfulam, Hong Kong
Department of Social Work and Social Administration, Pokfulam, Hong Kong

Sheying Chen Pace University, New York, NY, USA

Zhiyu Chen Department of Sociology, Georgia State University, Atlanta, GA, USA

Jacky Chau-kiu Cheung Department of Applied Social Studies, City University of Hong Kong, Kowloon Tong, Kowloon, Hong Kong

Shixun GUI Institute of Population Studies, East China Normal University, Shanghai, China

Rainbow Tin Hung HO Centre on Behavioral Health, The University of Hong Kong, Pokfulam, Hong Kong
Department of Social Work and Social Administration, Pokfulam, Hong Kong

Alex Yui-huen Kwan Department of Applied Social Studies, City University of Hong Kong, Kowloon Tong, Kowloon, Hong Kong

Pui Pamela Yu LEUNG Department of Social Work and Social Administration, University of Hong Kong, Pokfulam, Hong Kong

Bin Li School of Public Administration, Central-South University, Changsha, Hunan, China

Na Li Department of Public and Social Administration, City University of Hong Kong, Kowloon Tong, Kowloon, Hong Kong

Phyllis Hau Yan LO Centre on Behavioral Health, The University of Hong Kong, Pokfulam, Hong Kong

Vivian W.Q. LOU Department of Social Work and Social Administration,
Sau Po Centre on Ageing, University of Hong Kong, Pokfulam, Hong Kong

Baozhen Luo Department of Sociology, Western Washington University,
Bellingham, WA, USA

Guifen Luo Social Security Research Center of China,
Renmin University of China, Beijing, China

Hok Ka Carol MA Lingnan University, Tuen Mun, Hong Kong

Barbara R. McIntosh School of Business Administration,
University of Vermont Burlington, VT, USA

Center on Aging, University of Vermont, Burlington, VT, USA

Julie A. Norstrand Graduate School of Social Work, Boston College,
Chestnut Hill, MA, USA

Xiaomei Pei Department of Sociology, Tsinghua University, Beijing, China

Jason L. Powell University of Lancashire, Lancashire, UK

Pui Yee Phoebe TANG Lingnan University, Tuen Mun, Hong Kong

Youcai Tang Department of Sociology, Tsinghua University, Beijing, China

Linda Wong Department of Public and Social Administration,
City University of Hong Kong, Kowloon Tong, Kowloon, Hong Kong

Qingwen Xu Tulane University School of Social Work, New Orleans, LA, USA

Yuebin XU School of Social Development and Public Policy,
Beijing Normal University, Beijing, China

Yining Yang Pace University, New York, NY, USA

Heying Jenny Zhan Department of Sociology, Georgia State University, Atlanta,
GA, USA

Chun Zhang School of Business Administration, University of Vermont,
Burlington, VT, USA

Xiulan ZHANG School of Social Development and Public Policy,
Beijing Normal University, Beijing, China

Chapter 1
Introduction: Social Policy of the Changing Economic State

Sheying Chen

The economic state was conceptualized at the beginning of the 1990s, which was published years later with the release of my first book in English (Chen, 1996). At that time, there was enormous confusion surrounding the seemingly minimal (or singularly residual) role of the Chinese state, along with the diminishing state-run *Danwei* (work unit) and vanishing People's Commune, in providing social welfare amid rapid social changes and mounting social needs. Outspoken critics (mostly from outside mainland China) could easily take a political stance against the happenings, who however might not realize the underlying reasons for the status quo or even backward moves, along with the various problems that served as either the causes or the corollaries. Without a clear social resolution in sight, the ramifications of such frustration and criticism could only thwart the new economic drive and reform that were crucial in putting the country on a fast track to become a major part of the world economy (as we have all witnessed by now).

Economic State and Social Policy Study

Being one of the first teaching social policy in reformist China and then pursuing the subject in British Hong Kong and in America, I wished to master the "state of the art" and apply the advanced knowledge to finding answers to China's problems. I learned good analytical tools and theoretical perspectives as expected. I was disappointed, however, by the limitation of Western social policy study that evolved around the development of the welfare state: its history and prehistory, its ideals and advantages, its struggle and problems/drawbacks, its future directions and global

S. Chen (✉)
Pace University, One Pace Plaza 18th Floor, New York, NY 10038, USA
e-mail: sheyingchen@yahoo.com

S. Chen and J.L. Powell (eds.), *Aging in China: Implications to Social Policy
of a Changing Economic State*, International Perspectives on Aging,
DOI 10.1007/978-1-4419-8351-0_1, © Springer Science+Business Media, LLC 2012

1

variations, etc. (Esping-Andersen, 1990; Hirschman, 1980; Johnson, 1987; Mishra, 1984; Titmuss, 1958, 1963; Wilensky & Lebeaux, 1965; Zimmern, 1934), notwithstanding great reflective and research efforts that helped to deal with social problems in Western societies. My observation and experience with the Chinese system failed to validate the story of the rise and fall of the welfare state as very relevant. I had to conclude that trying to include China and some other (former) socialist states or Third World countries as-is in welfare state or "welfare capitalism" studies was simply stretching the latter too far. Those systems were so different that they could hardly lend themselves to the existing models within the Western portions of social policy research (by assuming this, I began challenging welfare state-centered social policy study as a complete discipline).

Specifically, my analysis revealed that the Chinese government before economic reform was structurally distinct from a Western welfare state (advanced or reluctant) in that the former was dominated by a disproportionately large number of economic departments while social welfare constituted only a tiny portion of a barely visible Civil Affairs Administration. The socialist–communist principles that guided the establishment of the Chinese state in the mid-twentieth century prescribed economic administration as its main function, including micro-management/planning based on public ownership of industrial enterprises and, to a lesser extent, rural production as well. That was a fundamental departure from the assumptions of the Western welfare state where economic production was largely private business and regulated primarily by the market. In short, the Chinese socialist state was created as an economic state rather than a welfare state (Chen, 1996). Its socioeconomic system and government were different, not to mention the nation's unique culture and history. Holding the same expectations and using the same explanations for the two divergent types of states would lead to nowhere but endless confusion. When the lion's share of China's public spending had to be direct economic investment in the absence of private funds/capital, complaining about its neglect of social spending based on a comparison with the welfare state's budget proportions would be misleading (ibid.). Therefore, welfare state-based public/social policy study had to watch its boundaries and emphases/biases when dealing with such a case as socialist China to avoid misunderstanding and miscommunication due to the fundamental differences.

Establishing the economic state for expanding comparative social policy study as a discipline, however, was not so easy. To understand the essence of the welfare state, the economic state, their differences, and various complications, I resorted to Morris' (1985) framework of a policy system, especially his notion of a "general public policy" (GPP) (i.e., a guide to the priority aims and preferred means of a policy system). While Morris did not elaborate on it and the idea seemed ignored by others, I found it important since social policy as one of the "sectoral policies" required an adequate understanding of the entire policy system to make good sense. To assess the real possibilities in aging undertaking, for example, I had to relate to economic and political issues at higher levels and within broader contexts (Chen, 2010b). I found GPP, instead of various traditional, pure ideological "-isms," to be the key to a germane interpretation of specific policy measures in greatly changed times, much more helpful than such a misused and outdated paradigm as state socialism (Chen, 2002a). In addition, the Chinese were used to an articulate and firm "basic line" deemed extremely important for their government, which was more serious and politically strict than the usually short-term and drifting agenda/ goal setting in Western welfare states. In my view, the study of GPP could incorporate the strengths of both while avoiding their drawbacks. It would help to illuminate the historical-cultural and economic-political-social context so that major social policy issues including aging could be better understood. By revealing developmental GPP patterns vs. non- and anti-developmental ones, such research in general should benefit the world's peaceful development (Chen, 2004). The Chinese economic state was essentially a unique GPP pattern, in contrast with the typical Western welfare state as another, suggesting that the latter was a particular rather than a universal model for all. Both the economic state and the welfare state were developmental GPPs (as opposed to such non- and anti-developmental ones as the war or warfare states) and thus deserve great research attention (I've challenged a popular use of the term "developmental state" for excluding the welfare state, possibly also others, and failing to adequately contrasting with truly non- and anti-developmental states, cf. Chen, 2004).

Politicization and Depoliticization

For China study, a major complication stems from what I have called politicization of the economic state, which interfered with China's development since the founding of the People's Republic (until it took a dramatic turn in 1978) (Chen, 2002b). Like the former Soviet Union and some other socialist-communist states, the People's Republic of China was conceived and born from wars and faced serious confrontations with Western capitalist states from the beginning. Internally, unlike the post-World War II movement toward a more harmonious welfare state (as opposed to the war or warfare states) in Western societies, the new born Chinese socialist–communist state remained highly militant after overthrowing so-called "three big mountains" (i.e., imperialism, feudalism, and bureaucrat-capitalism). That was unavoidable in view of even a "red scare" within the United States at the time (Wiegand & Wiegand, 2007). The saga of an internal "class struggle" in China, however, did not fade away but kept being intensified from time to time. The official fight with some real but more imagined "class enemies" took the form of frequent political movements that diverted the economic state from its original mission. It was by no accident that the country's economic system saw a pattern of administrative, political, or even military-type (command) operation. The early history of the economic state became so ironic that economic production had to give way to an officially avowed "continuing revolution under proletarian dictatorship." Such politicization culminated in a ruthless "Great Proletarian Cultural Revolution," which not only oppressed numerous arbitrarily decided "class enemies" but also brought the political or politicized economic state near bankruptcy and kept most Chinese people in poverty.

The change of Chinese top leadership in the late 1970s finally opened an opportunity window for the economic state to fully realize itself. The strategic decision by the post-Mao leadership under Deng Xiaoping in 1978 was often regarded as most important in terms of the "open door" and reform policies that changed China's course. In my analysis, however, the most crucial change was the depoliticization of the economic state. "Open door" and reform were only the corollaries, which would be impossible without replacing "continuing revolution under proletarian dictatorship" with economic construction as the government's top preoccupation. This fundamental change was at the root of a series of *bo luan fan zheng* (cleaning up chaos and bringing things back to order) measures to deal with the aftermaths of a disastrous decade of the "Cultural Revolution." The decision was so firm and with such urgency that the Chinese government swiftly repositioned itself as an out-and-out economic state by abandoning "class struggle as the key link" and keeping its economic focus for nearly three decades that ensued. A "strategic transfer of work emphasis" was the language used to indicate the shift of China's GPP without too much ideological debate about the Communist Party's basic line, reflecting the post-Mao leadership's practical/pragmatic approach and political wisdom. The Chinese state simply could no longer afford remaining economically backward and politically oppressive in view of such potential consequences as the fall of the Soviet Union. The de-politicized economic state, now not just a form of government but a resolute new ideology, resulted in an economic miracle that impressed the rest of the world. On the other hand, there were mounting social problems that often met with blind eyes and lip services under the out-and-out economic state, with GDP growth as the primary performance indicator for government leadership at all levels. The Chinese were almost sacrificing everything else for the development of economy during a period when "everything was looked at for money" and all kinds of people were "jumping into the sea" (going to business).

Such "economicization" of Chinese society following the depoliticization of the economic state did not bode well for social services. The nation's social security or safety net before the reform was actually quite advanced by any developing or poor country's standard. When the economic reform broke up its iron rice bowl-like occupation- or *Danwei*-based welfare system, the "community" was largely unprepared to take over the huge responsibility as expected (Chen, 1996). There were "serialized reforms" in social security etc., though they understandably lagged behind even if not deemed next in importance. Many scholars have noted the lack of social policy studies in China during this period, yet lack of funding or poor planning was not necessarily the reason for a major retreat in social protection. The depoliticized Chinese state's almost exclusive focus on the economy, no matter how negative it had been on the provision of social welfare etc., was mandated by the economic state's mission and driven by a desperate need to seize the last chance to prove itself and survive the serious challenges in the post-Mao era. Here, a broad historical context and a deep understanding of the system are required to decipher the overall development strategy with a realistic assessment of the possibility for advancing various causes between different sectors within a specific time period.

Deeconomicization and Balanced Development

"Open door" and reform were made possible by the Chinese government's decision to de-politicize itself, or to re-focus the state's work in terms of a strategic transfer of its emphasis onto economic construction. The finally realized, fully committed economic state, however, was immediately eroded by the changes that ensued. Economic reform and "open door" led to organizational modifications (or "institutional adjustment"), of which the "de-economicization" of the state was the most significant. Economic reform made many economic departments in the government unnecessary, undermining it as an economic state structurally and functionally. Economic growth lifted the standard of living and also increased government revenue for redistribution. Politically, advocates for disadvantaged groups including social policy researchers and social welfare professionals kept pushing for a new transfer of the government's work emphasis with more attention to widespread social problems and mounting social needs. While depoliticization implied economicization of the entire society since 1978, the economic state at the same time entered yet another process of de-economicizing itself as a result of reform (in this sense, the socialist economic state never had a chance to stay and be tested in full).

Due to the painful lesson of distraction in the past, the Chinese state was determined to stick to its mission as an economic state before substantial improvement of the nation's economic situation despite all the structural changes and political pressures. Nevertheless, widespread social problems including unemployment at the turn of this century and later greatly improved public finance over the last decade have gradually opened an opportunity window for a new realignment of the country's GPP (Chen, 2010a). The economic state appears increasingly out of tune with the most pressing social demands. It has also lost much of direct control over the economy with the exception of the banking sector, etc. (Morrison, 2011), especially at the microlevel and thus become increasingly outdated. Faced with all these, the Chinese government finally started modifying its basic guidelines, from an exclusive focus on economy/efficiency to more serious attention to social needs/justice, especially over the past few years when government revenue enjoyed considerable growth.

The economic state era in Chinese history seems to be near an ending, with remarkable achievements as well as controversies (Frazier, 2006). Economic construction is likely to remain as the most important foundation, though no longer as the center of everything. History tends to repeat itself at new levels, and the goal of more balanced development will require restoration and expansion of social protection at a higher standard. China is learning from Western welfare states, if not returning to the pre-1978 social/*Danwei* provision that was regarded as the major "superiority of socialism" and a model for developing countries (e.g., health care). It would be interesting to see how much "Chinese characteristics" will be kept vs. how much a convergence theory will be at work in this process.

Aging in China at the Crossroads

Individual and population aging is one of the most significant social policy issues facing China today. Due to the complexity of the subject, any major solution to related social needs will require an adequate understanding of the policy system (including its GPP) of the rapidly changing economic state (or a would-be welfare state, or welfare pluralism in a similar vein). On the other hand, aging research in China offers a great opportunity to understand and exemplify the country's social policy and its dynamic change process. My own social policy study began with aging research (Chen, 1996). As one of the sectoral policies (Morris, 1985), unanswered questions regarding aging policy led me to pursue those fundamental issues in the Chinese system and in the social policy discipline. Due to the dramatic changes of Chinese society over the past few decades, social issues and policy responses (including welfare, health care, etc.) may look very differently over different time periods. In view of a new GPP with greater societal attention to such social issues, it is an opportune time and of great interest to re-examine aging in China and their implications to the social policy of the changing economic state toward more balanced development (as the new norm). I am pleased, therefore, to introduce a group of scholars whose research and writings may prove to be important contributions to the understanding of the policy implications.

Jason L. Powell, my co-editor, wrote Chap. 2 for this book. He deals with bio-medicalization of aging which helps to expand the framework which my above discussion may have implied for studying aging in China. This is important in view of the historical trend of medicalization as seen in American society and other developed countries. He illuminates the need to move beyond the bio-medicalization of aging and how this (still) seeps into ambivalence toward older people in settings such as in China. While resources do need to be targeted on the vulnerable older people, he argues, the presumption that older people as a whole are an economic and social burden must be questioned. By juxtaposing vulnerable aging with active aging, the other chapters then come next to illustrate the varying ways aging has been positioned and shaped by various social, cultural, and political discourses.

In Chap. 3, Barbara R. McIntosh and Chun Zhang deal with an intertwined and interesting topic of aging and business in terms of the role of work and workplace implications as well as changing expectations in both the USA and China. Recent economic and demographic trends have shown important changes in individuals' expectations about work beyond the traditional retirement age and in employers' need for older workers in both countries. To help policy makers and others to adjust to this change in the workplace, the chapter offers several recommendations including planning strategically for these human capital shifts, managing the multigenerational workplace, ensuring the transfer of both explicit and tacit knowledge as aging workers exit the workplace, and providing flexibility in both hours and location of work to bolster recruiting and retention efforts. The authors emphasize that creating a supportive organizational culture is a key underpinning to making all the changes associated with the aging of the workplace.

XU Yuebin and ZHANG Xiulan devote their writing in Chap. 4 to income security with an overview of China's two polarized systems: social insurance and social assistance. They provide a detailed account of different categories of contributory as well as government funded, means-test benefit programs. As the authors point out, for the most part since economic reform started in the early 1980s, major efforts of the Chinese government have been focused on establishing an insurance-based social security system, including pensions. Social assistance began to play an increasingly important role during the mid-1990s in response to emerging poverty in the cities and persistent poverty in the countryside. Despite remarkable expansion in recent years, levels of benefits vary considerably under different social insurance schemes while the absence of broad-based policy interventions has limited the effect of social assistance in reducing poverty among the elderly.

Chapter 5 by Xiaomei Pei and Youcai Tang pays special attention to old age support in rural China. They examine the potential for local communities to generate and allocate resources in the context of rural development to meet the challenges of decreasing family resources and inadequate state provision. Their research included in-depth interviews with elderly people, their families, community leaders, and government officials of selected villages in three Chinese provinces. The data show that it is possible for rural communities to generate and distribute resources for old age support, to offer community opportunities for social inclusion through fair flows of resources to promote social harmony and stability, and to accelerate economic growth. Based on the findings, the authors urge policymakers to link state plans for old age protection with rural community development and to encourage grassroots efforts for supporting older people.

Chapter 6 by Guifen Luo supplies additional insights into issues of social protection of rural residents including the rural elderly. Key features of the Chinese system are reviewed from a historical perspective and an urban bias is pointed out. The content, nature, and the problems of the rural pension pilot program and the rural cooperative medical scheme were discussed. The development of policy goals and its implications for both traditional welfare institutions and rural elderly care are further explored. The chapter offers an interpretation of the interaction among social policy, traditional family support pattern, and rural elder care.

In Chap. 7, Guifen Luo continues to deal with rural aging by focusing on the impact of rural–urban female labor migration that left dependent children and the elderly in the countryside. Based on interviews with rural–urban migrants in Anhui and Sichuan provinces, the chapter examines coping strategies adopted by rural–urban migrant families to deal with the tensions caused by changes in generational care chains. The grandparents took care of the grandchildren, while the young migrating couples reciprocated by giving their parents financial support, other material help, and promises of better support in the future. The chapter shows that this inter-generational support pattern helped to compensate for weakened family support in China's rural areas.

In Chap. 8, Jacky Chau-kiu Cheung and Alex Yui-huen Kwan examine the utility of enhancing filial piety for elder care in China by distinguishing between individual filial piety and filial piety as an aggregate social norm, and by assessing the latter's contribution to family elder care as opposed to state elder care. Their research used data from a survey of 1,219 older adults in six Chinese cities. Interestingly, the results generally show that the social norm of filial piety did not consistently strengthen the utility of family elder care, with its effect varying significantly among the six cities. The results suggest that the social norm of filial piety would not sustain individual filial piety or family elder care in ways favorable to older Chinese. This thought-provoking finding should prompt policy researchers to rethink about ways to enhance family support as long desired.

Chapter 9 by Qingwen Xu and Julie A. Norstrand examines social capital and health outcomes among older adults in China by highlighting gender differences. The authors used data from the Chinese General Social Survey (CGSS) of 2005, which included a representative sample of China's urban and rural households ($N = 1,854$) of older Chinese (60 years and over, 46.4% female). The authors found significant differences on some dimensions of social capital, as well as different associations between dimensions of social capital with health outcomes, all by gender. Their results suggest that accounting for gender may be important when developing interventions to maximize social capital in Chinese communities.

Chapter 10 by HO Tin Hung Rainbow, LO Hau Yan Phyllis, CHAN Lai Wan Cecilia, and LEUNG Pui Yu Pamela illuminates an east–west approach to mind-body health of Chinese elderly. This is important given the desire for independent and active aging and quality of life in the face of strained elder care provisions. The authors discuss how the interplay of Eastern health philosophies and the Western biomedical model can promote the physical and psychological of Chinese elderly and to flexibly adapt to their inevitable decline in health statuses. As the traditional Eastern models of health emphasize nurturing life, attaining balance, and an integrated view of the being, they also offer an ideal approach to applying scientific evidence on the mind-body connection to the development of therapies in treating disorders.

Chapter 11 by Vivian W. Q. LOU and Shixun GUI investigates urban family caregiving and examines its impact on caregiver mental health. The chapter includes a review of long-term care policy and service development with an empirical study on a missing element in the long-term care model in Shanghai, that is, family caregivers. Their findings show that about one-fifth of the caregivers were at risk of developing depressive symptoms. The recommendation of the chapter includes establishing a national plan to clarify the definition of long-term care and policy objectives and include the needs of family caregivers that range from finance to health and to related knowledge/skills.

Chapter 12 by Bin Li and Yining Yang extends the investigation of Li and Chen (2010) of a key aspect of aging in place with important impact on family care, that is, housing. There has been a disconnect between aging and housing in China research, and this chapter links the topic with the study of social stratification in urban China. Using national statistical and survey data, the chapter makes several recommendations for policy consideration in support of home- and community-based elder support, from improving housing provision to increasing the supply of geriatric nursing professionals.

Chapter 13 by Heying Jenny Zhan, Baozhen Luo, and Zhiyu Chen examines institutional care for the elderly in China. The development of institutional care is reviewed within the context of the Chinese elder care system including family support and social provisions. Elder care institutions in Nanjing and Tianjin and their residents are profiled to show specific examples. Empirical data from a survey conducted in Zhenjiang City, Jiangsu province, are also used to examine the willingness of elderly Chinese for institutional care. The chapter reveals a rapid growth of non-profit and non-governmental organizations in the provision of elder care services, however with a lack of governmental regulation and quality control leading to blind competition. With the unavoidability of this growth in several decades to come, the authors call for innovative approaches to quality control of elder care homes including professionalization and humanizing elder care institutions.

In Chap. 14, Linda Wong and Na Li examine changing welfare institution and the evolution of Chinese nonprofit organizations by focusing on elder care homes in urban Shanghai. Using selected elder care homes in the city as case examples, the authors attempted to trace the institutional change in the development of residential care as well as its impact on the evolution of non-state care agencies over the past decade. Their empirical findings suggest that the state has indeed strengthened its role in welfare planning, financing and provision through formal and informal institutional arrangement. In this process, welfare NPOs serves as agents of the state and were regulated and incorporated under a system of state dominance. The authors argue that even with the expansion of non-state welfare provision, state hegemony in welfare development remains unchallenged. And such state dominance will remain to be a long-term feature of the new welfare economy in China.

Finally, Chap. 15 by CHAN Cheung Ming Alfred, TANG Pui Yee Phoebe, and MA Hok Ka Carol presents an Aging Policy Appraisal Index (APAI) developed by the Asia-Pacific Institute of Aging Studies (APIAS) at Lingnan University in Hong Kong. This tool was created in response to UNESCAP's call to enact aging policies in reference to the Shanghai Implementation Strategy (SIS), which was developed in the footsteps of the Madrid International Plan of Action on Ageing (MIPAA). The APAI acts as a comprehensive indicator of policy implementation and a validation instrument for end users' appraisal of life and service quality, particularly in the Asia Pacific Region. A case study was conducted on Macao Special Administrative Region (Macao SAR), China, which is used in the chapter for explaining the development of APAI and its results and implications for the local government.

Readers of this book will find its unique strengths such as covering not only mainland China but also its SARs, overcoming a usual urban bias with inadequate attention to rural issues and needs (Hebel, 2003), and bearing on the same major topic with different perspectives and approaches. On the whole, this volume, coupled and compared with theoretical literature in the field of Chinese social policy over the past few decades, provides a timely testimony to the relevance of the "economic state in transition" and the significance of the GPP study in understanding past changes and future directions in the field of aging. In return, the rich empirical material and insightful theoretical thinking on issues of aging will undoubtedly help to further develop social policy as a discipline useful for all nations in the world.

References

Chen, S. (1996). *Social policy of the economic state and community care in Chinese culture: Aging, family, urban change, and the socialist welfare pluralism.* Brookfield, VT: Ashgate.

Chen, S. (2002a). State socialism and the welfare state: A critique of two conventional paradigms. *International Journal of Social Welfare, 11,* 228–242.

Chen, S. (2002b). Economic reform and social change in China: Past, present, and future of the economic state. *International Journal of Politics, Culture, and Society, 15*(4), 569–589.

Chen, S. (2010a, June 10). A new turning point of Chinese general public policy: From "economic construction as the center" to "balanced socioeconomic development based on economic growth." *Chinese Social Sciences Today.*

Chen, S. (2010b). Qualitative research and aging in context: Implications to social policy study in China. *Qualitative Sociology Review, 6*(1), 34–47. http://www.qualitativesociologyreview.org/ENG/Volume15/QSR_6_1_Chen.pdf.

Chen, A., Liu, G. G., & Zhang, K. H. (2004). *Urbanization and social welfare in China.* Burlington, VT: Ashgate.

Esping-Andersen, G. (1990). *The three worlds of welfare capitalism.* Princeton, NJ: Princeton University Press.

Frazier, M. W. (2006). *One country, three systems: The politics of welfare policy in China's authoritarian developmental state.* Paper presented at the conference on capitalism with Chinese characteristics: China's political economy in comparative and theoretical perspective, Indiana University, May 19–20. Retrieved August 15, 2011 from http://www.polsci.indiana.edu/china/papers/frazier.pdf.

Hebel, J. (2003). Social welfare in rural China. In P. Ho, J. Eyferth, & E. B. Vermeer (Eds.), *Rural development in transitional China: The new agriculture.* London: Frank Cass.

Hirschman, A. O. (1980). The welfare state in trouble: Systemic crisis or growing pains? *The American Economic Review, 70*(2), 113–116.

Johnson, N. (1987). *The welfare state in transition: The theory and practice of welfare pluralism.* Sussex: Wheatsheaf.

Li, B., & Chen, S. (2010). Aging and housing in urban China. In S. Chen & J. Powell (Eds.), *Aging in perspective and the case of China: Issues and approaches.* Hauppauge, NY: Nova Science.

Mishra, R. (1984). *The welfare state in crisis: Social thought and social change.* Sussex: Wheatsheaf.

Morris, R. (1985). *Social policy of the American welfare state: An introduction to policy analysis* (2nd ed.). New York: Longman.

Morrison, W. M. (2011). *China's economic conditions* (Congressional Research Service, 7-5700). Retrieved August 15, 2011 from http://www.fas.org/sgp/crs/row/RL33534.pdf.

Titmuss, R. M. (1958, 1963). Essays on "the welfare state." London: Allen & Unwin.

Wiegand, S. A., & Wiegand, W. A. (2007). *Books on trial: Red scare in the heartland.* Norman: University of Oklahoma Press.

Wilensky, H., & Lebeaux, C. (1965). *Industrial society and social welfare.* New York: Free Press.

Zimmern, A. (1934). *Quo Vadimus? A public lecture delivered on 5 February 1934.* London: Oxford University Press. Cited in: Hennessy, P. (1992). *Never again: Britain 1945–1951.* London: Cape.

Chapter 2
China and the Bio-Medicalization of Aging: Implications and Possibilities

Jason L. Powell

Abstract Policymakers, economists, social analysts around the globe are increasingly concerned about the rising number of older people in their society. There are worries about the inadequacy of pension funds, of growing pressures on welfare systems, and on the inability of shrinking numbers of younger people to carry the burden of their elders. This chapter focuses on such issues in China, where the older people have become a rapidly expanding proportion of the population. While resources do need to be targeted on the vulnerable older people, the presumption that older people as a whole are an economic and social burden must be questioned. This is an agist view that needs to be combated by locating how bio-medical views on aging seep into policy spaces in China that position negative perceptions of aging as both individual and populational problems. The chapter then moves to observe the implications of bio-medicine for older people in China in terms of "vulnerable" aging but deconstruct such "fixed" explanations by juxtaposing *active aging* as key narrative that epitomizes "declining to decline" as espoused by bio-medical sciences.

Introduction: From "Global Aging" to "Aging in China"

There is no doubt that in many societies around the globe older people are a growing proportion of the global population (Krug, 2002). The age structure of the global population is changing from a society in which younger people predominated to one in which people in later life constitute a substantial proportion of the total population. While the biological and psychological models of aging inscribe it as an

J.L. Powell (✉)
University of Lancashire, Livesey House LH 311, PR1 2HE Preston, Lancashire, UK
e-mail: jpowell1@uclan.ac.uk

S. Chen and J.L. Powell (eds.), *Aging in China: Implications to Social Policy of a Changing Economic State*, International Perspectives on Aging, DOI 10.1007/978-1-4419-8351-0_2, © Springer Science+Business Media, LLC 2012

"inevitable" and "universal" process, an aging population is neither (Phillipson, 1998). Transformations in the age profile of a population are a response to political and economic structures. A major concern for organizations such as the United Nations and World Bank focuses on the number of such "dependent" older people in world society (Krug, 2002).

Indeed, older people in particular constitute a large section of populations in global aging. In relation to public services that have to be paid for by "younger" working people, the percentage of the population has been used to signify such "burdensome" numbers. Not only are older people seen as dependent but also children under school leaving age and people over the retirement age. Dependency rates—that is, the number of dependents related to those of working age—altered little over the twentieth century and yet the notion of "burden" group retains its legitimacy. The reason there has been so little change during a period of so-called rapid aging populations is that there has been a fall in the total fertility rate (the average number of children that would be born to each woman if the current age-specific birth rates persisted throughout her childbearing life) (Phillipson, 1998).

Changes in the age structures of all societies also affect total levels of labor force participation in society, because the likelihood that an individual will be in the labor force varies systematically by age. Concurrently, global population aging is projected to lead to lower proportions of the population in the labor force in highly industrialized nations, threatening both productivity and the ability to support an aging population (Powell, 2009). Coupled with rapid growth in the young adult population in Third World countries, the World Bank (1994) foresees growing "threats" to international stability, pitting different demographic-economic regions against one another. That the views the relationship between aging populations and labor force participation with panic recognizes important policy challenges, including the need to reverse recent trends toward decreasing labor force participation of workers in late-middle and old age despite mandatory retirement in both Western and Eastern countries such as the UK (Jackson and Powell, 2001) and China (Chen and Powell, 2011).

Notwithstanding this, in China there is also an ongoing huge increase in the aging population that replicates global trends. It can be seen that the percentage of the aging population has increased from just over 4.4% to just under 7% from 1953 to 2000 (Cook & Murray, 2001). Note that the increase has not been constant, reflecting the negative impact of the Great Leap Forward and the successive famines upon the demographic profile, shown in the results of the 1964 Census.

Since that latter date, the percentage of the elderly has nearly doubled, while the actual numbers have more than tripled, being approximately 3.6 times that of 1964 by the year 2000. By 2000, there were 88 million Chinese aged 65 or over, compared to just under 25 million in 1964, an increase of 63 million plus in 36 years.

As is well known, China's population policy, most usually referred to as the Single Child Family Program (although some Chinese commentators regard the phrase as a misnomer, given that more than one child is possible within a number of situations), has led to a rapid deceleration in the birth rate, which was only

13.38 per thousand in 2001, compared to 21.06 in 1990 and 18.25 in 1978 (Cook & Murray, 2001).

This controversial policy, lambasted and praised in equal measure depending on perspective, has meant a rapid expansion in the number of single children only households at the very same time that the proportion of the elderly has also increased as a new era of prosperity has reached many households in China.

Estimates suggest that up to 300 million less people have been born as a result of this process of state intervention, but Murray (2004) has built on the earlier work of D.G. Johnson and others to suggest that improved living standards via modernization would have led to the same outcome voluntarily as growing numbers of urban dwellers in particular chose to reduce their family size. This last point hints at the important spatial dimension of demographic change in China. There is a marked contrast between urban and rural life in China and, closely related to this, a marked contrast between Eastern China and Western China. It is in the heavily urbanized "Gold Coast" of the Eastern Seaboard in which China's spatial transformation is most dramatic, with fast-expanding cities being especially concentrated in the Pearl River delta of the Southeast and the Yangtse Delta in Central East China (Cook & Murray, 2001).

Hence, it is not simply that the Chinese government has belatedly recognized "the graying" of populational constructions and policy implications, it is that they continue to look for knowledge of aging as the power to define old age as a social problem in terms of dualistic distinctions between deviancy and normality. An aging population, like that of an individual being studied by bio-medical models, is seen as a "burden" problem in terms of economic management of Eastern (and Western) economies.

It could be argued when looking at the effects of a so-called "demographic time bomb" across US, Europe, and Asia, it may have been grossly exaggerated.

Such old age, therefore, has been perceived negatively via a process of "agism"—stereotyping older people simply because of their chronological age. Agist stereotypes such as "aging populations" act to stigmatize and consequently marginalize older people and differentiate them from groups across the life-course who are not labeled "old".

One of the ways to interpret social aging such as being a categorization whether it be in individual or populational terms is through use of theorizing on what it means to age in society; that is, concerns and social issues associated with aging and the ways in which these themes are influenced and at the same time influence the society in which people live. Thus, to understand the process of aging, looking through the lens of the "sociological imagination" is not to see it as an individualized problem rather as a societal issue that is faced by both First World and Third World nations as a whole.

In supporting this latter view, there is a need to focus on how populational discourses of aging in China are influenced and reinforced by bio-medical models of aging that help drive perceptions of older people as a burdensome group.

Where Does the Ambivalence to Aging Come from?

There are important implications for how aging is viewed by not only in terms of global aging but more specifically to China and the arrangement of political and economic structures that create and sanction social policies grounded in knowledge bases of "burdensome" populations (cf. Powell, 2001). Such knowledge bases are focused on: one, "biological aging" which refers to the internal and external physiological changes that take place in the individual body; two, psychological aging which is understood as the developmental changes in mental functioning—emotional and cognitive capacities.

Bio-medical theories of aging can be distinguished from social construction of aging: (1) focusing on the bio-psychological constituent of aging, and (2) on how aging has been socially constructed. One perspective is driven from "within" and privileges the expression from inner to outer worlds. The other is much more concerned with the power of external structures that shape individuality. In essence, this social constructionism poses the problem from the perspective of an observer looking in, whilst the bio-medical model takes the stance of inside the individual looking out (Powell & Biggs, 2000). There has long been a tendency in matters of aging and old age to reduce the social experience of aging to its biological dimension from which are derived a set of normative "stages" which over-determine the experience of aging.

Accordingly being "old," for example, would primarily be an individualized experience of adaptation to inevitable physical and mental decline and of preparation for death. The paradox of course is that the homogenizing of the experience of old age which the reliance on the biological dimension of old age entails is in fact one of the key elements of the dominant discourse on aging and old age.

It is interesting that comparative historical research on aging in Eastern culture highlights an alternative perception of aging; in eighteenth century China has highlighted a rather different path as to the conceptualization of aging as a scientific process developed by Western rationality. For example, observe that traditional Chinese society placed older people on a pedestal. They were valued for their accumulated knowledge, their position within the extended family, and the sense of history and identity that they helped the family to develop (Murray, 1998). Respect for elderly people was an integral part of Confucian doctrine, especially for the family patriarch.

This was a view that was also prevalent in Ancient Greece with the notions of "respect" for older people especially regarding gendered issues of patriarchy. Prior to industrialization, in India, there was a bestowment that older people had responsible leadership roles and powerful decision-making positions because of their vast "experience," "wisdom," and "knowledge" (Katz, 1996).

It seems with the instigation of Western science and rationality, aging began to be viewed in a different more problematic context than to the Confucian doctrine of aging epitomized in China and issues of respect for aging in India. Martin Heidegger (1971) makes the similar point when he spoke of the Westernization of the world through the principles of Western science and language.

Indeed, the technological developments due to industrialization, Westernization, and urbanization—under the purview of distorted form of modernity—have neglected these statuses of aging by downgrading its conceptualization. Part of understanding individuality in Western culture, the birth of "science" gave legitimate credibility to a range of bio-medical disciplines of whom were part of its umbrella. In particular, the bio-medical model has become one of the most controversial yet powerful of both disciplines and practice with regard to aging (Powell & Biggs, 2000).

The bio-medical model represents the contested terrain of decisions reflecting both normative claims and technological possibilities. Bio-medicine refers to medical techniques that privilege a biological and psychological understanding of the human condition and rely upon "scientific assumptions" that position attitudes to aging in society for their existence and practice. Hence, scientific medicine is based on the biological and psychological sciences. Some doctrines of the bio-medical model more closely reflect the basic sciences while others refer to the primary concern of medicine, namely diseases located in the human body. Most important is that these beliefs hold together, thereby reinforcing one another and forming a coherent orientation toward the mind and body. Indeed, the mind–body dualism had become the location of regimen and control for emergence of science in a positivist methodological search for objective "truth." The end product of this process in the West is the "bio-medical model."

In this sense, bio-medicine is based on the biological and psychological sciences. Some doctrines of the bio-medical model more closely reflect the basic sciences while others refer to the primary concern of medicine, namely diseases located in the human body. Most important is that these beliefs hold together, thereby reinforcing one another and forming a coherent whole (Powell, 2001). By developing an all-encompassing range of bio-medical discourses, many forms of social injustice could be justified as "natural," inevitable, and necessary for the successful equilibrium of the social whole such as mandatory retirement and allocation of pensions (Phillipson, 1998).

Bio-medical gerontology is a fundamental domain where medical discourses on aging have become located and this is very powerful in articulating "truths" about aging. Under the guise of science and its perceived tenets of value-freedom, objectivity, and precision, bio-medical gerontology has a cloth of legitimacy. Biological and psychological characteristics associated with aging have been used to construct scientific representations of aging in modern society. The characteristics of biological aging as associated with loss of skin elasticity, wrinkled skin, hair loss, or physical frailty perpetuates powerful assumptions that help facilitate attitudes and perceptions of aging. It may be argued that rather than provide a scientific explanation of aging, such an approach homogenizes the experiences of aging by suggesting these characteristics are universal, natural, and inevitable. These assumptions are powerful in creating a knowledge base for health and social welfare professionals who work with older people in particular medical settings such as a hospital or general doctor's surgery and also for social workers (Powell, 2009).

These new forms of social regulation were also reflected in the family and the community. Hence, modern systems of social regulation have become increasingly blurred and wide-ranging (Powell, 2009). Increasingly, modern society regulated the populational construct by sanctioning the knowledge and practices of the new human sciences—particularly psychology and biology. These are called geronto-logical "epistemes" which are "the total set of relations that unite at a given period, the discursive practices that give rise to epistemological figures, sciences and pos-sibly formalized systems" (Foucault, 1972:191). The "psy" complex or bio-medical epistemes refers to the network of ideas about the "nature" of individuals, their perfectability, the reasons for their behavior, and the way they may be classified, selected, and controlled. It aims to manage and improve individuals by the manipu-lation of their qualities and attributes and is dependent upon scientific knowledge and professional interventions and expertise. Human qualities are seen as measur-able and calculable and thereby can be changed and improved. The new human sciences had as their central aim the prediction of future behavior (Powell, 2009).

Powell and Biggs (2000) suggest that a prevailing ideology of agism manifests itself in the bio-medical model via its suggestion that persons with such biological traits have entered a spiral of decay, decline, and deterioration. Along with this goes certain assumptions about the ways in which people with outward signs of aging are likely to think and behave. For example, there are assumptions that "older people are poor drivers" or that older people have little interest in relationships that involve sexual pleasure that are all explained away by "decline" and "deterioration" master narratives that comprise an aging culture. The effects of the "decline" and "decay" analogies can be most clearly seen in the dominance of medico-technical solutions to the problems that aging and even an "aging population" (Phillipson, 1998) is thought to pose. Here, the bio-medical model has both come to colonize notions of age and reinforce agist social prejudices to the extent that "decline" has come to stand for the process of aging itself (Powell, 2001).

Powell (2001) have used the expression "bio-medicalization of aging" which has two closely related narratives: (1) the social construction of aging as a medical prob-lem and (2) agist practices and policies growing out of thinking of aging as a medi-cal problem. They suggest:

> Equating old age with illness has encouraged society to think about aging as pathological or abnormal. The undesirability of conditions labeled as sickness or illness transfer to those who have these conditions, shaping the attitudes of the persons themselves and those of others towards them. Sick role expectations may result in such behaviors as social with-drawal, reduction in activity, increased dependency and the loss of effectiveness and per-sonal control—all of which may result in the social control of the elderly through medical definition, management and treatment

(Quoted in Powell, 2001:19).

These authors highlight how individual lives and physical and mental capacities that were thought to be determined solely by biological and psychological factors, are, in fact, heavily influenced by social environments in which people live. This remains invisible to the bio-medical approach because they stem from the societal interaction before becoming embedded and recognizable as an "illness" in the aging body of the person. For example, in the "sociology of emotions" the excursion of inquiry has proposed that "stress" is not only rooted in individualistic emotional responses but also regulated, classified, and shaped by social norms of Western culture (Powell & Biggs, 2000).

This type of research enables the scope of aging to be broadened beyond bio-medical individualistic accounts of the body. On this basis alone, sociology invites us to recognize that aging is not only a socially constructed problem by bio-medical sciences but also the symptomatic deep manifestation of underlying relations of power and inequality that cuts across and through age, class, gender, disability, and sexuality (Powell, 2001; Powell & Biggs, 2000).

At this level of analysis, sociology addresses bio-medicine as one of the elements of social control and domination legitimated through power/knowledge of "experts" (Powell & Biggs, 2000). Such expert formation has also been labeled as agist. Agism is where the assumptions made about old age are negative, which treats older people not as individuals but as a homogenous group, which can be discriminated against.

Chinese society uses age categories to divide this ongoing process into stages or segments of life. These life stages are socially constructed rather than inevitable. Aging, too, is a production of social category. At any point of life span, age simultaneously denotes a set of social constructs, defined by the norms specific to a given society at a specific point in history. Thus, a specific period of life: infancy, childhood, adolescence, adulthood, middle age, or old age is influenced by the structural entities of a given society. Therefore, aging is not to be considered the mere product of biological–psychological function, rather a consequence of sociocultural factors and subsequent life-chances. Indeed, society has a number of culturally and socially defined notions of what Phillipson (1998) calls the "stages of life." However, a fundamental question is how bio-medical gerontology has stabilized itself with a positivist discourse that not only reflects history but also the total preoccupation with the "problems" of aging that has important implications for older people and health lifestyles in China.

Bio-Medicine, Family Care, and Aging: Implications for China

The dominant bio-medical discourse of aging in China dwells on the processes of physical deterioration associated with becoming older. In this perspective, the aging body has to deal with increased levels of incapacity, both physical and mental, and becomes increasingly dependent on younger people for sustenance and survival; it is the family through informal care that has to provide care for older people who may have illnesses, according to the Law on the Rights and Interests of the Elderly, introduced by the PRC in 1996. The bio-medical problematization of aging has secreted wider questions of power and inequality; especially influential is occidental modernity. A powerful discourse is thus developed which follows that of the West, via notions of "social inclusion" and "family care," and the all-important role of the consumer in buying products for the elderly, from disability aids through to private pensions Powell and Cook (2000). The latter suggest that this process constructs the aging body as a site of surveillance by the Chinese state, constructing them as, following Foucault, objects of power and knowledge in which "it's your age" is the prevailing authority response to the elderly "customer."

Powell and Cook (2000) have further noted that older people will be increasingly probed for social, psychological, and economic factors such as "frailty" or "expected level of supervision." "There are indications, for example, that where care homes are provided, these are for the more active elderly, rather than those in greatest need" (Powell and Cook, 2000:7).

This Foucauldian point has been borne out via an article on Shanghai (China Daily, 2004) with "most nursing homes in Shanghai have entry criteria that target a narrow minority of elderly people. Some admit only those who are capable of independent living while others accept only bedridden patients. While dementia is a common condition among the elderly, those afflicted by it are generally excluded by the criteria." Even more seriously, if a patient's condition changes according to these criteria, he or she is forced to leave the home. "The lack of a continuum of care creates devastating situations for the patients and their families." Further, nursing homes have only minimal level of medical support available, and patients are transferred to hospitals too readily if they have an ailment much beyond the common cold. The patient can then lose their place in the nursing home if their bed is transferred and thus be subject to further stress. "Preventive care, physical therapy, and spiritual care, which are crucial components of care for the elderly, are generally overlooked. Many nursing homes do not provide such services out of concerns for cost or accident liabilities." The Shanghai article also notes, damningly, that:

> The financial burden of long-term care accumulates on an elderly population already enduring tighter budget constraints because of retirement and unemployment. In the absence of government subsidy, the higher fees charged by self-sustained nursing homes deprive elderly people with limited financial means of their access to care
>
> (China Daily, 2004).

In the light of these and other issues the PRC government is attempting to change the ways in which the elderly are perceived, via campaign slogans such as "respect the elderly" and "people first." The former campaign seeks to encourage younger people to visit the elderly on a regular basis in order to reduce the sense of isolation that the elderly can feel, to look out for their needs out on the streets and to generally raise awareness of the situation of older people. There is a resonance here with ancient Confucian tradition in which the elderly were venerated. There have also been attempts to encourage younger people to think about "healthy aging," but this is meant in terms of ensuring that they themselves have adequate financial provision as they age. "People first" is the attempt to recognize that "aging is an individual-specific process" and that "a functional healthcare system for the elderly should integrate all aspects of care ... emphasizing and fulfilling individual needs and preferences" (China Daily, 2004).

Older people in rural areas are more likely to have to face emigration of their children to the cities as China's urbanization proceeds apace. This can leave them physically and socially isolated in a remote rural area, no longer able to rely upon their family to look after them in their old age, as was once the tradition.

Indeed, as social norms and values change, younger people may no longer be willing, even when they are able, to support elderly parents, and in recent years the law has been used to take children to court in order to force them to support their parents. For example, a new law came into force on October 1, 1996 on "The Rights and Interests of the Elderly" that explicitly states that: "the elderly shall be provided for mainly by their families, and their family members shall care for and look after them" (Du & Tu, 2000).

Notwithstanding the legal process, at its most extreme, the concerns of elderly people are expressed via suicide—gerontocide. The elderly can struggle in the face of the massive social changes that China is facing, and the abandonment of the tradition in which they themselves would have looked after their own parents and grandparents. They may feel so stressed and alienated that suicide seems the preferred option.

For example, a 76-year-old man blew himself up in a courtroom in protest during a case against his family, who had offered only 350 Yuan a month to support him when 600 Yuan was required (8 Yuan = 1 USD) (Cook & Murray, 2001). In a society of rapid transformation, older people in particular may be vulnerable to the sense of abandonment within a more materialistic and selfish new world epitomized by the forces of global capitalism and seeping impingement into day-to-day living of older people in China.

Declining to Decline: Active Aging?

At the other end of the scale is the active elderly, probably the ideal state for all elderly people. Briefly, older people traditionally were more likely to be active within rural areas of China, in part because they had to be in order to maintain their livelihood. This particular tradition continues today; official data for the 1990s showed that 26% of people in rural areas still depended on their own labor earnings compared to only 7% in urban areas.

This is not about advocating that elderly people should have to work to continue to earn a living, but we are suggesting that an active lifestyle be promoted where possible. In the cities of China today, it is heartening to see the colonization of open space throughout the hours of daylight and even into the evening by the elderly who are engaged in a wide variety of activities, from the traditional (such as taijiquan and qigong), walking one's pet bird, traditional dance, or poetry writing using brushes dipped in water only, illustrating the ephemeral and passing nature of life itself. Hopefully, the increased pollution which China's cities face will not erode the potential gains from these and other activities for older people. An alternative discourse on aging can point to ways in which the elderly can be *deproblematized* as a negative medical, economic, political, and social category.

This must begin with appreciation of older people first of all as people rather than as a category. Older people share, however, apart from their longevity, a wide and deep experience of life itself, and thus of life situations. In China, older people used to be venerated because they were, almost literally, the founts of wisdom, the holders of accumulated knowledge far in advance of the younger members of the family and community. Today, knowledge is far less likely to be oral, and far less dependent on accumulation by the individual. Instead, it is increasingly available at the touch of a button via an Internet search engine, even if there are some restrictions on web provision in the PRC. But, it can be suggested, there should still be a major role for the accumulated wisdom of older people's experiences as carriers of historical wisdom.

In China, there is a growing awareness of the need to have a sophisticated, multidimensional policy to respond to the needs of older people. But there is still a strong bio-medical emphasis on surveillance and control. It could be suggested that policy needs to be driven by the elderly themselves wherever possible. They should be encouraged to define and state their own needs, provided with support when this is required, but the overriding emphasis should be on providing support that fosters active lifestyle and independent living wherever and whenever possible. This means encouragement to the elderly to share their accumulated experience, to provide their oral histories and their views on the momentous changes that they, along with China itself, have lived through. Older people should be valued, and involved in wider society on terms that they themselves desire, recognizing that a wish for privacy and seclusion might be their preference.

Conclusion

Finally, it is worth noting that cases exist where narratives and micro-histories coexist and play a role in producing and strengthening social exclusion. The medical "gaze" refers here to discourses, languages, and ways of seeing that shape the understanding of aging into questions that center on, and increase the power of, the State, and restrict or delegitimize other possibilities. A consequence is that areas of policy may at first seem tangential to the medical project come to be reflected in its particular distorting mirror. The Chinese "Doumin"—a subcategory of a wider population officially cataloged as fallen people, beggars, or ruined households—were seen as inferior and condemned to bear low status on account of a number of beliefs prevailing among mainstream society. The narrative concerns a creational myth that asserts that the Doumin were closely related to Chinese ethnic minorities like the She and Yao and that all these groups shared the belief in Pan-hu, a common dog ancestor. As to micro-histories, we run into different stories which state that the Doumin were either: (1) descendants of Song Dynasty traitors, deserters, or prisoners; (2) remnants of antique non-Chinese ethnic groups; (3) foreigners who adopted the customs of Chinese lower social strata; and (4) descendants of domestic slaves. Yet in all cases the Doumins' excluded and outcaste status came about as a punishment society bestowed upon them. They were reduced to performing polluting occupations (ox head lanterns making, ironwork, barbers, caretakers, frog-catching, entertaining, among others) and limited to live in segregated quarters outside town. Furthermore, the Doumin were not allowed to study or take public office, nor serve as officers and were obliged to marry among themselves. This suggests that their social identity was a result of the legal status imposed on them and not the other way around. As Hansson (1996:87) expresses it: "Once fallen people had been labeled as beggars, they had little choice but to conform to the behavior expected from people who had the social identity associated with their legal status." For older people if they are regarded as an inferior category then their behavior will begin to mimic this categorization. "Dependent is as dependent does" is a major danger in the continuation of the dominance of the bio-medical model of aging.

A key point here is that the notion of the "bio-medical gaze," not only draws our attention to the ways that aging has become "medicalized" as a social issue in China, it also highlights the way in which older people are encouraged for as long as possible to "work on themselves" as active subjects. Thus, as has pointed out, older citizens are encouraged to take greater personal responsibility for their health and for extending this period of their active aging. Those who are defined in relation to their health then discover themselves transformed into passive objects of medical power in China. How do we go beyond this in managing old age?

We could suggest macro-social practices have become translated into particular ways of growing old that not only shape what it is to age successfully, but are also adopted by older adults, modified to fit their own life circumstances and then fed back into wider narratives of aging well. Coins this as the "Illderly" and "Wellderly" and managing aging experiences is about resistance to dominant discourses of bio-medicine. According to Frank (1998), the personal experience of illness is mediated by bio-medical procedures that shape and contribute to how the older people recognize their own process of ill health and recover. Katz (2000) notes that the maxim of "activity for its own sake" as a means of managing later life not only reflects wider social values concerning work and non-work, it also provides personal means of control and acts as grounds for resistance. In addition, Phillipson (1998) argues that changes in Westernized policies has occurred from seeing old age as a burden to seeing it as an opportunity to promote productive aging. This reflects an attempt to shape acceptable forms of aging whilst encouraging older adults to self-monitor their own success at conforming to the challenging paradigm to hegemony of bio-medicine and its neglect of the agency of older people.

References

Chen, S., & Powell, J. L. (Eds.). (2011). *Aging in perspective and the case of China*. New York: Nova Science.
China Daily. (2004, April 15). Tailoring health care policies for the elderly.
Cook, I. G., & Murray, G. (2001). *China's third revolution: Tensions in the transition to post-communism*. London: Curzon.

Du, P., & Tu, P. (2000). Population ageing and old-age security. In W. Z. Peng & Z. G. Guo (Eds.), *The changing population of China*. Oxford: Blackwell.

Foucault, M. (1972). *The archeology of knowledge*. London: Tavistock.

Frank, A. W. (1998). Stories of illness as care of the self: A Foucauldian dialogue. *Health, 2*(3), 329–348.

Hansson, A. (1996). *Chinese outcasts: Discrimination and emancipation in late imperial China*. Leiden: E.J. Brill.

Heidegger, M. (1971). *Poetry, language, thought*. New York: Harper & Row.

Katz, S. (1996). *Disciplining old age: The formation of gerontological knowledge*. Charlottesville: The University Press of Virginia.

Katz, S. (2000). Busy bodies: Activity, aging and the management of everyday life. *Journal of Aging Studies, 14*(2), 135–152.

Krug, E. G. (2002). *World report on violence and health*. Geneva: World Health Organization.

Murray, G. (1998). *China: The next superpower*. London: China Library.

Murray, G. (2004). *China's population control policy: A socio-economic reassessment*. PhD thesis, Liverpool John Moores University, Liverpool.

Phillipson, C. (1998). *Reconstructing old age*. London: Sage.

Powell, J. (2001). Theorizing gerontology: The case of old age, professional power and social policy in the United Kingdom. *Journal of Aging and Identity, 6*(3), 117–135.

Powell, J. L. (2009). Social theory, aging, and health and welfare professionals: A Foucauldian "toolkit". *Journal of Applied Gerontology, 28*(6), 669–682.

Powell, J., & Biggs, S. (2000). Managing old age: The disciplinary web of power, surveillance and normalisation. *Journal of Aging and Identity, 5*(1), 3–13.

Powell, J., & Cook, I. G. (2000). "A tiger behind and coming up fast": Governmentality and the politics of population control in China. *Journal of Aging and Identity, 5*(2), 79–90.

World Bank. (1994). *Averting the old-age crisis*. Oxford: Oxford University Press.

Chapter 3
Aging: The Role of Work and Changing Expectations in the United States and China

Barbara R. McIntosh and Chun Zhang

Abstract Economics and demographics are driving change in individuals' expectations about work beyond the traditional retirement age and in employers' need for older workers in both the United States and China. Adjusting to this change in the workplace, now and in the future, includes planning strategically for these human capital shifts, managing the multigenerational workplace, ensuring the transfer of both explicit and tacit knowledge as aging workers exit the workplace, and providing flexibility in both hours and location of work to bolster recruiting and retention efforts. Creating a supportive organizational culture is a key underpinning to making all the changes associated with the aging of the workplace.

Keywords: Aging workforce • Older workers • China • Multigenerational relations • Knowledge transfer • Workplace flexibility

Introduction

The United States and China are both experiencing dramatic changes in their labor force composition. A significant portion of each nation's workforce is approaching their traditional retirement age. At the same time, demographic projections point to future labor shortages. There will not be enough young workers to replace those eligible to retire. This chapter addresses the implications of these changes.

B.R. McIntosh (✉)
School of Business Administration, University of Vermont,
Burlington, VT, USA

Center on Aging, University of Vermont, Burlington, VT, USA
e-mail: mcintosh@bsad.uvm.edu

C. Zhang
School of Business Administration, University of Vermont, Burlington, VT, USA

S. Chen and J.L. Powell (eds.), *Aging in China: Implications to Social Policy of a Changing Economic State*, International Perspectives on Aging,
DOI 10.1007/978-1-4419-8351-0_3, © Springer Science+Business Media, LLC 2012

First, the discussion covers the growing evidence regarding extended work–life behaviors from the individual's perspective. Then, this chapter offers an analysis of the changes that employers will need to make in response to pressures stemming from these shifts in workforce composition. The reasons for change are slightly different in these two economies, but there is increasing pressure for aging workers to remain employed and for employers to retain this older cohort in both instances.

From the individual perspective, traditional models of retirement in the United States have virtually disappeared, and they are slowly eroding in China. In both countries, individuals are responding to economic pressures and longer life-expectancies by working longer. The 2007–2009 US recession accelerated a change in extended employment behavior among older-age cohorts. Aging workers recognized that their presumed retirement years may not be as financially secure as they originally thought. In addition, there is increasing recognition that continuing to work is important, given healthy, longer life spans; and it offers a viable option as the US shifts to jobs that are less physically demanding. In China, changing dependency ratios (the ratio between those working and those supported in retirement) are also altering labor force participation expectations.

From an employer's perspective, significant changes need to be made in human capital and human resource planning. In China, negative growth in the total working-age population is predicted to occur as early as 2013 (China Daily, 2011). In the United States, there are already shortages in healthcare, particularly nursing, and in some industries such as oil and gas. This chapter specifically addresses three areas that employers will need to proactively confront. First, managing the multi-generational workforce has not been an issue in the past. Now, there could be four generations in the workplace. In addition to gender and race differences, each generation brings different expectations into the workplace. The traditional cookie-cutter approach, treating every worker the same, will not meet the needs of this age-diverse workforce. Second, improving knowledge transfer will become a necessity. With so many potential organizational exits, knowledge retention will be key to the sustainability of an organization. Finally, this chapter analyzes alternative ways to integrate flexibility into the workplace. Whether this flexibility is in location or hours of work, it will be essential to retain younger and older workers alike and to maximize the competitive return on technological investment.

The Individual Perspective: United States

Even before the recession, there was evidence that aging Americans have been working longer. Between 1998 and 2008, the number of workers aged 55–64 increased by 64%; those aged 65–74 increased by 57%; and those 75 and older increased by 88% (Toossi, 2009). Aging workers are healthier and better able to work than earlier generations (Manton, 2008). At the same time, increased longevity adds to the risk of outliving one's retirement resources so, to some extent, continued employment is a necessity. The economic downturn called dramatic attention to precarious retirement finances when retirement accounts lost 40% or more. In excess of $2.1 trillion disappeared from 401K and IRA assets in 2008 alone! (Coombes, 2009).

In addition to being physically able to work and recognizing an economic need to do so, individuals are in a better position to work longer because the work itself has changed. The shift from manufacturing to service occupations means that jobs are not as physically demanding. Between 2008 and 2018 the total number of jobs is expected to increase by 10%, but service occupations and professionally related occupations are expected to increase by 14 and 17%, respectively (Lacey & Wright, 2009). There are now also more opportunities for flexible work arrangements which are attractive in the context of work–life balance for workers of all ages, and technological advances make it easier for older workers to continue working.

Equally important are the self-generated pressures to continue working. Members of the Baby Boom generation, in particular, identify with their jobs and the importance of contributing to society. Remaining productive is built into their value system. Indeed, survey results published by the Employee Benefit Research Institute in 2010 report that the major reason 60% of retirees gave for finding new paid work was that they wanted to stay active. An additional 32% said this was a minor reason. Fifty-nine percent reported that they enjoyed working, and an additional 28% reported this as a minor reason (Helman, Mathew Greenwald and Associates, & Copeland, 2010). It is also clear that a number of public and nonprofit organizations have formed to capture the productive energy of aging Americans. The federal government's Partnership for Public Service is one example, as is Experience Corps, which puts volunteers of 55 and over into public schools. Civic Ventures is a national organization promoting encore careers; and other growing grass-roots organizations helping aging Americans negotiate work and volunteering in the third stage of life include Coming of Aging, The Transition Network, and Discovering What's Next.

Continuing to be productive and contribute to society appear to be important to the individual, but it is important to recognize that, from the individual's perspective, personal finances and risk are the primary drivers. The percentage of workers very confident about having enough money for a comfortable retirement was measured at a 20-year low of 13% in 2009. This confidence level went up slightly in 2010 to 16%, but older American workers are still not prepared financially. More than half of workers (54%) report that the total value of their household's savings and investments, excluding the value of their primary home and any defined benefit plans, is less than $25,000 (Helman et al., 2010). In summary, Americans expect to be working longer. Indeed, the percentage of workers who expect to retire after age 65 has increased over time, from 11% in 1991 to 14% in 1995; 19% in 2000; 24% in 2005; and 33% in 2010 (Helman et al., 2010).

While there is clear evidence that the intention is to work longer, in the United States there are multiple forces restraining continued employment, including high unemployment, a slow growth economy, a change in the skill mix demanded, a youth culture, inflation in labor costs, hiring and retention practices in some firms, and negative stereotypes about older workers. These pressures cannot be underestimated. Jobs are not as readily available, and those older workers laid off during the recession have been facing longer periods of unemployment. In March 2010, unemployed workers age 55 and older had been out of work for an average of 35.5 weeks, compared to 30.3 weeks for job seekers 25–54. Meanwhile, job seekers between ages 16 and 24 have been jobless for an average of 23.3 weeks (Challenger, Grey & Christmas, Inc., 2010). In December 2009, the unemployment rate among those 55 and up reached 7.2%, the highest level for this age group in Bureau of Labor Statistics records going back to 1948. Since then, it has fallen slightly to 6.9% as of March 2010 (Challenger et al., 2010). Only 23.3% of unemployed youths were jobless for 27 weeks or more in 2009, compared with 33.4% of prime-age workers and 39.4% of older workers. Furthermore, the share of long-term unemployed made up of youths declined between 2007 and 2009, while the share made up of older workers rose. It is important to note that, in part, these statistics reflect the fact that facing economic uncertainty, unemployed prime-age workers and older workers became substantially less inclined to drop out of the labor force between 2007 and 2009. This created upward pressure on both their overall unemployment rates and long-term unemployment rates (Ilg, 2010). The barriers faced by older job seekers are also reflected in the increasing number of age discrimination claims filed with the US Equal Employment Opportunity Commission. The number of claims filed in 2008 rose to 24,582, the largest number of age discrimination claims filed annually since the agency started keeping records in 1992, and the number dropped only slightly in 2010 to 23,264 (US Equal Employment Opportunity Commission, 2010).

The Individual Perspective: China

An aging population and workforce are also presenting a significant challenge to the Chinese economy. A series of social changes and economic reforms in China from the 1970s to 1990s have resulted in not only an increasingly larger percent of the population being older but also an increasing number of retirees seeking to reenter the workforce.

The first social change affecting the aging population and their employment situation in China is the one-child policy implemented in 1979. In the hope of slowing population growth and ensuring economic stability, China adopted a one-child policy in 1979 that mandates the majority of Chinese couples to have only one child. With fewer children and improved living standards due to economic reforms, the percent of elderly has grown significantly over the last 30 years. Data released at the end of April 2011 by the National Bureau of Statistics showed that the proportion of the population aged between 0 and 14 fell from 22.9% in 2000 to 16.6% in 2010. Meanwhile, the number of people aged 60 and above grew from 10.3% to 13.3% over the same time period (China Daily, 2011). The Asian Development Bank forecasts that the proportion of those aged 60 and above is expected to rise to 33% by 2050. That would make China's population the same age as Denmark's, and older than that of the United States (26%) (Xinhuanet, 2011).

Traditionally, the elderly in China have been cared for by their children according to the traditional cultural beliefs based on the Confucius teaching of filial piety or "Xiao." Confucianism proposes that human beings are fundamentally relationship-oriented and proposes that social and economic order can be achieved through building a strong and orderly hierarchy of relations (Luo, 1997). Influenced by this philosophy most Chinese recognize the importance of building relations by reciprocating favors (Gu, Hung, & Tse, 2008). Of all the favors reciprocated, those in a parent–child relation are considered the most important and long-term (Park & Luo, 2001). Chinese parents are willing to give up whatever resources they have to provide a good future for their children. In return, the adult children "owe aging parents respect, obedience, loyalty, material provision, and physical care" (Zhan, 2011, p. 162). As cited in Zhan's study, intergenerational coresidence remains popular in China, with 54% of older men and 60% of older women in urban areas and 62% of older men and 75% of older women in rural China living with their adult children's families (Zeng, Liu, Zhang, & Xiao, 2004).

This family care system for the aging population has been challenged by the one-child policy. This policy reduced family size and the number of people born after 1979/1980 in China. As a result, many families currently have a "4-2-1" family structure (four aging parents need to be cared for by a couple who have one child) (Xinhua News, 2010). As the number of children available to support aging parents has decreased, the need and desire for the aging population to fend for themselves have increased. Financial necessity is a key reason for continued employment beyond the normal retirement age.

The second force that motivates the aging population to continue work is the pension scheme reform passed in the 1980s and 1990s. Before the reform, the majority of the aged population or retirees in urban areas lived on pension income paid for by their employers. Employees did not contribute to this pension, and the payments were covered by the government (Pei & Xu, 2011). This pension payment was approximately equal to the employee's last month's is wage before retirement, and the amount increased as the wages of the employees on the job increased (Tian, 2008). This retirement pension scheme remained essentially the same for the employees of the government offices and public institutions after the pension reform. However, for those employed by the state-owned firms, there was a change in the 1990s from a pension to an old-age insurance program (similar to the 401K plan used in the United States). The primary purpose of the change was to relieve the financial burden on state-owned firms making the pension payments (Pei & Xu, 2011). Old-age insurance was established to combine contributions by the government, employers, and employees. As a consequence, the retirement pensions for state-owned firm employees have been adjusted to target a replacement rate of 60%, significantly lower than the pensions for the retirees of the government offices and public institutions (Pei & Xu, 2011).

The pension reform, accompanied by increasing costs of living, created different financial needs for retirees from different sectors and also different motivations for them to reenter the workforce. When an employee reaches the mandatory retirement age (about 60 for men, and about 55 for women in China), the retirees who have a good pension plan and health insurance coverage are interested in working for nonfinancial reasons, while the retirees from state-owned firms and other sectors tend to seek additional employment to meet the financial needs of their families.

Furthermore, health insurance and health care costs are also a concern for the majority of the population. Universal insurance coverage used to be given to almost all rural and urban residents in China (Yip & Mahal, 2008). After the economic reforms in the 1980s, a city-based social health insurance scheme was adopted to replace the health insurance provided by employers and cooperatives (Today's Research on Aging, 2010). A large number of rural and urban workers were no longer insured with only 56% urban residents and 21% of rural residents having health insurance in 2003, according to the National Health Services Survey (Today's Research on Aging, 2010). Health insurance coverage also varies across income classes. According to the statistics reported in Today's Research on Aging (2010), "In urban areas, about 80% people in the highest income quintile have health insurance, while in rural areas, just 32% of people in the highest income quintile have it." (p. 3). Consequently, an increasing number of retirees reenter the workforce to earn additional income to cover their health care costs.

Expectations about working as one grows older in China are clearly changing. Traditionally, the Chinese considered the primary responsibilities of older people (above age 60) to be enjoying family life with their children and grandchildren, taking good care of their health, and passing the last stage of their lives in contentment and peace. With the change in these traditional beliefs about aging, the improvement of living standards, and the need for financial security, the aging population is becoming more active in managing their careers compared to the past. They are called the "gray hair troops" in the Chinese labor market, and in some special fields, they are more employable than younger people (Xinhua News, 2010).

With the increasing aging population, more and more retirees choose to reenter the job market. These older employees can be broadly classified as one of three types (Chang, 2010):

1. The knowledge-rich. These retirees typically were employed by the government offices and public institutions (hospitals, universities, and research institutions). They are well-educated and have valuable managerial experiences and/or technical expertise. They receive good pensions (continued salary payment and a health insurance plan) and have little financial need to work. Their main motivation for working after retirement is to remain active and contribute to society. Because of the valuable experience and expertise possessed by these older employees, some companies often target hiring them and recruit them immediately after their retirement.
2. The older entrepreneurs. These retirees also do not have significant financial needs and have saved some money. They no longer need to take care of their children, so they start their own businesses. These people are energetic. They run small businesses or clinics, schools, Internet cafés, etc.
3. The helpless older workers. These older workers are not educated and do not have expertise that is in demand. Their children also tend to have relatively low

education levels and low incomes. The major motivation for them to work is financial need. They have to work in order to support their family and themselves.

In China, most of the older employees belong to the first category, and there is a growing demand for the first group of people (Chang, 2010). Employers find the first group of employees to be patient and responsible, have a good work ethic, and tend to be committed to the company (Chang, 2010). Also, they don't have high salary expectations since they are financially comfortable. In addition, many older workers have rich experiences and strong connections with resourceful people. They can secure opportunities and ROI for the company within a short period of time. Furthermore, most of these older workers have good health insurance from their previous employers. The new company doesn't have to offer them benefits, which greatly reduces the cost of hiring.

In addition, some of these older, educated people are interested in contributing to society. For example, the national government organized the "gray bell in action" movement to support the inland, underdeveloped regions (Chelder, 2009). Many older experts take the minimum living expense compensation to volunteer. According to the statistics of China's Older Talents Center, between 2003 and 2009, more than 1,000 highly educated, older experts have participated in the "gray bell in action" movement. Their professions include doctors, teachers, scientists, entertainers, etc. (Chelder, 2009).

On the other hand, if an older worker does not a have good education and expertise, they face significant discrimination in the job market (Chang, 2010). Labor market regulations covering age discrimination remain underdeveloped in China, and few legal protections exist for older workers. In addition, labor law and labor contracts have not covered issues related to retirees reentering the job market (Chang, 2010).

The Business Perspective

Human Capital and Planning for Potential Labor Shortages

The underlying social dynamics are different, but both Chinese and American employers face aging workforce issues and human capital planning will be of growing importance. Whether it is a necessity because of financial uncertainty or a personal preference because they enjoy working, the trend among aging workers to extend their years of employment is actually an advantage from the business perspective. In fact, US demographics suggest that employers may need to turn to older workers out of necessity given potential labor shortages in the future. Projections by the US Government Accountability Office (2007) show a dramatic decrease in labor force growth between 1970 and 2080 as shown in Fig. 2.1.

In terms of age, it has been projected that between 2010 and 2030 the number of Americans between the ages of 20 and 64 will increase by only 10%; at the same time the number of those 65 and over will increase by almost 80% (from 13 to 19%)

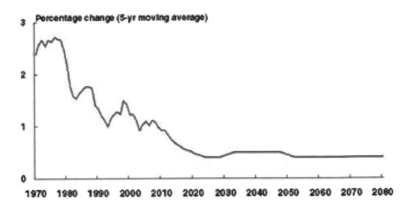

Fig. 2.1 US Labor Force Growth, 1970–2080 (Government Accountability Office, 2007). Source: GAO analysis of data from the Office of the Chief Actuary, Social Security Administration. Note: Percentage change is calculated as a centered 5-year moving average of projections based on the intermediate assumptions of The 2008 Annual Report of the Board of Trustees of the Federal Old-Age and Survivors Insurance and Federal Disability Insurance Trust Funds

(US Census Bureau, 2010). Longer labor-force participation may take the edge off projected labor shortages because if Baby Boomers, in fact, retire at the same rate as older workers have before them, and the economy recovers, there may not be enough workers to fill the available jobs (Bluestone & Melnick, 2010). Bluestone argues "Not only will there be jobs for these experienced workers to fill, but the nation will absolutely need older workers to step up and take them." This is particularly true in the social sector, which includes healthcare, education, social assistance, and nonprofit organizations. Bluestone and Melnick further note, "Based on government projections of how much each worker will contribute to the nation's Gross Domestic Product in 2018, the labor shortage could cost the economy $3 trillion over the 5-year period from 2018 to 2022." Fortunately, there is evidence that the labor force participation of adults over 50 is continuing to increase. According to a 2011 analysis conducted by the Urban Institute, "adults age 50 and older made up 31% of the labor force in 2010, up from 20% in 1995" (Johnson & Mommaerts, 2011, p. vii).

As previously discussed, China is also projecting a negative growth in their labor force, primarily because the overall population increased only 0.57% a year between 2000 and 2010 due to the one-child policy. China's population is expected to start to decline as early as 2015, according to the government forecasts (MacKinnon & Wheeler, 2011). The pressures are projected to be extraordinary, as the number of Chinese aged 60 and over is expected to be 200 million by 2015; and, by 2025, 20% of people living in urban areas will be 60 and older (MacKinnon & Wheeler, 2011).

While older workers' revised expectations about working additional years may ease potential shortages in general, there may be issues for any one organization. With this in mind, it is imperative that employers, both US and Chinese, take a hard look at their human capital and associated planning models. Specifically, it is important to do an age audit, and ask the following questions:

What is the age profile of my workforce?

What is the distribution of older workers throughout the organization? (Are they lumped in one or two areas?)

How many years of experience specific to this organization do they represent?

What has been the investment in training these workers?

Is the knowledge level and expertise replaceable?

What is the productivity of workers in this organization by age?

Are some of the older workers underemployed? (They have not been moved/promoted to more challenging jobs)

Answers to these and related questions should give a more complete picture about needed retention and/or rehiring efforts, as well as efforts that need to be directed toward improving multigenerational relations, knowledge transfer efforts, or implementing flexibility policies.

There is already some recognition that these planning efforts are important but still lacking. According to a 2009 US study conducted by the Sloan Center on Aging and Work, "four of every 10 employers (40%) surveyed anticipate the aging of the workforce will have a "negative/very negative" impact on their business over the next three years" (Pitt-Catsouphes, Sweet, Lynch, & Whalley, 2009, p. 5). This same study also found, however, that almost 3/4 of the employers surveyed (77%) had not analyzed projections about the retirement rates of their employees (Pitt-Catsouphes et al., 2009). Unfortunately, a 2009 MetLife study also reported that, while "37% of employers acknowledge that an aging workforce will have a significant impact on their companies, only 17% of employers offer resources or programs, such as staged or phased retirement, geared toward the aging workforce" (Metlife 2009a, 2009b, p. 34). Clearly, employers need to plan sooner rather than later in order to remain competitive.

Multigenerational Relations

Advance human resource planning is not the only issue. One key factor in the ability to retain older workers and increase the productivity of all employees may rest in the culture of the organization. Perceptions about management and coworker support contribute to retention (DelCampo, 2010; Jenkins, 2008), and the changing age profile within the organization suggests more attention needs to be directed toward managing generational diversity. Until now there has been very little research on the management implications of the multigenerational workforce. It has been suggested that there is a potential crisis stemming from intergenerational conflict (Dychtwald, Erickson, & Morison, 2006; Lancaster & Stillman, 2002). Others have suggested that this may not be widespread (Pitt-Catsouphes & Smyer, 2007; Sweet & Moen, 2006). Certainly, the recession may have acerbated this potential.

What is clear is that each generation brings a different set of values and priorities into the workplace (Raines, 2003), and each generational cohort views the others differently. Research reported by James, Swanberg, and McKechnie (2007), for example, shows that each succeeding younger cohort has increasingly more negative perceptions of workers 55 and over. According to a 2009 survey conducted for The Conference Board, "when the 2009 survey results are compared with those [from their comparable survey] in 1987, the largest decline in overall job satisfaction—from 70.8% in 1987 to 43.4% now—occurred among workers aged 65 and over.... Workers under the age of 25 had the second greatest decline, dropping from 55.7% in 1987 to 35.7% in 2009" (Franco, Gibbons, & Barrington, 2010, p. 9).

In a recent poll conducted by the Society for Human Resource Management (SHRM), more than half of the older workers raised concerns about younger workers' inappropriate dress (55%) and poor work ethic (54%). Additional issues included excessively informal language and/or behavior (38%), need for supervision (38%), inappropriate use of or excessive reliance on technology (38%), and lack of respect for authority (36%). By contrast, the top three concerns or complaints raised by younger workers about older managers are resistance to change (47%), low recognition of workers' efforts (45%), and micromanaging (44%). These issues were followed by rigid expectations of following authority/chain of command (38%), aversion to technology (31%), and low respect for workers' work–life balance (31%) (SHRM, 2011).

This multigenerational split in perceptions of others is further complicated by authority relationships. According to a 2010 CareerBuilder survey of workers aged 18 and over, "43% of workers ages 35 and older said they currently work for someone younger than themselves. Breaking down age groups, more than half (53%) of workers ages 45 and up said they have a boss younger than them, followed by 69% of workers ages 55 and up. According to the 2010 CareerBuilder survey, "16% of workers ages 25–34 said they find it difficult to take direction from a boss younger than them, while 13% of workers ages 35–44 said the same. Only 7% of workers ages 45–54 and 5% of workers ages 55 and up indicated they had difficulty taking direction from a younger boss (CareerBuilder, 2010).

Not surprisingly, the majority of younger workers does not intend to stay with their current employer and, as a result, may not be as invested in developing good working relationships across the generations. According to a 2010 Pew Research Center survey, "about two-thirds of all employed Millennials say it is very likely (39%) or somewhat likely (27%) they will switch careers sometime in their working life, compared with 55% of Gen Xers and 31% of Baby Boomers" (Pew Research Center, 2010, p. 46). Regardless of employees' intentions to remain in the organization, it is essential for top level managers to establish a framework of values and expectations about attitudes and behaviors in the workplace. In the context of the US workplace, these elements must include respect and valuing others' contributions to the work environment. Sensitivity to others' perspectives is increased with honest communication and openness to change; and top management must role-model these behaviors to reinforce employee expectations. Specific ways to improve multigenerational relations in the United States include openly discussing multiple perspectives and reinforcing the point that differences in ideas are not only respected but also rewarded. Mentoring, shadowing, and creating age-diverse teams also contribute to multigenerational recognition and support (see Zemke, Raines, & Filipczak, 2000).

In China, establishing a multigenerational culture of respect is far less of an issue, but there are differences in value orientations. The work values of different age groups in China have been significantly influenced by the social and economic reforms since the late 1970s. The group-focused, conformity oriented, Confucius values are more reflected by the work values of the baby boomer generation in China (born between 1946 and 1964). These employees are more loyal to their employers and willing to work with others. They tend to value group harmony and tolerance. In addition, they view the job based on external work values, such as reputation and income, while the younger workers (especially those born after the 1980s) are more interested in internal work values such as self-development and self-realization (Huang, 1993). In general, more experienced and older employees are less likely to leave the organization than younger employees (Mowday, Steers, & Porter, 1979, Li, Liu, & Wan, 2008), and older workers may have a stronger work ethic than younger workers (Li et al., 2008). Given these characteristics of work values of older employees in China, we can infer that the older workers are likely to respect hierarchy in organizations and cooperate with supervisors, regardless of their age.

Traditional Chinese values consider that old age commands respect due to the work experience and wisdom accumulated by the elderly over time. In the centralized economy (before the 1980s), seniority was a key determinant of promotion in state-owned firms, government offices, and public institutions. With the economic reform, however, seniority at the workplace has been gradually replaced by an expertise-based hierarchy. Chinese employees generally value hierarchy based on job titles. This work value enables supervisors to command respect, regardless of their age (Alas & Wei, 2008; Li, Liu, & Wan, 2008).

Alas and Wei (2008) conducted a study of 29 Chinese companies in 2005–2006 with 1,303 respondents in two cities (Beijing and Jihan). They hypothesized that the older people who started their working life in pre-1978 China value hierarchy more than the younger workers; however, they found no statistically significant differences between younger and older groups along this dimension. The employees in China, in general, emphasize a long-term orientation and management by objectives. They have a high level of respect for hierarchy and place great importance on supervision. As such, age-related value differences may not present a significant difficulty for managing the multigenerational workforce in China.

Knowledge Transfer

A related issue of increasing importance given the aging of the workforce and potential organizational exits is knowledge transfer. Certainly, the quality of multigenerational relations contributes to the organization's ability to have successful knowledge transfer programs, and such programs already appear to be under consideration in many organizations. According to a 2009 MetLife survey of 240 large US employers, "among those employers concerned primarily with the knowledge drain, 62% believe that the retirement of their older employees has a moderate to major negative impact on their organization's productivity. Interestingly, this cuts across all industries, including manufacturing (62%), finance (59%), and services (69%)" (Metlife 2009a, 2009b, p. 10). This same report also stated that among the steps being considered to manage the knowledge drain, "four in ten are considering introducing technology as a way to transfer knowledge to younger employees (41%), implementing part-time work programs as a way to help employees ease into retirement (39%), and offering pension benefits to partially retired/partially in-service employees (32%)" (Metlife 2009a, 2009b, p. 14). Additionally, a 2008 survey of small business owners reported that 28% of the small-business owners stated they "have planned for knowledge transfer from experienced older workers to other workers" (National Association of Professional Employee Organizations, 2008)

What is equally important for knowledge transfer to actually take place is the participation of employees in the process. In a 2008 survey of over 2,000 US employees, 69% of Baby Boomers rate as important the trait "readily shares knowledge with coworkers", and 71% report that the trait describes themselves. In contrast, 53% of GenY employees rate that trait as important, and 56% feel that it describes themselves. Among GenX employees, the rates are 62% important and 63% describes themselves. Matures have the highest rates on this trait, with 69% rating it as important, and 83% reporting that it describes themselves (Randstad, 2008, Fig. 20, p. 24).

This recognition and associated knowledge transfer planning are critical because it is argued that knowledge loss is one of the most expensive problems organizations face (Parise, Cross, & Davenport, 2006). Secondly, it is typically one of the most ignored problems (Leonard, 2004). It has been recognized that knowledge transfer and retention based solely on IT is usually not very successful (Leibold & Voelpel, 2006; Parise et al., 2006; Kluge, Stein, & Licht, 2001). A large literature supports the idea that establishing the right environment and culture is crucial to successfully implement a knowledge transfer or retention strategy (Kluge et al., 2001). The right culture is one implemented throughout the organization as a whole, where people are intrinsically motivated to share knowledge. This is particularly important in order to transfer both explicit and tacit knowledge.

What organizations are most concerned about retaining is *deep smarts*, a type of knowledge that is referred to by Leonard-Barton and Swap (2005) as being the engine of an organization. They elaborate that

> knowledge that provides a distinctive advantage, both for organizations and for managers as individuals, is what we term deep smarts." "Deep smarts are a potent form of expertise based on firsthand life experiences, providing insights drawn from tacit knowledge, and shaped by beliefs and social forces. Deep smarts are as close as we get to wisdom. They are based on know-how more than know-what the ability to comprehend complex, interactive relationships and make swift, expert decisions based on that system level comprehension but also the ability, when necessary, to dive into component parts of that system and understand the details. Deep smarts cannot be attained through formal education alone-but can be deliberately nourished and grown and, with dedication, transferred or recreated (Leonard-Barton & Swap, 2005).

The transfer of *deep smarts* in China may be less problematic, given the traditional respect accorded to the organizational hierarchy and elders in society in general. Similar to United States, however, there are generational differences in the values associated with knowledge transfer. Alas and Wei (2008), for example, reported that older Chinese prize specialty-related values more than younger Chinese. Specialty-related values are defined as values associated with "an employee's professional education and acknowledgement of an employee as a specialist in his or her field." Driven by this value, older employees are more likely to be willing to share their knowledge and expertise with the rest of the workers when receiving recognition and respect. This is further evidenced by the fact that the "knowledge-rich" retirees are motivated to be reemployed with limited financial compensation or volunteer their expertise to contribute to society.

Another consideration in knowledge transfer is that individual scientific domains have become increasingly specialized and complex, which requires high-quality knowledge workers (Leibold & Voelpel, 2006). There is literature that refers to this development as the knowledge era or knowledge economy, which is "characterized by a new competitive landscape driven by globalization, technology, deregulation, and democratization" (Halal & Taylor, 1999). *Deep smarts,* as discussed earlier, is an important component of this knowledge economy and, since high-quality knowledge workers develop *deep smarts* mainly through experience and practice, this quality is largely held by the older worker cohort (Leonard-Barton & Swap, 2005). Losing *deep smarts* can have very serious consequences for an organization, and that is why a major challenge will be to capture and transfer the knowledge of the people with *deep smarts* (Leonard, 2004).

Leibold and Voelpel (2006) list several strategies to capture and transfer knowledge. Strategies based on information technology include document mining, which is basically documented knowledge that can empower and support the worker; and knowledge mapping, which is a strategy that maps out the organization's intellectual capital. This knowledge is usually accessible on the Intranet or Internet and captures the organization's format and structures, employees' roles and responsibilities, knowledge that employees will need to do their jobs, etc. Leonard (2004) argues, just as Kluge et al. (2001) and Parise et al. (2006), that technology can help in transferring knowledge, but that before investing in databases, managers need to realize that tacit knowledge is hard to transfer. Databases may not be used or might not serve their purpose with respect to really transferring knowledge. Leonard (2004) points out that the only way to transfer tacit knowledge and *deep smarts* is through experience. Since this is a slow process, organizations could speed up this process by facilitating guided experience(s). Organizations also need to motivate and train their employees with *deep smarts* to act as coaches and pass their wisdom along through guided practice, guided observation, guided problem solving, and guided experimentation.

According to DeLong (2004), there are four strategies for transferring and retaining *deep smarts*. These strategies include storytelling; organizing communities of practices which are basically networks of people who share the same practice, language, problems, values, etc.; conducting After Action Reviews, which are based on the assumption that, whenever a project or an assignment fails, a lot of knowledge is wasted; and finally, incorporating mentoring and coaching, which DeLong (2004) refers to as being among the most effective ways of transferring knowledge. Usually a more experienced worker mentors a less experienced worker, which insures a greater range of different types of knowledge (human knowledge, social knowledge, and cultural knowledge) being transferred.

De Long and Mann (2003) discuss two more knowledge retention strategies: Phased retirement and utilizing retirees. Rather than letting experienced employees completely separate from the organization, it has also been suggested that retirees should remain "on call" or used as consultants (Thilmany, 2008). These flexibility strategies fit with US older workers' long professed desire to work in "retirement" (AARP, 2002; AARP 2007; Watson Wyatt Worldwide 2004; Harris Interactive & Dychtwald, 2005). In China, it may be useful to create retiree pools and consciously seek out former employees to rehire "post retirement." This may involve consciously tapping into the "gray hair troops."

Flexibility

There is growing evidence that flexible scheduling helps not only with older worker retention but retention across the generations. Happiness at the workplace is higher (Randstad, 2008); there are higher levels of organizational commitment (Eaton, 2003); and higher levels of reported well-being (Moen & Kelly, 2007). In addition, employees on flexible schedules report lower work overload and greater satisfaction with work–family balance (Pitt-Catsouphes et al., 2009).

Flexibility can be implemented in many different ways. There are flexible employment contracts (consulting, leaves of absence), hours of work (work schedule, compressed work week, part-time work, part-year, seasonal), exit programs (phased retirement), location (telecommuting/telework), work assignment (job sharing), and work redesign. The complexity of possible arrangements is overshadowed, however, by research demonstrating significantly higher job satisfaction, engagement, and commitment to the employer in firms offering workplace flexibility (Baltes, Briggs, Huff, Wright, & Neuman, 1999; Bond, Thompson, Galinsky, & Prottas, 2002; WorldatWork, 2011). In addition, flexibility policies benefit employers by increasing productivity (Eaton, 2003); reducing absenteeism (Dalton & Mesch, 1990); serving as a supplement or tradeoff for higher wages (Baughman, DiNardi, & Holtz-Eakin, 2003); and helping with retention (Pavalko & Henderson, 2006; WorldatWork, 2011). According to a 2010 industry sector report from the Sloan Center on Aging & Work, "40% of the retail organizations, and 45% of the organizations in other sectors, reported that workplace flexibility somewhat/significantly increases business effectiveness" (Sweet, Pitt-Catsouphes, Besen, Hovhannisyan, & Pasha, 2010, p. 37).

Unfortunately, according to a 2011 report on a survey of managers from over 500 different business organizations, the ROI of workplace flexibility is not being

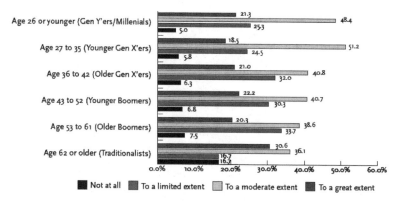

Fig. 2.2 Extent to which Workplace has Linked Flexibility with Overall Business Effectiveness by Generation

measured. Only 7% of organizations attempt to quantify the ROI of flexibility programs by measuring productivity, customer satisfaction, product quality, etc." (WorldatWork, 2011, Fig. 22, p. 8).

In a 2009 study of access to flexible work options, "nearly two-thirds (64.2%) of the respondents agreed that their work team links workplace flexibility with overall business effectiveness to a "moderate/great extent." Employees in the Traditionalist generational group were more likely to report that their organizations have linked flexibility to business effectiveness to a great extent (but are also most likely to report "not at all") (Pitt-Catsouphes et al., 2009, Fig. 9, p. 9) (Fig. 2.2).

It is a common practice in China that when an employee is approaching retirement age, they will be moved from the "first line" to the "second line" in their workplace. This enables them to have a significantly lower workload, fewer responsibilities, and for the employer to arrange for transitions of responsibilities to take place. The Chinese employees, however, do not seem to value flexible working hours or vacation times, regardless of age (Xu, 2009). According to Xu's (2009) study on aging in China, among all employment expectations and job rewards, only 5.7% of Chinese employees mentioned that good working hours were an important aspect of work, and only 5.2% of Chinese employees consider flexible working hours as important. Younger employees in China appear to be more aware of these factors as indicators of job quality than older workers; however, Xu (2009) found that the Chinese employees' general emphasis on these factors is low. More interestingly, ratings of good working hours and generous holidays (indicators of workplace flexibility) decreased between 1995 and 2001 in China. In 1995, 45.7 and 14.5% of Chinese employees reported that good hours and generous holidays are important job aspects, respectively. In 2001, the percentages dropped to 33 and 11.1%, respectively (Xu, 2009). Flexible working hours could be even less important for an aging workforce, given that they are motivated by external work values such as reputation and income, and they do not value aspects of the job that promote one's comfort and enjoyment.

While workplace flexibility has been used in China primarily as a transition tool and is not currently valued by the employees themselves, the evidence in the United States is clear that implementing flexibility in a variety of ways is an effective strategic tool for addressing workforce change.

Conclusion

In both the United States and China economics and demographics are driving change in the composition and structure of the workplace. Recent losses in pension accounts and retirement savings due to the US recession have underscored the uncertainty facing older Americans in their retirement years. At the same time, Americans are living longer, healthier lives and anticipate a much longer period of "retirement." Working longer is the expected response whether the driver is economic necessity or simply a desire to remain productive. In China, the employment opportunity structure for the aging population is highly segmented. The "knowledge-rich" retirees are in high demand, while the "hopeless older worker" risks being discriminated against in the highly competitive labor market in China.

In both countries, projected labor shortages suggest that employers will need to review the role of older workers. In China, the emergence of negative growth in the total working-age population, which some demographers predict will happen as early as 2013, is likely to contribute to slower economic growth and higher inflation (China Daily, 2011). Wage increases are expected as a result of this shortage, and in 2010 minimum wages increased by more than 20% on average. In fact, the government has vowed to double workers' pay over the next 5 years (China Daily, 2011). These wage increases coupled with increased uncertainty following pension changes and smaller family support structures, will make continuing to work more attractive to aging Chinese. Employers, facing a shrinking supply of workers, will also be looking toward experienced older workers.

Employers facing labor shortages and the possible availability of older workers will need to explicitly address several new workplace issues. The emerging multigenerational workforce necessitates reinforcing the importance of respect and valuing the contributions of all coworkers. A culture of open, honest communication supports not only multigenerational relations but also knowledge transfer. Systematic knowledge transfer systems including mentoring, shadowing, etc. are required to stop the brain drain inevitable with an aging/retiring workforce. Finally, workplace flexibility, whether it is in hours of work or location, is desired by workers of all ages in the United States; and this flexibility is now supported by technology. While flexibility is not currently an issue in China, the attractive workplace in the future will explicitly recognize work–life balance issues across the generations.

References

AARP. (2002). *Staying ahead of the curve*. Washington, DC: AARP.

AARP. (2007). *Staying ahead of the curve*. Washington DC: AARP. Retrived on Nov 2, 2011, http://www.aarp.org/research/surveys/stats/surveys/public/articles/2007_Staying_Ahead_of_the_Curve.html.

Alas, R., & Wei, S. (2008). Institutional impact on work-related values in Chinese organizations. *Journal of Business Ethics., 83*, 297–306.

Baltes, B., Briggs, T., Huff, J. W., Wright, J., & Neuman, G. A. (1999). Flexible and compressed workweek schedules: A meta-analysis of their effects on work-related criteria. *Journal of Applied Psychology., 84*, 496–513.

Baughman, R., DiNardi, D., & Holtz-Eakin, D. (2003). Productivity and wage effects of family fringe benefit benefits. *International Journal of Manpower., 24*, 247–259.

Bluestone, B., & Melnick, M. (2010). *After the recovery: Help needed*. Metlife and civic ventures. Reterived on Feb 26, 2010, from www.northeastern.edu/news/stories/2010/03/dukakis_center_report.html.

Bond, J., Thompson, T. C., Galinsky, E., & Prottas, D. (2002). *Highlights of the national study of the changing workforce: Executive Summary* (Vol. 3). New York: Families and Work Institute.

Careerbuilder. (2010). *More than four-in-ten workers over the age of 35 currently work for a younger boss, finds new CareerBuilder survey*. Reterived on February 18, 2010, from http://www.careerbuilder.com/share/aboutus/pressreleasesdetail.aspx?id=pr554&sd=2/17/2010&ed=12/31/2010&siteid=cbpr&sc_cmp1=cb_pr554_.

Challenger, Grey & Christmas, Inc. (2010). *Older workers join, succeed in job search*. Retrieved on April 29, 2011, from http://www.challengergray.com/press/PressRelease.aspx?PressUid=135.

Chang, L. (2010). *New problem presented to the older Chinese: Reemployment*. Beijing: Banyuetan News.

Chelder. (2009). *Diverse types of reemployment opportunities for the older Chinese*. Reterived on April 2, 2009, from Chelder.com.cn.

China Daily. (2011). http://news.xinhuanet.com/english2010/china/2011-05/03/c_13856468.htm.

Coombes. (2009). *MarketWatch*, Sept 21, 2009.

Dalton, D. R., & Mesch, D. J. (1990). The impact of flexible scheduling on employee attendance and turnover. *Administrative Science Quarterly, 35*, 370–387.

DelCampo, R. G. (2010). *Managing the multi-generational workforce from the GI generation to the millennials*. New York: Gower.

Delong, D. (2004). *Lost knowledge*. London: Oxford University Press.

De Long, D. W., & Mann, T. O. (2003). Stemming the brain drain. *Knowledge Management., 1*, 39–43.

Dychtwald, K., Erickson, T. J., & Morison, R. (2006). *Workforce crisis*. Boston: Harvard Business School Press.

Eaton, S. C. (2003). If you can use them: Flexibility policies, organizational commitment and perceived performance. *Industrial Relations, 42*(2), 145–167.

EEOC. (2010). Charge statistics FY 1997 through FY 2010. Retrieved on Nov 2, 2011, http://www.eeoc.gov/eeoc/statistics/enforcement/charges.cfm.

Franco, L., Gibbons, J., & Barrington, L. (2010). *I can't get no ... job satisfaction, that is: America's unhappy workers* (Conference Board Research Report No. 1459-09-RR). New York: The Conference Board.

Gu, F., Hung, K., & Tse, D. (2008). When does guanxi matter: Issues of capitalization and its dark sides. *Journal of Marketing, 72*(4), 12–28.

Harris Interactive & Dychtwald, K. (2005). The Merrill Lynch new retirement survey: A perspective from the baby boom generation. Reterived from Nov 2, 2011, http://www.totalmerrill.com/retirement.

Halal, W. E., & Taylor, K. B. (Eds.). (1999). *Twenty-first century economics: Perspectives of socio-economics for a changing world*. New York: Macmillan.

Helman, R., Mathew Greenwald and Associates, Copeland, C. & VanDerhei, J. (2010). The 2010 retirement confidence survey: Confidence stabilizing, but preparations continue to erode. *Issue Brief, March 2010*. No. 340 Employee Benefit Research Institute. Retrieved on Nov 2, 2011, http://www.ebri.org/pdf/briefspdf/EBRI_IB_03-2010_No340_RCS.pdf.

Huang, T. C. (1993). The study of the youth work values and organizational centripetal force (Paper in Chinese), *Research report for the youth of Taibei*. Taibei Youth Instruction Committee.

Ilg, R. (2010, June). Long-term unemployment experience of the jobless, *Issues in labor statistics summary 10–05*. Bureau of Labor Statistics. Accessed on http://www.bls.gov/opub/ils/summary_10_05/long_term_unemployment.htm.

James, J.B., Swanberg,J.E., & McKechnie, S.P. (2007). Generational differences in perceptions of older workers' capabilities. Issue Brief 12. Chestnut Hill, MA: The Center on Aging and Work/Workplace Flexibility, Boston College.

Jenkins, J. (2008). Strategies for managing talent in a multigenerational workforce. *Employment Relations Today, 34*(4), 19–26. Winter 2008.

Johnson, R. W., & Mommaerts, C. (2011). Age differences in job loss, job search, and reemployment. Program on Retirement Policy, Discussion Paper No. 11–01. Washington, DC: Urban Institute. Retrieved on Nov 2, 2011, http://www.urban.org/UploadedPDF/412284-Age-Differences.pdf.

Kluge, J., Stein, W., & Licht, T. (2001). *Knowledge unplugged: The McInsey & Company global survey on knowledge management*. New York: Palgrave.

Lacey, T. A., & Wright, B. (2009). Occupational employment projections to 2018. *Monthly Labor Review., 132*(11), 82–123.

Lancaster, L., & Stillman, D. (2002). *When generations collide*. New York: Harper.

Leibold & Voelpel. (2006). *Managing the aging workforce*. Germany: Publicis Corporate Publishing and Wiley.

Leonard, D. (2004). *Why don't we know more about knowledge? MIT Sloan Management Review.14-18*.

Leonard-Barton, D., & Swap, W. C. (2005). *Deep smarts: How to cultivate and transfer enduring business wisdom*. Boston: Harvard Business School Publishing Corporation.

Li, W., Liu, X., & Wan, W. (2008). Demographic effects of work values and their management implications. *Journal of Business Ethics., 81*, 875–885.

Luo, Y. (1997). Guanxi: Principles, philosophies, and implications. *Human Systems Management, 16*, 43–51.

MacKinnon, M. &Wheeler, C. (2011, April 29). China's future: Growing old before it grows rich. *The Globe and Mail*.

Manton, K. G. (2008). Recent declines in chronic disability in the elderly US population. *Annual Review of Public Health., 29*, 91–113.

Metlife. (2009). *MetLife emerging retirement model study: A survey of plan sponsors*. New York: Metlife. Accessed on http://www.metlife.com/assets/institutional/services/cbf/retirement/EmergRetireModel-Study.pdf.

MetLife. (2009). *Study of employee benefits trends: Findings from the 7th annual national survey of employers and employees*. New York, NY: Metropolitan Life Insurance Company. Retrieved on Nov 2, 2011, http://whymetlife.com/trends/downloads/MetLife_EBTS09.pdf.

Moen, P., & Kelly, E. L. (2007). *Flexible work and well-being study: Final report*. Minneapolis, MN: University of Minnesota. Retrieved on Nov 2, 2011, http://www.flexiblework.umn.edu/FWWB_Fall07.pdf.

Mowday, R. T., Steers, R., & Porter, L. (1979). The measurement of organizational commitment. *Journal of Vocational Behavior, 32*, 92–111.

National Association of Professional Employee Organizations. (2008). *Report on the February 2008 survey of small businesses: Small businesses outpace larger ones in planning for impact of aging workforce*. Alexandria, VA: NAPEO. Retrieved on Nov 2, 2011, http://www.napeo.org/newscenter/research.cfm.

Park, S., & Luo, Y. (2001). Guanxi and organizational dynamics: Organizational networking in Chinese firms. *Strategic Management Journal, 22*, 455–477.

Parise, S., Cross, R., & Davenport, T. H. (2006). Strategies for preventing a knowledge-loss crisis. *MIT Sloan Management Review., 47*(4), 31–38.

Pavalko, E. K., & Henderson, K. A. (2006). Combining care work and paid work: Do workplace policies make a difference? *Research on Aging., 28*, 359–379.

Pei, X., & Xu, Q. (2011). Old age security, inequality, and poverty. In S. Chen, J. L. Powell, S. Chen, & J. L. Powell (Eds.), *Aging in perspective and the case of china: Issues and approaches* (pp. 133–149). New York: Nova Science.

Pew Research Center. (2010). *Millennials: A portrait of generation next: Confident. Connected. Open to change.* Washington, DC: The Pew Research Center. Reterived on Nov 2, 2011, http://pewsocialtrends.org/assets/pdf/millennials-confident-connected-open-to-change.pdf.

Pitt-Catsouphes, M. &Smyer, M. (2007). *The 21st century multigenerational workplace.* (Issue Brief 9). Chestnut Hill, MA: The Center on Aging and Work/Workplace Flexibility, Boston College.

Pitt-Catsouphes, M., Sweet, S., Lynch, K., & Whalley, E. (2009a). *Talent management study: The pressures of talent management* (Issue Brief No. 23). Chestnut Hill, MA: Sloan Center on Aging and Work at Boston College. Reterived on May 7, 2011, http://agingandwork.bc.edu/documents/IB23_TalentMangmntStudy_2009-10-23.pdf.

Raines, C. (2003). *Connecting Generations.* Menlo Park, CA.: Crisp Publications.

Ranstad (2008). *The world of work 2008.* Rochester, NY: Harris Interactive, Inc. Reterived on Nov 2, 2011, http://www.us.randstad.com/2008WorldofWork.pdf.

SHRM. (2011). SHRM Poll: Intergenerational Conflict in the Workplace. *Society for Human Resource Management, April, 29,* 2011.

Sweet, S., & Moen, P. (2006). Advancing a career focus on work and family: Insights from the life course perspective. In M. Pitt-Catsouphes, E. E. Kossek, & S. Sweet (Eds.), *The work and family handbook: Multi-disciplinary perspectives, methods, and approaches* (pp. 189–208). Mahwah, N.J.: Lawrence Erlbaum.

Sweet, S., Pitt-Catsouphes, M., Besen, E., Hovhannisyan, S., & Pasha, F. (2010). *Talent pressures and the aging workforce: Responsive action steps for the retail trade sector* (Industry Sector Report No. 3.1.0). Chestnut Hill, MA: Sloan Center on Aging & Work at Boston College. Reterived on Nov 2, 2011, http://www.bc.edu/research/agingandwork/meta-elements/pdf/publications/TMISR03_Retail.pdf.

Thilmany, J. (2008). Passing on know-how: Knowledge retention strategies can keep employees' workplace-acquired wisdom from walking out the door when they retire. *HR Magazine,* June. Reterived on Nov 2, 2011, http://findarticles.com/p/articles/mi_m3495/is_6_53/ai_n27503950/.

Tian, C. (Ed.). (2008). *China's social security system.* Beijing: Foreign Language Press.

Today's Research on Aging. (2010). China's rapidly aging population, *Population Reference Bureau, No. 20.* July.

Toossi, M. (2009). Labor force projections to 2018: Older workers staying more active. *Monthly Labor Review., 132*(11), 30–51.

US Census Bureau (2010). The next four decades: The older population in the United States, 2010–2050. *Current Population Reports 25–1138.*

U.S. Government Accountability Office (2007). *The challenges and opportunities of demographic change in America.* GAO-07-1061CG. Washington DC: US Government Accountability Office. Reterived on 2, 2011, http://www.gao.gov/highrisk/agency/dol/.

Watson Wyatt Worldwide. (2004). *Phased retirement: Aligning employer programs with worker preferences.* Washington DC: Watson Wyatt Worldwide.

WorldatWork. (2011). *Survey on workplace flexibility.* Reterived on Feb 28, 2011, http://www.worldatwork.org/waw/adimLink?id=48160.

News, X. (2010). Gray hair reemployment shakes the Chinese elders' traditional lifestyle. *Nov, 28,* 2011.

Xinhuanet. (2011. May 3). *Economy threatened by aging demographic in China.*

Xu, Q. (2009). *Mind the gap China.* Boston College: The Sloan Center on Aging and Work.

Yip, W., & Mahal, A. (2008). The health care systems of china and India: Performance and future challenges. *Health Affairs, 27*(4), 921–932.

Zemke, R., Raines, C., & Filipczak, B. (2000). *Generations at work: Managing the clash of veterans, boomers, xers and nexters in your workplace*. NY: Amacom.

Zeng, Y., Liu, Y., Zhang, C., & Xiao, Z. (2004). *Analysis on determinants of healthy longevity in China*. Beijing: Peiking University Press.

Zhan, H. J. (2011). Elder care in China. In S. Chen & J. L. Powell (Eds.), *Aging in perspective and the case of China: Issues and approaches* (pp. 161–173). New York: Nova.

Chapter 4
Pensions and Social Assistance: The Development of Income Security Policies for Old People in China

Yuebin XU and Xiulan ZHANG

Abstract Current income security policy in China includes two polarized systems: social insurance and social assistance. Social insurance consists of five categories of contributory social insurance programs in areas such as old-age pensions, unemployment, medical care, workers' injury and compensation, and maternity benefits. Social assistance is composed of a series of government funded, means-test benefit programs, providing assistance to individuals and households falling into absolute poverty for the maintenance of a basic living, medical care, education, housing, or other needs. For most of the time since economic reforms started in the early 1980, the major efforts of the Chinese government have been focused on establishing an insurance-based social security system. While social insurance has been extended substantially in recent years, covering people in both formal and informal sectors and the rural population, levels of benefits vary markedly between people under different schemes. Social assistance began to play an increasingly important role during the mid-1990s in response to emerging poverty in the cities and persistent poverty among the rural population. However, due to the absence of more broad-based policy interventions particularly in health and education, the effects of social assistance in reducing elderly poverty tends to be limited. This chapter described the development of income security policies for old people in China, including pensions and social assistance.

Keywords Old-age pensions • Social assistance • China

Y. XU (✉) • X. ZHANG
School of Social Development and Public Policy, Beijing Normal University, Beijing, China
e-mail: xuyuebin@bnu.edu.cn

S. Chen and J.L. Powell (eds.), *Aging in China: Implications to Social Policy of a Changing Economic State*, International Perspectives on Aging,
DOI 10.1007/978-1-4419-8351-0_4, © Springer Science+Business Media, LLC 2012

Introduction

Since China's economic reform started in the early 1980s, social insurance has been seen by both the government and academics as the most appropriate tool to finance social security for old people in China. On the one hand, social insurance is viewed as matching the spirit of individual responsibility generally thought to be compatible with a market economy that China's reform has been moving toward (Chow & Xu, 2001). On the other hand, it was also necessitated by the then urgent need to facilitate the restructuring of state-owned enterprises, which has been a crucial part of China's reform. As such, for most of the time following the inception of the reform, the major efforts of government in reforming the social security system have been focused on establishing various employment-related social insurance schemes, including old-age pensions, unemployment benefits, medical insurance, and workers' injury and compensation benefits. Programs were first set up for the urban and formal sector and then gradually expanded to the rural and informal sector. In terms of the old-age pension system, currently it consists of four separate schemes, catering respectively for enterprises, government organs and public institutions, rural population, and urban residents outside the workplace. More recently, an increasing number of cities have developed retirement pension schemes for rural–urban migrant workers, either attached to urban employee pension schemes or created as separate pools. Pension reforms have been focused mainly on the one for enterprises in the cities and old-age pensions for the rural population, while others were launched only recently on an experimental basis.

Entering the mid-1990s, a shift of emphasis was made from social insurance to social assistance in response to emerging poverty in the cities and persistent poverty among the rural population. A means-tested social assistance scheme, the Minimum Living Standard Guarantee System (dibao), was first established nationwide in the cities in 1999 and extended to the countryside in 2007. The scheme provides cash assistance to families with incomes falling below the local poverty line. While it is not a program designed specifically for old people, it has turned out that a substantial number of beneficiaries are old people, mostly those out of coverage of the pension system. In the rural areas, a traditional social assistance program, named the "five guarantee system" (wubao), has remained a major program for old people deprived of family support. In the cities, in recent years there have been an increasing number of cities setting up universal social pensions for old people reaching a certain age. As such, current formal income security policies for old people in China include two polarized systems: old-age pensions and social assistance for households living in poverty. While the former covers mostly the urban and formal sector, the latter is a safety-net program to catch those people who are deprived of social insurance coverage and fall into absolute poverty. In 2009, China had a total number of 167 million old people aged 60 years and above, occupying 12.5% of the national population (National Bureau of Statistics, 2011). Among them, roughly 89 million lived in the countryside and 78 million in the urban areas.[1] In 2010, around 92 million or 55% of old people received pensions, including 63 million urban and 29 million rural pensioners (Ministry of Human Resources and Social Security, 2010). On the other end of the social security system, in 2009 there were about 15.35 million or 10% of old people aged 60 years and above receiving dibao, including 3.35 million or 5% of nonagricultural and 12 million or 12% of agricultural elderly population, respectively[2] (Ministry of Civil Affairs, 2011). In terms of payment rates, average pension payment increased from less than 600 Yuan per month in 2001 to about 1,300 Yuan per month in 2009, with replacement rates dropping from around 60 to 50%.[3] For old people living on social assistance, the monthly benefit rates were 165 Yuan for the urban elderly and 64 Yuan for the rural ones in 2009 (Ministry of Civil Affairs).

The following sections of this chapter will first give an account of the reform processes in China's pension systems and the resulting products, first for the urban and then for the rural population. This is followed by description of the development of social assistance, focusing on its relation with and impact upon the aged. The chapter concludes with an assessment of the policy-making process and economic security policies for old people in China and suggestions for further efforts from the government.

Pension Reforms for the Urban Population

The first nationwide pension system in China was established in 1951 when State Council issued Regulations on Labor Insurance, which applied to all types of workunits including state-owned enterprises (SOEs), government organizations, public institutions, and social organizations all over China. Benefits covered pensions, medical care, workers' injury and compensation, maternity benefits, and a number of temporary relief programs. In 1955 a separate system was created for employees in government organs and public institutions. In 1958 the two systems were combined into a single system, and in 1978 it was divided again into two systems, one for employees in enterprises and another for those in government organs and public institutions. This divide has continued into the present period as the major pension systems in China. Both systems were urban-based and operated as pay-as-you-go, and defined benefit schemes, covering mainly the public sector.[4] One difference

---

[1] Figure for rural elderly population was calculated based on *China Population Statistical Yearbook - 2010*.

[2] Including roughly seven million dibao and five million "five guarantees" recipients (Ministry of Civil Affairs, 2011).

[3] Calculated based on *China Labor Statistical Yearbook – 2010*.

[4] The public-ownership economy of China during the plan era consisted of two subeconomic systems: state-owned and collectively owned economy.

between the two schemes was that the pension system for government organs and public institutions was financed with central avenues and administered by individual work-units while that for enterprises was both financed and administered by individual enterprises—insurance by the work-units. Being a noncontributory benefit system in nature, employees from government organs and public institutions are not required to make individual contributions during the years of employment. Upon retirement, he/she is paid with different levels of benefits based on his/her wage levels and structure prior to retirement and the number of years of employment. Usually, a retiree from a government organ can get 90% of the basic wage if he/she has worked for 30 years. In 2009, the system covered 15.24 million employees and 4.6 million retirees from government organs and public institutions (China Labour Statistical Yearbook, 2010). In 2004 average annual pension benefits were 9,715 Yuan, which were 16,532 Yuan for retirees from government organs, 14,911 Yuan for retirees from public institutions, and 8,081 Yuan for retirees from enterprises (China Labour Statistical Yearbook, 2005). While sources of official data after 2005 are unavailable, it is widely acknowledged that the disparities between the different systems in pension benefits have been increasing. In recent years, there have been wide claims from the general public as well as attempts of the government for reforming the pension system for government organs and public institutions, but progress has been slow.

As such, reform of the pension system in China has been focused on the one for employees in enterprises, which started with in the early 1980s when rapid increases in the number of retirees relative to that of employees in SOEs had resulted in huge pension burdens in the state sector, leading to a "pension crisis" for most SOEs. Between 1978 and 1985 the number of retirees increased fivefold, and pension costs rose from 2.8% of total wages for urban employees to 10.6% (World Bank, 1997). As such, early pension reforms were implemented in the form of local experiments, which were intended to relieve SOEs of pension burdens through social pooling across enterprises. By the mid-1980s a variety of locally administered pension schemes were set up. The schemes were mostly based on counties or districts, differing widely in both design and administration. Because pooling was limited mainly within the state sector, its effects in reducing pension burdens of SOEs were limited. In addition, wide diversity in pension benefits and contribution rates as well as in pension burdens across enterprises frequently led to disincentives for well-off enterprises to pay contributions to the pools while the poor ones were unable to pay contributions (Chow & Xu, 2001).

Nationwide reform of the pension system for enterprises started in the early 1990s and was completed by the mid-2000s. Between 1991 and 2005, the central government issued a number of documents, which gradually brought the system into the current shape. Specifically, State Council Document 33 of 1991, entitled Resolutions on Reform of the Pension System for Enterprises, called for the establishment of a provincially unified social pooling system covering all types of employees and enterprises. Different from the previous pooling across enterprises, Document 33 required workers to make individual contributions to the pools. This was followed by State Council Document 6 of 1995, which decided to turn the pay-as-you-go system into a defined-contribution and partially funded one: a social pooling plus a funded individual account. This change was made based on several widely assumed advantages of the funded approach, including efficiency, equity as well as effectiveness in combating population aging (Chow & Xu, 2002). It was held that the previous pay-as-you-go system was designed mainly to meet the needs of employees in the state sector. If employees in the non-state sector are included in such a system, inequity will result, as they would become net contributors. On the other hand, a funded approach is viewed to be not only being able to reflect the market economy principle of self-reliance and individual enterprise that China has been encouraging since the economic reforms, but also would motivate employees from outside the public sector to participate in the system.

In practice, State Council Document 6 proposed two plans in which social pooling and individual accounts could be combined with different emphasis, and allowed local governments to choose between them or design their own plans based on their local situation. This led to wide variations in contribution rates, the size of the individual account, and benefits across the nation, depending on local ideologies about or interpretations of the principle of combining efficiency and equity. Such a widely variegated national pension system was not only difficult to administer by central government, it also posed problems of portability and gaps in benefits across and between localities. To correct the problem, State Council issued Document 26 in 1997, Decisions on the Establishment of a Unified Pension System for Employees in Enterprises, proposing to make three "unifications" to the system in the years to come: First, the contribution rates were unified to be 20% of the total wage for enterprises and 8% of the wage for individual contributions. Second, the size of the individual account was unified to be 11% of a worker's wage, into which individuals contribute 8% and enterprises 3%. Finally, the method of providing benefits was unified. Retirement benefits were to consist of a basic pension and an individual account pension, taking into consideration both the wage and the accumulated savings in the individual account. A worker who reached retirement age after contributing for at least 15 years was to get a monthly basic pension equivalent to 20% of the average local wage in the year prior to his/her retirement, plus a monthly individual account pension equivalent to the funds in the individual account divided by 120 (months).

In the subsequent years, reform of the urban enterprise pension system was mostly focused on adjusting the benefit structure and size of the individual account. In 2000, State Council revised the pension scheme again through its Document 42, which contained three major changes: First, the size of individual account was reduced from 11% of the wage to 8% to be based solely on individual contributions. Second, funds in individual accounts were to be managed separately from social pooling funds. This separation was intended, on the one hand, to make individual accounts into fully funded funds to be managed through the capital market, and on the other hand to prevent social pooling from borrowing money from individual accounts. Third, to compensate for benefits reduced due to reduction in the size of individual accounts, the level of the basic pension was raised by relating it more closely to the years of contribution in excess of 15 years. A worker reaching retirement age after contributing 15 years would get a monthly basic pension equivalent to 20% of the average local wage. If the worker has contributed more than 15 years, he/she will get another benefit from the basic pension for each additional year of contribution until the total benefit reaches 30% of the average wage. After several years of experimentation of the revisions in three provinces in the northeast of China where pension burdens were particularly high, the benefit structure and the size of the individual account were confirmed in 2005 in State Council's Decisions on Perfecting the Pension System for Employees in Enterprises, which required nation-wide adoption of the revised system starting from 2006 on. The Decisions also made some modifications in the benefit formula. The amount of pension from the individual account is determined with reference to life expectancy, retirement age, and interest accumulated in the account. A retiree is able to receive a monthly payment equivalent to the funds in the individual account divided by 139 (life expectance at age 60) instead of being divided by 120 (months). The amount of the basic pension was made to take into account both the average and individual wage prior to retirement as well as the number of years of individual contribution. That is, a worker reaching retirement age can get a monthly payment of 1% of the mean of the average wage plus the indexed individual wage for each year of individual contribution. Since then, the major efforts of the government have been to expand the coverage to the informal sector and turn the individual accounts from notional to fully funded accounts.

The pension system for urban residents began in local trials in 2009 and 2010 in a number of cities such as Beijing, Chongqing, Xi'an, and Kunming. The experiment was proposed to expand to cover 60% of the cities all over China in 2011, expecting for nationwide adoption by 2012. According to State Council 2011 *Guidelines for Establishing Pensions for Urban Resident*, eligible participants include all urban residents aged 16 or above other than current students, who are not engaged in wage-earning activities and are covered by urban pension schemes for employees. The system operates in the same approach as the New Rural Pension System (NRPS), except that individual contribution rates were divided into ten levels ranging from an annual amount of 100 to 1,000 Yuan for participants to select. Both central and local governments provide subsidies for the schemes. The central government bears full funding for the basic tier for counties in the middle and western regions and 50% for those in the east region, and local governments are required to subsidize at least 30 Yuan for each participant in 2011. Benefits also include two tiers—a basic pension and individual account pension. The basic pension is funded with subsidies from the central government and matching funds by local governments, while individual contribution and subsidies by local governments for each participant both goes into the individual account. A participant reaching 60 years after contributing 15 years can get a monthly minimum of 55 Yuan from the basic pension plus a monthly sum equal to 1/139 (based on life expectance at age 60) of the total funds accumulated in the individual account. For currently old people aged 60 years and above, benefit from the basic pension is a universal benefit and does not require individual contribution. Local governments are allowed to determine the structure of both individual contributions and benefits according to the local context.

Pension Reforms for the Rural Population

In the countryside, concern for rebuilding the rural social security system came to the attention of the government in the mid-1980s. In the "Seventh Five-year Plan" passed in 1986, the government first proposed to build "a socialist social security system with Chinese characteristics." Population aging and the implementation of the single-child family policy, which also contributed to the acceleration of population aging, were among the widely acknowledged imperatives for building the rural social security system. Again, priority was given to the establishment of old-age pension schemes. This was also facilitated by the fact that an increasing number of rural labor force were employed in nonagricultural production particularly in the mushrooming village and township enterprises (VTEs). In many of these places, most of the labor force worked in VTEs, and their family farm plots were often managed as "collective farms" by village committees. As such, the mid-1980s began to see an increasing number of locally initiated rural pension schemes in the wealthier localities, which were largely a modified version of the urban schemes.

The implementation of the rural pension schemes also went through a process of experimentation and promotion, as was the case with other reform policies. As early as in the early 1980s a variety of community-sponsored and managed rural pension pools appeared in the rich regions where employment in VTEs was the dominant source for farmers to earn income. The schemes differed widely in terms of funding, management, coverage, and benefits. In 1987, the Ministry of Civil Affairs was put in charge of investigating the schemes and "exploring rural grass-roots social security" by State Council (Ministry of Civil Affairs, 1987). After several years of experiment, MCA began to promote the establishment of "county-level unified rural old-age pension scheme" in economically developed areas in the early 1990s, and most of the former community-based schemes were merged into the new county-level schemes.

The funding of the rural pension schemes was based on the same principle as that of the urban enterprise pensions. That is, individuals, collectives, and the government shared the responsibility for financing. According to the Basic Plan of Rural County-based Pensions issued by the Ministry of Civil Affairs (1992), the schemes were run on a voluntary basis. Eligible participants included all farmers aged between 20 and 59 years old, who can decide both the rates of individual contributions (ranging from a monthly sum of 2 to 20 Yuan) and the method of contributing them (monthly or in lump sum). Collective economic entities or villages were required to subsidize the scheme with a maximum limit of 50% of individual contribution out of the collective income. However, the government's share of the financing responsibility was in the form of preferential taxation policies—it allowed collectives to use profits for subsidizing the scheme before taxation. Finally, fund reserves were required to invest in government bonds. Special sections were set up within the local civil affairs department to manage the schemes, with 3% of the funds being allowed to be used for administrative costs.

With the enthusiastic promotion of the Ministry of Civil Affairs, the county pension schemes were rolled out across the country rapidly in the following years. By the end of 1997, over 90% of the counties had implemented the schemes, with 74.5 million farmers participating in the schemes and 0.61 million old people receiving pensions (Ministry of Civil Affairs, 1998). Following the establishment of the Ministry of Labor and Social Security in 1998, the management of the rural old-age pension schemes was handed to the new ministry, which was given the responsibility of designing and managing social insurance schemes for both urban and rural areas. Afterwards, the number of counties implementing the schemes began to decline steadily following the 1998 Asian Financial Crisis, which led to a negative assessment of the government of the financial viability of the community pools as a means to provide rural old-age security. Most schemes particularly in the less-developed regions came to a halt. By the end of 2006, the number of participants declined to 53.74 million, with 3.55 million old people receiving pensions (National Bureau of Statistics, 2006).

The resumption of the rural pension schemes began cautiously in the early 2000s. In the 2002 CCPC 16th Plenary Session the central government encouraged only those localities with sound economic conditions to explore the establishment of rural old-age pensions, medical insurance schemes, and minimum living standard guarantee schemes. A number of wealthier localities took the lead to establish new rural pension schemes subsidized with government revenues. Nationwide experiment of this new model began in 2007 following the CCPC 17th Plenary Session which encouraged all local governments to experiment on government subsidized rural pensions. The experiments finally led to the launch of the NRPS in 2009 when State Council issued Document 32 "Guidelines for Experiment of the New Rural Pension System." The new scheme adopted the partially funded approach similar to the urban enterprise pensions, which consists of two tiers of benefits: a basic pension and an individual account. The basic pension is set at a minimum of 55 Yuan per person per month. For current old people aged 60 and above, this tier of benefit is a universal benefit and does not require individual contribution. The benefit from the individual account is a monthly sum equal to 1/139 (based on life expectance at age 60) of the total funds accumulated in the individual account. A major departure from the old rural pension scheme of the 1990s is that the central government shares a substantial proportion of the funding: It provides cash transfers to subsidize local governments for the basic benefit tier. In particular, the central government bears full funding for the basic tier for counties in the middle and western regions and 50% for those in the east region. Local governments are required to subsidize at least 30 Yuan for each participant in 2009. The document also proposed that the new schemes cover 10% of rural counties in 2009 and gradually cover all counties by 2020. Central transfers have provided not only incentives for local governments to establish the new schemes, but also necessary financial support for localities in the middle and western regions which are mostly economically constrained. Between 2009 and 2010, the number of counties implementing the new schemes increased from 320 to 840, participants from 87 million to 102.77 million and pensioners from 15.56 million to 28.63 million (Ministry of Human Resources and Social Security, 2009, 2010).

Experiment for Pensions for Rural–Urban Migrant Workers

This patch of pension system caters specifically to rural–urban migrant workers. Local trials such as those in Guangdong, Shenzhen, Beijing, and Shanghai where rural migrant workers have been concentrated, started mostly around the turn of the 1990s following the passing of Labor Law in 1994 and State Council 1999 *Temporary Regulations on the Collection of Social Insurance Premiums*. While the labor law stipulated that all workers including contract and self-employed ones participate in local pension schemes where they work, the *Regulations* provided specific methods for the inclusion of migrant workers in the enterprise pension system. In the subsequent years, various policy guidelines and methods were provided by the central government to channel the efforts of local schemes, which varied considerably across cities. In 2009, the Ministry of Human Resources and Social Security issued *Methods of Rural Migrant Workers Participating in the Pension System*, which contained the basic blueprints for the implementation of the system. Funding is shared between individual participants and enterprises. Enterprises contribute 12% of the wage, and individuals contribute 4–8% of the wage. Similar to enterprise pensions, pension benefits consist of a basic pension and an individual account pension. If a worker has contributed for 15 years or above, he/she is able to receive a monthly payment of both basic pension and individual account pension. In case the worker has contributed for less than 15 years, there are two methods: If he has participated in NRPS, his account is to be transferred to NRPS; if not, he will be paid with the money accumulated in the individual account in a lump sum. Similarly, local governments are allowed to determine the structure and levels of both individual contributions and benefits according to the local context. By 2010, the schemes covered a total number of 32.84 million rural residents (Ministry of Human Resources and Social Security, 2010).

Social Assistance for the Urban Population

During the planned economy era, China had a government-funded social relief program which was established in the early 1950s for childless old people outside a work-unit named "Three Nos" households, that is, people without work ability, family caregivers, and sources of income. The program provided cash and/or in-kind help in order for them to maintain a basic living. Because most residents had a work-unit to take care of their needs, few people were in need of this support, making social assistance a negligible element in the social security system.

Since the mid-1990s, poverty became an increasingly disturbing issue in the cities. There were many factors leading to the phenomenon of what was termed "new urban poverty"—increasing numbers of urban residents falling into poverty. Apart from rapid increases in the number of unemployed or laid-off employees due to the restructuring of SOEs, the incomes of current employees and retirees were also affected due to generally poor economic performance in most SOEs. Many SOEs had difficulty in delivering pensions for retirees and financial assistance for laid-off employees.[5] In some SOEs even wages for current employees could not be guaranteed. People may fall into poverty even though they had a job, pension, or other benefits. Furthermore, as increasing numbers of urban labor force shifted from the state to non-state sector where social security policies were virtually absent, this rendered the coverage of the ongoing social insurance schemes extremely limited. Finally, due to increased commodification of many basic social services such as education and health care previously provided by work-units to their employees and family members free of charge, urban families were faced with huge financial burdens, leading to the phenomenon of "poverty due to high medical or educational expenses." As most of the newly poor came from among laid-off employees from SOEs, social stability was threatened.

Obviously, the social insurance schemes failed to function as was expected. Thus, toward the end of the twentieth century the attention of the central government was drawn to social assistance as a tool to deal with the problem of urban poverty. This led to a general shift of emphasis in social security reform from social insurance to social assistance. In the following years, a series of social assistance programs were set up throughout the country, first in the cities and then expanded to the rural population. The schemes include dibao and a number of supplementary assistance schemes for medical care, education, housing, and a variety of other preferential policies such as in heating. Among them, dibao is the most important program which also provides the basis for eligibility for other benefits. The scheme was first introduced in 1993 in Shanghai, followed by voluntary trials in a number of wealthier cities, and extended nationwide in 1999 following State Council's Regulations on the Establishment of a Minimum Living Standard Guarantee System for Urban Residents (urban dibao), which required all cities to establish dibao for its urban residents before the end of 1999 and spelled out some specific requirements with regard to the eligibility and financing of the scheme. In brief, dibao was designed as a means-tested benefit program financed out of government revenues. It provides cash assistance for households with per capita incomes falling below local poverty lines (or social assistance lines), which are determined mostly at city or county levels based on the standard budget methods and also taking into consideration local financial capacity. An eligible household receives the difference between the total benefits eligible (the assistance line times the number of persons within the household) and the total household incomes. The objective of dibao is to provide a "safety-net" to households falling into absolute poverty. In practice, its benefits cover mainly food, clothing and a few daily necessities such as fuel, electricity, and water.

---

[5] Assistance for laid-off employees was a temporary benefit program jointly funded by SOEs, local governments, and unemployment insurance funds. It was provided to them to maintain a basic living for a maximum period of 3 years. In 2003 this benefit was merged into the unemployment insurance scheme, and the laid-off employees began to be treated as the unemployed.

The implementation of urban dibao witnessed the increasing role of the government in delivering social welfare in the past decade. The earlier trial schemes were set up by local governments based on local conditions, varying markedly across cities in terms of financing, levels of benefit rates, and eligibility criteria. For various reasons, local governments were generally reluctant to take up the burden of financing benefits for poor people emanating from enterprises. The early schemes, therefore, operated mostly in the form of "taking home your own child" approach—dibao provided assistance only to the "three nos" households who were the targets of publicly funded social relief administered by the Ministry of Civil Affairs. For those households affiliated with a work-unit, the role of civil affairs was only to determine eligibility while applicants were required to submit application to the work-units. Upon approval, the work-units were supposed to deliver benefits to them together with their wages. As such, the receipt of dibao benefits for the non-civil affairs targets depended on the financial ability of the enterprises.

State Council 1999 Regulations put an end to local trials by providing a nationally unified basic structure for the program. One important change brought about by the Regulation is the extension of eligibilities from previously only a few categories of people to all urban households with incomes falling below the poverty line. The Regulation, however, continued with the principle that financing of dibao was primarily the responsibility of local governments, and that funding from the central government would only be available when local finance was difficult.[6] This was also changed in 1999 when the central government for the first time allocated 400 million Yuan to subsidize local governments for setting up dibao. In the following years, accompanying increases in the amount of central transfers, the number of dibao recipients also increased rapidly from 2.66 million in 1999 to 4.03 million in 2000 and further to 11.7 million in 2001 (Ministry of Civil Affairs, 2002). It seemed that more money from the central government would result in more people eligible for benefits. Baffled by the figures, in early 2002 the Ministry of Civil Affairs made a national investigation of poor households, which identified a total number of 19.38 million poor people eligible for assistance. It was also realized that reliance on local governments for funding was insufficient, a main reason for the heretofore generally low benefit levels and small coverage across cities. In the following years, central transfers increased substantially. By 2008, total expenditure for dibao amounted to around 40 billion Yuan, of which transfers from the central government reached over 60% of the total costs (Ministry of Civil Affairs, 2008). Along with the increases of central transfers, the number of dibao recipients also increased rapidly from 22 million in 2002 and further to 23.48 million in 11.42 million households in 2009,

---

[6] Starting from 1980, financial responsibilities were decentralized by dividing both revenues and expenditures between the central government and localities. Since then the financing of social programs have been the responsibility of individual localities, including provincial governments, municipalities and counties. Local governments were not motivated in setting up social programs, and they tended to decide on the number of recipients to be covered based on how much money they could afford for the task.

accounting for about 6% of the urban population at an average cash allowance of
CNY 165 per person per month. Among the recipients, around 3.35 million were
people over 60 years old, representing over 14% of the recipients or around 5% of
the urban elderly population (Ministry of Civil Affairs, 2010).

Social Assistance for the Rural Population

During the collective economy period, rural collectives functioned as a welfare
mechanism for the rural population in a similar way to that of the work-units in the
cities. Community-based welfare provisions were made into an integral part of
the collective economic mechanism, covering the basic needs of the commune
members in an all-inclusive way. The number of people who might fall out of the
collective protection was also negligible. Old people deprived of family support in
their later years were taken care of by the collectives through a social relief program
named "Five Guarantees" program (wubao) which was for old people without
family caregivers and sources of income. The reform of the planned economy and
collective safety-net resulted in the absence of social protection for most of the rural
population. Assistance by the government steps in only when family resources are
exhausted. Main efforts of the government were focused on reducing development-
related poverty through a series of geographically targeted antipoverty measures
named Development-based Poverty Alleviation Policies.[7] Although the number of
rural poor fell from 250 million in 1981 to 35.97 million in 2009, based on the
official poverty line of an annual per capita net income of 1,196 Yuan,[8] the remaining
poor, which mostly consist of the elderly, children, the sick, and disabled people,
can hardly benefit from the Development-based Poverty Alleviation Policies, leading
to vast unmet needs for the rural population. Entering the twenty-first century,
increased attention of the government was given to the setting up of rural social
assistance programs targeting the households. Among them, the major ones include
wubao and dibao.

Wubao provides income support and services to old people without family
caregivers and sources of income. Those eligible for Wubao benefits include orphans
aged below 16 and old people who are unable to provide for themselves through
labor and cannot avail themselves of the support of family caregivers, as defined by
State Council Regulations on the Work of Rural Five Guarantees in 2006 (Ministry

---

[7] The project started in the mid-1980s which consists of a number of county-based programs. Poor
counties were chosen by either the national or local governments as targets for poverty reduction.
The main focus was to create income-generating sources and improve living conditions for rural
people, by such measures as improving rural infrastructure, providing employment opportunities
through public works or enterprises funded by state poverty alleviation funds, and organizing the
migration of poor people to well-off places.

[8] "Situation of China's Human Resources", http://news.xinhuanet.com/politics/2010-09/10/c_
12540033_3.htm.

of Civil Affairs, 2006). The program was established in the early 1950s as a collectively financed safety-net for the rural elderly and continued into the reform era as a major welfare policy for the aged. Provisions included five categories: food and fuel; clothing, bedding articles, and pocket money; housing with the basic necessities; medical care, and for those elderly unable to perform the daily activities, a person was to be arranged to provide nursing care; and an adequate burial funeral. While the basic features of the program such as eligibilities and types of benefits remain unchanged, it has undergone several revisions in response to changes in the rural social and economic context following the reforms. In the beginning years of the rural reform, the main efforts of government were to ensure continued provision by villages and townships for wubao households (Ministry of Civil Affairs, 1982). This is mainly because the dismantling of the collective economy following rural economic reforms that started in the early 1980s lead to an abrupt collapse of the former collective welfare programs in many parts of the rural areas. Another reason was that these services tended to be neglected by local government authorities who were preoccupied with the economy (Chow, 1988; Leung, 1990; Tian & Hu, 1991).

One major measure concerning wubao during the 1990s was State Council's Regulations on the Work of Rural Five Guarantees passed in 1994, which laid down the basic principles with regard to eligibility, financing, types and levels of benefits, and methods of providing the service. Wubao was still defined as a collective welfare, but the township government was put into charge of its administration. Sources of funding were reaffirmed as to include the rural collective economy and the "township unified collections and village retainings," whichever was suitable for the local situation. Eligibility for the benefit included a definition of legitimate caregivers which were those persons who bore the legal duty of family support in accordance with the Marriage Law of 1980. Provisions remained unchanged but levels of benefits were left to the discretion of the local government. The only requirement of the central government was that it should not be below the general living standards of the local villagers. The 1994 Regulations met with some difficulties in financing the scheme. The "township unified collections and village retainings," which were originally intended to cover village administrative costs and a limited number of communal welfare items, including care for the needy elderly, soon became abused by the local government as new sources of incomes. Both the items and amount of collections climbed rapidly in the following years, which became an increasingly heavy burden on the farmers as well as a major source of social disturbance in the countryside. For these reasons, throughout the 1990s the central government took various measures to "reduce peasant's burdens." This led to the implementation of the "Fee-to-Taxation" reform policies in 1998, by which the previous practice of collecting fees from villagers was removed and replaced by levying an agricultural tax on the farmers. Further in 2003, agricultural taxation was also abolished. Thus, townships and villages were no longer able to rely on "unified collections" for sources of funding, and the financing of the Five Guarantees program again became a problem.

This led to the second revision of the *Regulations on the Work of Rural Five Guarantees* in 2006, which eventually turned wubao from a collective to state-run scheme. The financing responsibility was taken over by the government in the form of a block grant by the central government for township and village administration, which includes money for financing benefits for the elderly cared separately in the villages. Meanwhile, county government was held to be responsible for the funding and management of the "Homes for the Aged." Thus, wubao was divided into two separate programs with the central and local government sharing the financing responsibility. The Regulations also redefined the benefit levels for the Five Guarantees targets as the average standards of local residents, which are regarded as a progress over the previous vague definition of "general" standards. It also provided a definition of legitimate caregivers, which includes parents, children, and grandparents who bore the legal duty of family support as set by the Marriage Law of 2001. Benefits are grouped into five categories, including food and fuel; clothing, bedding articles, and pocket money; housing with the basic necessities; medical care and a person to be arranged to provide nursing care for those elderly unable to perform daily activities; and an adequate burial funeral. Levels of benefits were left to the discretion of local governments. Under this program, wubao households can be cared for either separately in villages or collectively in "homes for the aged," depending on the physical condition of the elderly or the availability of beds in the homes. In 2009, the program covered 5.53 million rural elderly people, of which 1.72 million were cared for collectively in the homes for the aged and 3.82 million separately in villages, with an average benefit rate of 2,600 Yuan for those in the homes and 1,840 Yuan for the separately cared (Ministry of Civil Affairs, 2009).

The most important cash benefit program for the rural poor population was dibao. Being also a means-tested benefit, rural dibao began in local trials in the mid-1990s in a few economically developed provinces. With the promotion of the Ministry of Civil Affairs, it was adopted by an increasing number of counties in the less-developed regions in the following years. Due to economic constraints, however, most of them dropped out of the dibao experiment and moved to a temporary assistance scheme named Tekun, which provided temporary relief to rural households impoverished by major illness or loss of family labor. Since 2006, the central government began to pay increasing attention to social assistance programs focusing on household-level interventions. Following the 2006 CCPC Sixth Plenary Session, local governments were all encouraged to set up rural dibao. The decision to establish rural dibao nationwide was restated in Premier Wen Jiabao's Government Work Report to the March 2007 National People's Congress, and was announced further in the State Council Circular on Establishing Rural Dibao Nationwide in July 2007, which proposed that dibao would be established throughout the country by the end of 2007.

Inspired by the widely acknowledged success of urban dibao, rural *dibao* was rapidly extended to most counties by the end of 2007. One major stimulus for local governments to adopt the schemes was the availability of central transfers for the program. Before 2007, dibao schemes were almost financed exclusively by local governments, which often shared the funds among different administrative levels (e.g., provincial, prefecture, municipal, county and township governments) and village collectives. In 2007, for the first time, the central government allocated 3,000 million Yuan to subsidize local governments for the schemes, an amount that increased substantially in the following years. The design of the rural schemes is a modified version of the urban ones. In general practice, a poverty line or standard of benefits would first be determined by the county government. Households with a per capita income falling below the poverty line would be eligible for benefits. By 2009, the program covered 48 million rural residents, accounting for 5.5% of the agricultural population at an average cash allowance of CNY 64 per person per month. Among the recipients, around seven million were old people, accounting for 34% of the total recipients (Ministry of Civil Affairs, 2010). If the five million elderly "five guarantee" households were included, the number of rural old people living on social assistance may amount to about 12% of the rural elderly population.

In fact, the proportion of the old population covered by the *dibao* underestimates the magnitude of the elderly poverty. In most places, old people tend to be systematically excluded from eligibility for *dibao* due to the integration of family obligations into social assistance. Based on the household as the benefit receiving unit, social assistance in China functions to substitute family support, which is identified as the main source of support for individuals in need. The general practice is that if old people have adult children, either living separately or under the same roof, the incomes of children would be taken into account in measuring the incomes of their parent household. They are eligible for *dibao* only if their children are proven through a means-test to be economically unable to support them. As current old people tend to have more than one adult child, it is possible for them to have at least one child who has the financial ability to support them. There are also cases that adult children are reluctant to be means-tested even though providing the support to their elderly dependent is a financial burden. This would lead to the denial of the elderly for social assistance. This practice particularly disfavors the rural elderly, who are mostly uninsured but tend to have more children for sources of family support. It is expected that the rapid population aging trends in China in the coming decades,[9] the increasing mobility of family, the ongoing social transformation, and the "4-2-1" family structure resulting from the one child policy will substantially affect the family support pattern in the future. Recently, some localities have circumvented the eligibility criteria to *dibao* based on family and means-testing support by separating the household registration of the elderly from that of their children and thus enable them to apply for *dibao* without considering their adult children' incomes. Many provinces established categorical social assistance or social pension schemes where the *dibao* benefits are paid to old people and people with disabilities on an individual basis instead of the household as the receiving unit.

---

[9] It is estimated that the population aged over 60 in China will reach 248 million or 17.17% of the total population by 2020, growing at an annual average rate of 3.28%, which represents five times the national population growth rate (0.66%). As a result, by 2050, the population over 60 of age will reach a peak with over 430 million or 31% of the total population.

Conclusion

Policy making and implementation in China typifies the Chinese approach to delivering social protection programs. Largely a crisis management process, the development of the social security system has been mainly for the purpose of resolving the urgent problem of the socially threatening consequences of restructuring SOEs (Chow & Xu, 2001). Issues of stability have played the most important role in shaping China's post-reform social security system into the current form. In the course of reforming the pension system, the government has experiment first and adjustments later based on a learning-by-doing approach and advices from international organizations particularly the World Bank. The role of the central government has been limited for the most part to establishing broad guiding principles, while local governments are encouraged to experiment with different solutions or models based on local circumstances and financial capacity (Leung, 2003, 2006). While social insurance has covered mainly those people in the formal labor market particularly those in the state sector, social assistance has been both stringent and insufficient (Leung, 2006). It has proved that the role of the central government in financing social programs is the most important factor in the establishment of an effective social security system. Although the establishment of a social security system has long figured on the agenda of the central government since the mid-1980s, due to the practice of fiscal and welfare decentralization, most of the responsibility for financing social programs was transferred to local governments—a factor which has contributed greatly to the absence of social protection programs in China. Indeed, overreliance on local governments for financing social protection measures for the rural population has not only left vast unmet needs in the poor areas, but has increased regional and rural–urban disparity in social and economic development. Poverty among the elderly is still a significant phenomenon. In 2009, nationally 15.35 million or 10% of old people aged 60 years and above lived on dibao benefit.

With regard to the pension system in particular, the transition from a pay-as-you-go system to a partially funded system could not fail to be a challenging issue for all nations undertaking such a change, because it requires substantial increases in the amount of payroll taxes taken from the current labor force which will further increase costs for enterprises. In addition, the transition has also resulted in considerable liabilities in respect of both old and new workers. This all required that the government should substantially increase its responsibility for covering such deficits. Recent years have seen an important shift of China's developmental priorities from predominantly economic growth to social development, and the central government has taken an increasing role and responsibility for the funding of social programs through central transfers. This is a good indication of progress.

References

China Labour Statistical Yearbook. (2005). Retrieved July 13, 2011, from, http://w1.mohrss.gov. cn/images/2006-11/16/27110316153762520791.pdf

China Labour Statistical Yearbook. (2010). (pp. 428–430). Beijing: China Statistics Press.

Chow, N. W. S. (1988). *The administration and financing of social security in China*. Hong Kong: Centre of Asian Studies, University of Hong Kong.

Chow, N., & Xu, Y. (2001). *Socialist welfare in a market economy: Social security reforms in Guangzhou*. China: Ashgate.

Chow, N., & Xu, Y. (2002). Pension reforms in China. In C. J. Finer (Ed.), *Social policy reforms in China: Views from home and abroad* (pp. 129–142). Aldershot: Ashgate.

Leung, J. (1990). The community-based social welfare system in rural China: mutual help and self-protection. *Hong Kong Journal of Social Work, 25*(3), 196–205.

Leung, J. (2003). Social security reforms in China: Issues and prospects. *International Journal of Social Welfare, 12*(2), 73–85.

Leung, J. (2006). The emergence of social assistance in China. *International Journal of Social Welfare, 15*, 188–198.

Ministry of Civil Affairs. (1982). *A compilation of civil affairs documents*, Unpublished document of the Ministry of Civil Affairs.

Ministry of Civil Affairs. (1987). *Exploring rural social security schemes*. Changsha: Hunan University Press.

Ministry of Civil Affairs. (1992). Trial plan for rural county-based pensions. Retrieved June 4, 2011, from http://w1.mohrss.gov.cn/gb/ywzn/2006-02/15/content_106552.htm

Ministry of Civil Affairs. (1998). Statistical report on the development of civil affairs undertakings in 1997. Retrieved June 4, 2011, from http://cws.mca.gov.cn/article/tjbg/200801/20080100009420. shtml

Ministry of Civil Affairs. (2002). Situation of Dibao in 2002, Unpublished documents by the Department of Disaster and Social Relief.

Ministry of Civil Affairs. (2006). Regulations on the work of rural five guarantees. Retrieved June 4, 2011, at http://www.mca.gov.cn/article/zwgk/fvfg/zdshbz/200711/20071110003488.shtml

Ministry of Civil Affairs. (2008). Situation of Dibao in 2008, Unpublished documents by the Department of Disaster and Social Relief.

Ministry of Civil Affairs. (2009). Statistical report on the development of civil affairs undertakings in 1997. Retrieved June 13, 2011, from http://cws.mca.gov.cn/article/tjbg/201006/20100600081422. shtml

Ministry of Civil Affairs. (2010). *China civil affairs statistical yearbook – 2010*. Beijing: China Statistics Press.

Ministry of Civil Affairs. (2011). Situation of Dibao in 2010, unpublished official documents.

Ministry of Human Resources and Social Security. (2009). Statistical communiqué on human resources and social security undertakings in 2009. Retrieved June 4, 2011, from http:// w1.mohrss.gov.cn/gb/zwxx/2010-05/21/content_382330.htm

Ministry of Human Resources and Social Security. (2010). Statistical communiqué on human resources and social security undertakings in 2010. Retrieved May 25, 2011, from http:// w1.mohrss.gov.cn/gb/zwxx/2011-05/24/content_391125.htm

National Bureau of Statistics. (2006). Statistical communiqué on labor and social security undertakings in 2006. Retrieved May 25, 2011, from http://w1.mohrss.gov.cn/gb/news/2007-05/18/ content_178167.htm

National Bureau of Statistics. (2011). Statistical communiqué on the Third Census in 2010. Accessed June 30, 2011, from http://www.stats.gov.cn/tjgb/rkpcgb/qgrkpcgb/t20110428_402722232.htm

Tian, X. Y., & Hu, W. L. (Eds.). (1991). *The economy of Chinese aged population*. Beijing: The Economic Press.

World Bank. (1997). *Sharing rising incomes: Disparities in China*. Washington: World Bank.

Chapter 5
Rural Old Age Support in Transitional China: Efforts Between Family and State

Xiaomei Pei and Youcai Tang

Abstract This chapter reviews the transformation of old age support in rural China and examines the potential for rural communities to generate and allocate resources for rural old age support in the context of rural development to meet the challenges of decreasing family resources and inadequate state provision. In-depth interviews with elderly people, their families, community leaders, and government officials of three villages, respectively, located in three provinces, provide us with clear insights into existing local institutional arrangements for rural old age support and the role of both government and communities in organizing such programs. They confirm that rural communities are able to generate and distribute resources for old age support, to offer community opportunities for social inclusion through fair flows of resources to promote social harmony and stability, and to accelerate economic growth. The findings of the study imply that there is a need for policymakers to link state efforts toward old age protection to rural community development, and to encourage grassroots efforts in old age support.

Keywords Old age support • Rural development • Resource distribution • Community organization

Introduction

Old age support as a social issue has increasingly become a focus of public attention in contemporary societies as a result of the rapid rise in proportion of elderly people in many societies. However, the issue presents different challenges for developed

X. Pei (✉) • Y. Tang
Department of Sociology, Tsinghua University, Beijing, China
e-mail: peixm@mail.tsinghua.edu.cn

S. Chen and J.L. Powell (eds.), *Aging in China: Implications to Social Policy of a Changing Economic State*, International Perspectives on Aging,
DOI 10.1007/978-1-4419-8351-0_5, © Springer Science+Business Media, LLC 2012

societies and developing societies. While in developed societies old people are usually covered with relatively comprehensive social security programs, the majority of old people in developing societies still rely on family resources, which are significantly undermined by economic transition locally as well as globally.

One difference in terms of providing social support for the aged between the already industrialized societies and the societies that are still in the process of industrialization lies in the difficulty in organizing public programs of social insurance. In industrialized societies where the majority of the population are employed in sectors paying wages and salaries, social insurance based on contributions of both employee and employer serves as an effective approach to public provision for old age. However, social insurance can hardly be organized for people in societies in transition, where the majority of the population is still rural and does not earn a stable income. As the elderly segment of the population increases rapidly in some transitional societies, it is necessary to explore ways of providing social support for the rural elderly.

China has been challenged by the same difficulties of providing social protection for the rural aged population that are often found in developing societies. It has experienced a dramatic demographic transition, indicated by the rapid increase in the number of the aged since the 1980s. By the end of 2008, the total number of those aged 60 and above reached 169 million, making this group 12.8% of the total population (National Bureau of Statistics of the People's Republic of China, 2009). Of the elderly in China, nearly 70% currently reside in rural areas, and very few are covered by existing state old age insurance programs (Xinhua News Net, 2009). At the same time, rural communities in China have experienced a profound social and economic transition, during which family resources for old age support have been substantially undermined.

China's old age security system is fragmented, and the benefits are delivered unevenly between rural and urban residents. While old age insurance schemes for urban residents are compulsory, employment-based, and state-subsidized, the schemes for the rural residents are voluntary, contribution-based, and until 2010 as discussed below received no subsidy from the state (Xinhua News Agency, 2006). A national survey in 2005 indicated that, despite the existence of a national and some local social insurance programs, the distribution of insurance coverage among the Chinese elderly is geographically highly uneven. In urban areas, 57.5% of the elderly are covered by old age insurance or retirement pensions, while in township areas, the proportion covered is 24.8%. In contrast, beneficiaries in rural areas make up only 4.6% of the rural aged. At present, 54.1% of the rural elderly have to rely on their family and 37.9% of them still work to support themselves (National Bureau of Statistics of the People's Republic of China, 2005).

The Chinese government first tried to organize a voluntary old age insurance program for the rural elderly at county level in the early 1990s. However, the lack of a stable income was a barrier to farmers' continued contributions. As a result, a very limited number of such programs in rural areas have been sustained throughout the rapid industrial transition. By the end of 2000, only 61 million rural people, 11% of the rural population, were still participating. The number of elderly benefiting from this insurance was even smaller, and the average level of yearly benefits from the insurance was even lower than that received from social relief programs for people in poverty. As a result, at the beginning of the new century the program was stagnating (Ding & Chen, 2005).

Several reasons have been identified for the failure of this rural old age insurance program. First, it was based on the idea of commercial insurance, with a flat rate contribution required from individual peasants. However, farmers did not have a stable income like urban workers. Moreover, their income was so low that they had difficulty finding extra money to contribute to the old age insurance fund. Even if they joined the program, they tended to keep the insurance at the minimum level. Second, the program was organized so that only those who were wealthier and younger could be included, and those who were relatively poor and older tended to be excluded. Third, even in places where farmers could make contributions, there was a lack of experience among program organizers in managing the funds, and problems of inappropriate fund management were widespread. Last but not least was the lack of a legislative basis for organizing such a program, which led to competition among government sectors for control over the funds, without effective systems for monitoring (Gao, 2003; Huang, 2004; Qiao, 1998a, 1998b).

In the face of these unresolved problems, the Chinese government has recently made efforts to establish a national rural pension supported by public finance, with the aim of meeting the needs of the rural aged population for income protection. The scheme was piloted in 2009 and is being introduced progressively as a component of the nation's old age security system starting from 2010. The scheme provides rural residents of age 60 and above with a minimum monthly income of ¥55.[1] Although it is designed as an insurance scheme, this basic minimum is guaranteed by the state, with contributions from both central and local government.[2] It is still early to evaluate the impact of this scheme on the rural elderly. However, existing studies of the experience of social pension schemes in some developing countries uniformly confirm the positive impact of such programs on poverty reduction among the aged (Barrientos, 2005; HelpAge International, 2006; Johnson & Williamson, 2006; Samson, van Niekerk, & MacQuene, 2006). Public opinion also regards the scheme as a positive step toward institutionalized rural old age security.

On the other hand, the low rate of benefit makes it reasonable to question the program's adequacy. Although old age insurance schemes are established for the purpose of income provision, old age support is also about coverage of health care and opportunities for social inclusion. There is an increasing public awareness of the existence of multiple needs of the rural aged for social assistance and the importance of a fair redistribution of resources at local and community level to supplement the state's efforts. The encouragement of the involvement of old people and organizations in the community becomes increasingly critical for the effectiveness and sustainability of any program of support.

---

[1] The average income of a rural resident in the year 2009 was ¥5,153. See National Bureau of Statistics of the People's Republic of China (2010).

[2] A fuller description of the scheme can be found on the website of The Central Government of the People's Republic of China (2009).

In searching for potential mechanisms for rural community involvement in the organization of support for the elderly, there is a need to answer the following questions. How does the local institutional and cultural environment affect the elderly? How are community resources mobilized and distributed? What is the relationship between various stakeholders in the community? How do the existing community programs for old age protection operate? Our study of three cases of rural aged support practices in different geographical locations in China, generates firsthand knowledge about the reality of rural old age protection in transitional China, and enhances our understanding of the above questions.

We will start with a brief review of the policy debates on the rural old age protection in transitional China. The review will be followed by a detailed description of the three cases, respectively, focusing on their efforts to provide community old age support and on how their practice is shaped by local social and economic transitions. Evidence drawn from the description demonstrates the influence of leadership, organizational legitimacy, shared values of fairness, the involvement of the aged themselves in community affairs, and the possibility of integrating community efforts with the local or state schemes on the levels and sustainability of community provision for the aged community members. The chapter will conclude with an exploration to the implications of the three cases for the development of social provision schemes for the rural aged in societies experiencing rapid economic development.

Rural Old Age Support: Family, State, and Community

Although family remains to be the main source of support for rural elderly in China, many researchers believe it is becoming a supplementary form rather than the mainstay along the changes in economic production and social institution (Li & Yang, 2005; Wang & Xia, 2001). As the ability of the family to provide care for their elderly members has been undermined by the shrinking of family size and tendency toward nuclear family type, families no longer possess the capacity to provide adequate care (Xia, 2003). In fact, family support has already become self-support in many places, turning the elderly into a socially disadvantaged group (Wang, 2004a, 2004b). Moreover, the commercialization of rural life made monetary resources increasingly critical for survival and rural families have been limited in their ability to provide enough cash for the elder members (Miao, 2005).

Research shows that most rural elderly are expecting social support as a main form of support in old age, although family support is still the reality (Cui, 2006). As more and more adult children migrated to the urban areas, there is a growing public expectation of the government to help both urban and rural families and to take action to reduce the risks to old age support (Duan & Zhang, 2007).

Fully recognizing the inadequate public support programs for the majority of the rural elderly, public attention has increasingly been turned to the responsibility of the government on providing universal relief subsidies to older people and their families (Tang, 2007; Zhang & Xu, 2003). Researchers in China usually agree on the need for social support for the rural elderly, but they disagree on whether it is feasible for China to adopt a universal social pension program at this time. Most of them believe that a nationwide pension program is not feasible but that local, flexible program would be (Liu, 1997; Qiao, 1998a, 1998b; Xi, Lu, & Hu, 1996; Yang & Du, 1997).

The experience of the countries like South Africa and Brazil in practice of social pension provides inspiration in thinking about the necessity of a universal noncontributory social program among some policy researchers in China (Lu, 2003, Tang, 1998). After all, Brazil and South Africa are large developing countries that share some similarities with China. Despite considerable political and financial barriers to be overcome to the development of such a program, it was noted that, as a developing country, China is at least in a position to choose a preferred multi-pillar social pension system almost from the start before these obstacles arise (James, 1998).

In an attempt to explore approaches to social protection for the rural elderly in a time of social transition, Long (2007) suggests the establishment of a multi-model system to solve fundamentally the problem of rural aged support. As the task of providing for the rural aged is huge, no single source of support could fund such an endeavor. Only a combination of resources from family, community, and the state could adequately provide welfare to the rural aged (Zhou, 2001). Furthermore, the growth of a market economy has also necessitated a comprehensive social security system for rural population (Jiang, 2003).

Some scholars have divided this three-in-one multi-model system into two subsystems based on a dichotomy of time or space. Some suggest that in developed areas community support should become the mainstay of the three-in-one rural aged support system, while in developing areas family support can remain as the mainstay for a period of time (Zheng & Wang, 2005). Others suggest that currently family support could remain as the mainstay of a rural aged support system with community support and social support as a supplement, while in the long run the social support should become the core system with community support as the main part and family support as supplement (An & Dong, 2002).

Along with the public discussion about community function for old age support, in some relatively developed areas of eastern China, community support activities have already become the mainstay of a three-in-one system and worked rather effectively in some villages. In the view of some researchers, community support in rural areas represents the collective effort of providing for the security of the rural elderly and the expedition to modernization and development of the rural community itself (Cheng & Zhao, 2006; Ye, 2005). With the development of a collective economy as well as the increase of the collective power, community support has a very bright future and much space for development (Liu & Li, 2006). Moreover, community efforts are viewed as a step toward the establishment of formal support systems (Jiang & Zhou, 2003; Yang & Wu, 2003). On the other hand, despite the optimistic vision of the future welfare system that covers the rural population and the general design for a model of multiple sources of support for the rural elderly, there is a lack of clear definition of responsibilities of the community, the family, and the state in transition.

Three Cases of Community Support for the Rural Aged

We introduce in this section three cases being studied to understand the principles and mechanisms of community resource mobilization and distribution, and the organization of community efforts for old age support in rural China during the economic transition. We refer to these villages as Case A, Case B, and Case C. The selection of villages was deliberate, and designed to take into account the different levels of development across the Chinese countryside. Among the three villages chosen, one is a case located in a more economically developed area where community life remains relatively collective. The second is located in a region where the level of development is about the national average, and the third case is located in a region that is relatively underdeveloped. As China has a large territory and is so geographically diverse, we do not want to make any generalization on the prevalence of community old age support in China based on the three cases. Instead, we want to understand the channels through which community resources flow to individual elderly persons, the effects of the flows on individuals and households, the relations of flows to community development, and the economic and social positions of the elderly in the family and community.

Although over half of all Chinese still live in rural areas, it is very difficult to make generalizations about their living standards because huge inequalities exist between those who live in relatively developed areas and those in underdeveloped areas, between those who live along coastal areas in the east and those living inland in the west. These differences are also reflected in the status of the elderly. We can see from Table 5.1 that demographic differences among the three cases are obvious. While the elderly make up 17% of the total population in Case A and 18% in Case B, respectively, the same age group makes up only 9% in Case C. Another obvious difference is in income status. In Case A, all the aged villagers are covered by local and community old age insurance programs. However, the majority of the aged in Case B and Case C are not protected by any old age insurance programs. In the absence of a stable income from social security, old people in these two villages had to rely on income generated from work, where they were capable of it, along with cash and in-kind provision from adult children. In Case B there were well-organized

Table 5.1 Status of the elderly in the villages studied

Case characteristics	A	B	C
Villagers aged 60 and over	142 persons (17% of the total population of 822)	479 persons (18% of the total population of 2,668)	166 persons (11% of the total population of 1,524)
Living arrangement	Almost all live independently	Most independently	Some independently; some with children
Income	Stable and multiple sources	Less stable but multiple sources	Unstable and limited sources
Labor participation	General participation in paid labor	General participation in paid labor	General participation in agricultural labor

paid labor opportunities for the old and subsidies were provided by the village's Association for Old Persons (AOP) to those in poverty. In Case C, there were no organized efforts to support the aged.

This study found that children's resources for supporting their aged parents were limited. They usually had no problem with providing food, clothes, and daily necessities to those aged parents in need of support. However, providing cash to pay for their parents' medical bills and other expenses for their social activities could become a burden hard to bear. This finding has also been confirmed by a survey of the New Rural Cooperative Medical Scheme (NRCMS) in China, where it was found that even with the NRCMS in place, about 20% of the elderly had to pay the bills out of pocket and nearly 38% of them had the NRCMS reimburse only small proportions of the bills, with reimbursement rates declining along with the increase in expenditures (Zhang, Yi, & Rozelle, 2010).

Despite these differences in the proportion of the aged and their income status, old persons across the three villages shared similarities in living arrangements and labor participation. Contrary to beliefs regarding the living arrangements of old rural people with large families, the elderly in these three villages tended to live in a separate household from their adult children. Moreover, whether they were protected by income maintenance programs or not, they tended to participate in productive labor as long as they were physically capable.

Characteristics of the Communities Studied

The three communities vary in terms of size and resources for old age provision. Table 5.2 illustrates the variations in community provision, coverage, funding, and resources generation for old age support. A review of the development of each case in the following sections is intended to provide a contextual background.

Table 5.2 Old age support in the three villages studied

Case support	A	B	C
Provisions	Pension/housing/job opportunities/medical care/long-term care subsidies	Paid job opportunities/ living subsidies	Occasional subsidies for organized entertainment activities
Coverage	All members of the village	Those physically capable; those in poverty	Those socially active
Funding	Budget by village committee	Budget by Association for Old Persons	No budget
Community resources	Profits from village-owned industrial output	Profits from collectively owned and NGO-managed businesses	Profits from farmland leasing

Case A: Integrating Old Age Provision into Community Development

Our first case was located in an area where manufacturing industry had developed rapidly over the past three decades. It was a village of 822 residents, 17% of them aged 60 and over. In the previous three decades, the village had quickly accumulated collective resources through the successful organization and operation of four collectively owned enterprises. In 2008, the village's total GDP was ¥200 million, the average annual income *per capita* was ¥19.3 thousand, and the total income of the village was ¥10 million.

With the funds generated from industrial production, the community leader and his decision-making group developed a collective welfare system that ensured a secure later life for villagers. The village welfare provisions for the aged included pension income, free medical care, a free and furnished retirement apartment, subsidies to families caring for frail elderly members, and opportunities for paid work for physically abled elderly persons. Villagers started to benefit from these village provisions at the age of 55. The elderly had a monthly pension income of ¥300–600 according to age. They could choose to live with children or independently in an apartment in the retirement housing area which was located within the same village. If they were healthy and willing to do some work, they would have the opportunity to get a paid job as a gardener, a street cleaner, or cafeteria helper. If they were sick, they could get free medical care at the village clinic and reimbursement for medical expenses on hospital visits.

Case A seems to confirm the argument that the community provisions for old age support only happen where substantial resources are accumulated through economic development. However, a further examination of the history of the village showed that *the development of community old age provisions started long before the resources were accumulated.* In the early 1980s, when the village's resources were very limited, the community leaders prioritized pension provision to aged community members in their development agenda. At that time, they believed that this was the right approach to the implementation of state policy on family planning. Unlike most of the rural villages in the area, where land and other community resources were privatized, the village had maintained collective production and organization. At first, the pension benefit level was small and coverage was relatively low. With economic growth and an increase in the amount of resources had come a gradual rise of pension benefit levels and the expansion of old age welfare coverage. This was based on giving priority to a fair distribution of the products of economic growth among the villagers. In 2007, community expenditure on pension payments reached ¥611,400, about 6% of the total collective income that year.

The focus on welfare provisions for the aged as part of the village development in Case A benefited from the political determination of the community leaders who played a critical role in constructing community welfare programs, to ensure "wealth for all." Since the 1980s, the village leader, a charismatic figure, has persistently held a belief in collective ownership of the land and other production materials as an approach to collective prosperity. Under his leadership, the village experienced rapid growth in industrial and agricultural production and a collective distribution of the wealth accumulated through economic development. As a result of this welfare provision, the distribution of income and material resources among villagers was relatively even.

Case A also demonstrated *a coordination of government and community support*. As mentioned above, the village is located in a relatively well-developed area compared with other parts of the country. In 1992, the local government started to organize a pension program for farmers, funded by insurance contributions from farmers and subsidies from the local government. The pension from the old age insurance provided elderly people with an income that was stable but relatively small. The organization of the community pension provision in Case A funded by resources generated from collective production effectively supplemented the local government insurance programs. As a result, the individual income of the aged persons in Case A was substantially higher than that of the aged in the surrounding villages, which enabled them to maintain an independent status in their community and families. When the state rural medical insurance program started in 2006, the contribution of the individual villagers to the medical insurance was paid by the collective fund.

Case B: Empowering the Elderly Through Community Organization

Case B illustrates a self-organized effort by elderly community members to create resources for community welfare provisions as part of their effort to contribute to the economic growth of the village. The village had a population of 2,668 people and the average income of the villagers in 2008 was ¥7,506, 20% higher than the average in the surrounding villages. Farmers there were generally involved in vegetable growing and generated income mainly from selling vegetables and other farm products. As a result, their income level had remained low and the community had very limited resources for old age support. Unlike other villages nearby, Case B had a senior citizen organization, the AOP, which sought opportunities for older villagers to live independently while contributing substantially to local economic development.[3] The AOP was initiated and led by a group of people who had retired either from teaching or from commercial institutions in 1992. The village AOP was owned collectively by its members and registered as a nongovernment organization to promote community welfare for the aged. In order to generate resources for community welfare provisions, the village AOP was involved in managing and regulating the village-based wholesale vegetable market. After 16 years of development, the market has become one of the largest vegetable trading centers in the region, providing commercial services to a population of over 100,000 vegetable

---

[3] Associations for Old Persons in rural China are village-based nongovernment organizations aiming at mutual assistance and community welfare provision for the aged. A fuller description of AOP can be found in Gan (2008).

growers from about 1,000 villages. The market also created about 1,400 jobs for local people, and an elderly person who was willing and physically capable in the village usually had an opportunity for a paid job.

The market became the major source of income for the community. Along with the development of the market, the AOP generated funds for starting up 13 service business entities such as vegetable storage, oil supply, inns, and restaurants, two trial programs for vegetable growers—vegetable planting and tropical fruit growing demonstrations—and three workfare programs particularly for old people involved in gardening, freshwater aquiculture, and courtyard production. By 2008, this self-organized association had accumulated assets of over ten million yuan, paid total tax of 17 million yuan to the state and contributed over five million yuan to the village collective. In 2008 alone, AOP generated an income of 13 million yuan.

Older people in Case B benefit from the efforts of the AOP in terms of opportunities for flexible employment and related income, which enhances their ability to be financially independent and secure. They also benefit from the village welfare programs funded by an increasingly sound collective income source. These programs include collective contributions to the state rural medical insurance, known as NRCMS, for all the villagers, welfare subsidy provision to the AOP's 528 members (including elderly people in the surrounding villages), the operation of the old age school and the seniors' centre. In general, the AOP acts as a significant resource for community old age support and other welfare provisions.

The overall increase in villagers' living standards and the specific empowerment of the elderly through organized efforts has led to changes in intergenerational relationships in village families. A villager manager described the gradual formation of a more positive intergenerational relationship within families since the 1990s. It has been noted that many elderly people still lived together with one of their adult children. However, more and more old people chose, and were able, to live independently of their children in a separate household. Like the elderly living in old age apartments in Case A, this independent living arrangement has promoted more positive interaction between generations, since it enables the elderly to relate to their children on a more equal basis.

In Case B, we see how old people can be empowered to maintain an independent life, how a community can develop with the resources contributed by its older members, and how the aged can benefit from their community. In the absence of pension coverage for the majority of elderly villagers, grassroots efforts to mobilize collective resources for old age support by the aged themselves appeared to be especially valuable and effective.

Case C: Creating an Opportunity for Independence

Case C presents a less organized effort by the rural elderly and a lack of community incentives for old age provision. Here, individual efforts were made to seek opportunities for economic independence and living autonomy. Case C is located along the Yellow River in the northwest of China, an area much less developed than coastal areas. Currently, its main economic output comes from agricultural produce, specifically asparagus and apples. The area was the largest base for asparagus growing in China and had eight asparagus processing businesses. This village had a population of 1,524 people, with an average income of ¥4,000 in 2008. It had very limited community welfare provision. About 60% of the elderly lived separately from their adult children and family support was also limited. The elderly generally made a living by working in the fields as long as they were capable.

The unique phenomenon about Case C was the choice of some aged villagers who lived and worked on the bank of the Yellow River about three miles from the village.[4] Encouraged by local policy, some villagers started to cultivate the land on the bank in the 1980s. The land was productive and income was better than working on the land in the village. They built houses on the river bank and lived there for over two decades. We noticed at the time of the study that these villagers were between the ages of 63 and 80. The elderly on the river bank believed that this was a more appropriate place for them to live. They said the air was cleaner and the water softer in comparison to the village. They lived in very modest houses of 20–30 m² with two or three furnished rooms. Some had a TV set while some did not. The houses tended to be surrounded by cultivated land or orchards. All the old people living on the river bank had remained physically and functionally independent. While living independently in their own houses, they have formed a neighborhood that allows for close communication.

Not all the elderly of the village lived on the river bank. About two-thirds still lived inside the village, and the majority of them were actively involved in farming production, including some 25 people aged 80 or over. As the income generated from agricultural production was unstable, the elderly villagers were living very modestly. Old couples tended to live independently in a house of their own. Those who had lost a spouse tended to continue living independently or with one of their children, depending on whether they had adult children in the village, and on their own financial situation and that of their children. Older people had to rely on children for a living when they were no longer able to work. There were problems for those who did not have any income, who were too old to work or whose children were not in a financially sound situation. Many of these people were elderly women. Even if they lived under the same roof, they tended to be neglected by other members of the family.

Community organizations, such as the village AOP, limited their role to mediating in family conflicts and liaising between individuals who were in need of help and the village administration. Unlike the AOP in Case B, which remained economically productive, the AOP in Case C engaged no efforts in economic production and, therefore, had no resources for welfare provisions for the elderly villagers.

---

[4] The shifts in the course of the Yellow River leave a strip of fertile soil along the banks.

Factors for Community Provision

The three cases above provide us with a sample picture of old age support in rural China. With variations in terms of geography, history, and level of development, the three cases show a clear inclination toward independent living among the majority of the aged population. A growing number of elderly choose to live separately from their children when they have a stable income of their own, especially when income can be maintained from social insurance programs. But this separate living within the community does not disengage the old people from their families. On the contrary, it gives both the elderly and the younger generation more free space and more equal status, promoting harmonious generational interaction.

It is well recognized that historically the ethic of *filial piety* has served as a moral basis for intergenerational relations in Chinese families. The tradition still functions in contemporary Chinese families, with adult children being obliged to provide material support to aged parents without resources. The government or community provision of old age income in Case A and B relieved part of the obligation from the children. In Case A, when the aged persons become independent with the support of community resources, there is a change of style in intergenerational contact. Along with an increase in emotional exchange and a decrease in financial exchange, was a change of expectations of their children for support. As one old person in Case A stated: "Filial piety means that children make frequent visits to the aged parents, spend time chatting with them, and provide care in times when they fall ill." In contrast, those who do not have support, either from governments or from the community, like the elderly in Case C, expressed clearly their expectation of financial support in old age from their children.

Although social security provision from the state and local government are critical to ensure independence for the elderly, the community's role is also important in enhancing their independence through generating and distributing resources, organizing welfare benefit delivery, creating opportunities for social and economic participation, and coordinating state and local efforts. Some evidence for the potential of communities for old age provision has been identified from the case studies above and is summarized in the following sections.

The Critical Role of Leadership

We see clearly that *grassroots organizations with trusted leadership are crucial for community old age provision.* Both Case A and Case B provide evidence for the impact of organized efforts on community welfare provision. While Case A relied heavily on a formal administrative organization in building a community welfare system, Case B was found to have an organization run by old people themselves. Both organizations worked effectively under leadership that was trusted by community members, and that was coherent so that organizational efforts could be sustained for a relatively long time. In contrast, a lack of organized old age provision in Case C has been found to be closely related to the dysfunction of the village administrative body as a result of distrust among the villagers. The distrust was caused directly by a lack of accountability of the community leaders in managing collective affairs and the difficulty of the villagers to participate in community decision-making processes. The comparison between these cases makes clear the important role of effective community organizations in mobilizing resources for old age provision.

Community Organization and Its Legitimacy

We have found a positive relationship between community old age provision and community development in Case A and Case B, which suggests that *resources for old age provision can be created and mobilized in communities where organizational efforts have been made.* The two villages started to work on community welfare provision when community resources were still very limited, in the early 1980s and 1990s, respectively. Collective provision for the elderly has been seen by village leaders as part of community development and as a way to share the products of growth. When old age provision is put on the community development agenda, it is in the context of the accumulation of community resources. The impact of old age provision on community development has been reflected in the active involvement and contribution of the protected aged population to the economic activities of the village, as valuable human resources. Thus, a positive interaction between the elderly and the community has been formed. In Case C, we see the opposite occurring: where there is a lack of community protection for the aged, there is also an absence of organized efforts for community development. Disengagement between the village administration and the villagers was obvious and the elderly villagers were left with little support from the community.

We have noticed that organized collective activities in rural communities were legitimized in contemporary China by the reality that land has been owned collectively, rather than privately, by villagers. Collective ownership provided a legal basis for the organization of collective production and consumption. In the last three decades, there has been a shift toward a market environment for rural communities. Our study shows that rural villages have responded to the change differently. Case A and B responded to the external market through active involvement as an integrated collective entity. Compared to the individuals, collective entities tended to be more powerful in market bargaining, and the profits generated by collective trading were higher than what individuals could achieve. The result is that they generated resources from their market activities and made welfare funding available to the villagers.

Here we see the benefit of collective organized action in coping with the economic transition. There is a positive relationship between resource accumulation and collective welfare development in which the aged are protected. On the other hand, we see from Case C that a lack of internal organization in responding to the market environment led not only to a relatively weak position of the individuals in the marketplace in accumulating private resources, but also to the inability of the village to accumulate collective resources for the welfare of the community members. These cases underline the importance of community organization for the welfare of the rural aged Chinese in a time of economic transition.

A Fair Distribution of Collective Resources

A comparison of the three cases has led us to see that a fair distribution of community resources is more important than the amount of resources available in the development of community provision for the aged. In fact, it is not difficult to find wealthy communities in contemporary China without social provision for the aged. Obviously the availability of resources does not necessarily lead to this provision. The cases studied here have identified community values regarding distribution as an important variable in old age provision. We see in Case A a principle of sharing in resource distribution. One of the community leaders said: 'Funds are always limited and there are always many places where funds are requested. We need to priorities enough funds for the most vulnerable, and the aged persons are among them.' In Case B, we also see that the contribution of the aged to the economic growth of the village has been well recognized, and such recognition justifies the community's efforts regarding old age provision. In contrast, the absence of shared values for fair distribution and negligence toward the aged among community members were found in Case C.

The Aged Being a Stakeholder in Community Development

We have found that *the aged are a valuable resource for rural development in general and for community old age provision in particular.* We see in Case A the active involvement of aged villagers in various productive positions and community political activities, such as participating in various meetings for community affairs. In Case B, major economic entities have been developed and managed by the aged under the leadership of the AOP. These village enterprises generated tremendous interest in the village collective and created job opportunities for both community members and the people from surrounding communities. Along with community development is the expansion of old age provision. In fact, we see here that old age support has been provided by the aged community members themselves. Instead of being a burden, the elderly are valuable contributors to the community. Even in Case C, where few organized efforts have been made, many individual old people maintain a productive life. It seems that, if effectively organized, these individual efforts could become a force in collective development.

Creating job opportunities for the aged in the community required a change of attitude among community members toward their aged members, especially among the community leaders. Previously the elderly tended to be treated as vulnerable and dependent. Continuing involvement in productive work proved the value of the elderly as a resource for the family and community. A close look at the phenomenon would reveal again the factor of community organization to the elderly's realization of their value for community development. In Case A and B, the village organization provided relatively appropriate conditions under which the elderly could find a job in a collective setting. As a result, the value of the elderly as resources and their awareness of contributing to the community tended to be maximized. In Case C where the elderly proved their value through independent work, lack of organizational support limited their ability to contribute to community development.

The Potential for Integration of Support

Evidence has been found that integration of community old age provision with state and local government efforts appears to be effective in meeting the needs of the rural aged. We see that in the area where Case A is located, the local government started to implement an old age insurance program for the local elderly in 1992. Nevertheless, the benefit level has been too low for beneficiaries to maintain a basic living. The provision of community pensions as an important supplement in Case A effectively addresses the inadequacy of the local program. A sense of security among the elderly villagers has increased along with their sources of income. In Case B, we see an active coordination between the AOP and the related district government offices, which leads to collaborative efforts of the community and the local government in building a nursing facility for those aged who need care.

The Personalized Cultural Environment for Community Provision

This study of three rural communities in contemporary China provides some clear evidence on the potential of community organizations in delivering social support for aged community members. The realization of this potential is conditioned not only by availability of collective resources, but also by shared principles for resource distribution and organizational leadership. In this sense, communities are having an increasing impact on local welfare provision. At the same time, we also see some limits to community efforts. There is generally a lack of an institutional basis for old age provision even in communities where current provision is relatively comprehensive. Where community decisions on old age provision rely heavily on charismatic leadership rather than on institutional procedures, the sustainability of this provision can be questioned.

While community old age provision reduces the level of dependency of the aged on their adult children, the practice intensifies the dependency of the elderly on community leaders. In Case A, we found that the elderly villagers were more likely than the younger people to express their gratitude to the leader of the village committee. They almost uniformly believed that they owed their welfare to the morality and ability of this leader. In Case B, we found a similar phenomenon of a high level of trust in one community leader by the aged population. Obviously, even in those relatively well organized rural communities, the connection between the elderly and community leaders was still based on traditional ethics and the moral standards of individual leaders. Thus, a charismatic leader, rather than institutional arrangements, had become the guarantee of welfare provision for the aged. As a result, in these villages, we have found a change in the position of the aged from relying on adult children to relying on community leaders for old age support.

If the community organization of provision for the aged found in Cases A and B could be viewed as a model for welfare provision, it appears to be interest-based, rather than rights-based. Decisions on provision were made by the leader, the aged persons as the beneficiaries then displayed gratitude and loyalty, thus strengthening the authority of the leader. In terms of community provision, the model is fragile when a change of leadership happens. Nevertheless, despite the question of sustainability of community provision resulting from a lack of formal institutional arrangements, one should keep in mind that relations in rural communities in contemporary China are still personal rather than institutional. As it takes time to build institutions, how to cope with the cultural environment in organizing community provision for the aged poses a challenge to the policymakers.

In summary, what we have found from the three cases are some obvious factors that affect community efforts to organize old age provision. A trusted community leadership and a well-functioning community organization seem to be crucial factors for community old age provision. Moreover, a shared sense of fair distribution is more important than the level of development in allocating resources for old age provision. It is important to realize that the aged themselves can also act as a valuable resource for community welfare provision, when they are given the opportunity to participate in resource production. In terms of the national organization of old age provision, there exists the potential to integrate community provision into local and state old age insurance schemes to make social support mechanisms more effective. On the other hand, we also identify an obvious lack of institutional arrangements under a legal framework, which challenges the sustainability of existing community efforts.

Policy Implications

This discussion of the role of the community in support of the rural aged has important implications for rural policy development in China. It leads to an understanding of the impact of the local institutional and cultural environment on the life of the rural aged. It also delineates the mechanisms for resource generation and distribution of old age support in rural communities, in which the relationship among some stakeholders is revealed. These findings show that the development of social policies in rural China in the future should at least include the coordination of available resources from various sources, the encouragement of community efforts through funding and information dissemination, and the institutional building of welfare provision to make old age security programs sustainable.

The Importance of a Fair Distribution of Resources

The importance of fair resource distribution for state policymaking is also revealed by our study. Recent decades have witnessed a rapid accumulation of resources by the state, along with rapid economic growth. If a fair distribution can occur at the community level, it should also be possible at the state level. Obviously the policy issue should be one of distributive justice, rather than that of the availability of resources.

In the cases where community provision for the aged is organized, we have found the existence of a relationship based on interest exchange between the elderly and community leaders. The lack of clarification of rights and obligations in the life of rural communities in China is prevalent. As a result, the distribution of collective resources to the aged is highly reliant on personal leadership and fair distribution could be problematic from time to time, as indicated in Case C.

A critical issue lies in the problem of ownership. It appears that community provision for the aged is based upon a shared ownership of community resources. Shared ownership is in fact indicated by the reality that all the villagers are owners in name, and community leaders are empowered to act as the agents of the shared owners to manage resources. In a sense, the individual villagers, although being the owners of collective resources, do not have rights in deciding the distribution of collective resources. They rely heavily on their agent in distribution, and whether it is a fair distribution or not depends entirely on the moral ethics of the agent. In pursuit of a fair distribution of collective resources, it is a matter of urgency to introduce the concept of rights-obligation-based relationships into the rural community.

Geographic Variation in Institutional Planning for Welfare Provision

Our study provides evidence of a lack of institutional basis for existing provision for the elderly in cases where community efforts have been made. This concern also leads to questions about the sustainability of existing community efforts. To make community efforts sustainable, there is a need for policymakers to build institutions for funding support and information dissemination.

This study confirms previous studies (Na, 2007; Wang, 2004a, 2004b; Wang, 2007; Xu, 2001) of the variations of local institutions in rural China. The living conditions of the rural elderly also vary greatly as a result of the differing impacts of local institutions on them. As the concept of institution may include a wide range of social arrangements, attention should be given to the impact of market development and of government financial systems. This approach would be helpful in institutional building for community old age provisions. The impact of market development on the welfare of the rural aged can be seen in the fact that the elderly in rural areas obtain direct opportunities for accumulating resources through participation in the market. At the same time, they can share the results of development through collective redistribution. Case A represents a situation of positive impact for the rural elderly. Located in a rich area along the east coast, the development of businesses provided the local people, including the aged, with employment opportunities. This significantly raised the living standards of the local elderly population. However, elderly people in general are marginalized in the market and the very old have lost their ability to participate in market activities. Perhaps the importance of market development lies not in the work opportunities for the elderly as indicated by Case B, but in the opportunities for organized efforts to accumulate resources for redistribution, as represented by Case A.

Geographic variation is also reflected in the local government financial system that is related to public funding for local old age provision, such as poverty relief and insurance schemes. As levels of economic development are so uneven, with some areas developing much faster than others, local governments vary different in their ability, willingness, and approach to fund local provision for the aged. In relatively underdeveloped areas, such as the village where Case C is located, the local government provides only minimum funds for assisting the poor. The area where Case B is located represents an average level of development. Local government is believed to be in a better shape to generate funds for old age support. However, there is no action from the local government in monetary terms of organizing such provision. In the absence of local funding, community provision for the aged in Case B is organized in a relatively fragmented manner, supportive as it seems to be. In the area where relatively rapid economic growth has been achieved, old age provision is more easily accommodated on the local government agenda. In fact, community provision alone is not enough to secure a financially sound status for the aged in Case A, and the local old age insurance scheme makes a critical contribution to it. Generally speaking, local institutional arrangements have an impact on the wellbeing of the rural aged. Institutional planning for rural community welfare provision should take market and government forces into consideration.

Coordination of Resources for Welfare Provision in Plans for Development

Finally, the findings of this study also echo the argument for the importance of welfare provision as an integral part of rural development (Hall & Midgley, 2004). It appears that rural communities are as active as the state in searching for ways to meet the challenge of old age support, and some local efforts predate state action. The experience accumulated by some rural communities is extremely valuable to state and local governments in building welfare for the aged. To meet the growing support needs of the rural aged, the state should incorporate community organized efforts into its plans. For example, the state social pension program is expected to extend coverage to all the aged in rural China by the end of 2010. The monthly income of ¥55 for an elderly person would be very helpful but not enough to maintain a decent living in many areas in China. Community provision would not only provide the aged person with more sources of income, but also serve as a supplement to share responsibility with the state.

In recent years, the state has been transferring funds to local governments to help the economic transition of rural communities through a movement called "New Rural Construction."[5] Policymakers should seriously consider supporting rural communities to incorporate old age programs into development through funding and information dissemination so that community provisions can be sustained and developed.

The findings from this study also have implications for other developing societies undergoing rapid demographic transition. It is well known that old age support as a policy issue has increasingly attracted public attention as a result of the rapid aging of many societies. However, the issue presents different challenges for developed and less developed societies. While in the former the elderly are usually covered by relatively comprehensive social security programs, the majority of the aged in developing societies still rely on family resources, which are significantly undermined by economic transitions locally as well as globally. While social pensions offered in some transitional societies are viewed as an effective way to address the need for old age support, relatively low levels of benefits and sustainability of funding are always issues in the adequacy of such provision. If this evidence from community organized efforts in rural China can offer any suggestions for policy development for old age provision in developing countries, it is that they can form a relatively effective and supplemental approach to government action.

Acknowledgments We acknowledge the funding of Social Protection in Asia (SPA) policy-research and network-building program which is funded by the Ford Foundation and IDRC and managed by the IHD New Delhi, India, and IDS Brighton, UK, for the study. Part of the content of this chapter was published in *IDS Bulletin*, July Issue 2010, under the title "Old Age Protection in the Context of Rural Development."

References

An, Z. L., & Dong, Y. G. (2002). The discussion about the option of the pattern of providing for the aged in Chinese countryside. *Journal of Northwest Agriculture and Forestry University (Social Science Edition) [Xibei Nonglin Keji Daxue Xuebao]*, 7, 59–63.

---

[5] A political movement in which the state provides funding to rural villages to build infrastructures for the communities. See Xinhua News Agency (2007).

Barrientos, A. (2005). *Non-contributory pension and poverty reduction in Brazil and South Africa, mimeo*. Manchester: Institute for Development Policy and Management (IDPM), University of Manchester.

Cheng, X. L., & Zhao, H. L. (2006). An analytical study of underdeveloped rural areas. *Journal of Social Work [Shehui Gongzuo], 7*, 58–60.

Cui, Y. G. (2006). Empirical research on the rural support for the aged and people's corresponding preference. *Journal of Nanjing School of Demography Management [Nanjing Renkou Guangli Ganbu Xueyuan Xuebao], 7*, 28–31.

Ding, S. J., & Chen, C. (2005). *Security for the aged in rural China during the economic transition period*. Beijing: China Finance and Economics Press.

Duan, S. J., & Zhang, L. Q. (2007). Analysis on the support for the old people of one-child family in rural area. *Xibei Demography Learned Journal [Xibei Renkou], 3*, 108–111.

Gan, M. T. (2008). Rural grassroot organisations and community public life: Rural associations for old persons in Fujian. *Journal of Fujian College of public Administration [Fujian Xingzheng Xueyuan Xuebao], 1*, 17–22.

Gao, H. R. (2003). Why the project "social supporting for elderly in county level" does not work in rural China. *Market and Demography Analysis [Shichang yu Renkou], 9*, 17–21.

Hall, A., & Midgley, J. (2004). *Social policy for development*. London: Sage.

HelpAge International (2006) Why social pensions are needed now. *HelpAge International Briefing Paper*. Retrieved March 9, 2007 from www.helpage.org/Researchandpolicy/PensionWatch/Feasibility.

Huang, X. L. (2004). Rural aged insurance in institutional transition: problems and strategies. *Journal of Shayang Teachers College [Shayang Shifan Gaodeng Zhuanke Xuexiao Xuebao], 6*, 59–62.

James, E. (1998). New models for old-age security: Experiments, evidence, and unanswered questions. *The World Bank Research Observer, 13*(2), 271–301.

Jiang, M. Z. (2003). Model selection of rural aged support system in transitional China. *Journal of Jiangxi Agriculture University (Humanity and Social Science Version) [Xibei Nongye Daxue Xuebao], 3*, 114–117.

Jiang, L., & Zhou, C. L. (2003). An analytical study of rural aged support in China. *Guangxi Journal of Social Sciences [Guangxi Shehui Kexue], 2*, 163–164.

Johnson, J. K., & Williamson, J. B. (2006). Do universal non-contributory old-age pensions make sense for rural areas in low-income countries? *International Social Security Review, 59*(4), 47–65.

Li, M., & Yang, H. (2005). Mode choice of the elderly support in less developed areas in China. *Zhongzhou Learned Journal [Zhongzhou Xuekan], 3*, 74–77.

Liu, W. (1997). Views and suggestions on social insurance for the rural aged. *Journal of Sociological Research [Shehuixue Yanjiu], 4*, 61–63.

Liu, N., & Li, R. (2006). The major approaches of aging security in rural area and the development of rural community service. *Journal of Jining Normal Institute [Jining Shifan Zhuanke Xuexiao Xuebao], 4*, 67–70.

Long, F. (2007). On improving rural family-based support models for the aged. *Agriculture Economy [Nongcun Jingj], 5*, 3–6.

Lu, C. H. (2003). Building a System of Rural old-age Insurance for the Development of a Harmonious Society. *Journal of Guangxi College of Education [Guangxi Jiaoyu Xueyuan Xuebeo], 5*, 105–109.

Miao, H. J. (2005). Analysis and policy suggestion of social support for the rural aged in northeast of China. *Theory Circle [Lilunjie], 2*, 178–181.

Na, W. (2007). How well the elderly in your family: A review of the condition of the rural elderly. *Research on Aging [Laoling Wenti Yanjiu], 1*, 45–47.

National Bureau of Statistics of the People's Republic of China. (2005). *National survey on a sample of 1 per cent of the total population*. Beijing: China Statistics Press.

National Bureau of Statistics of the People's Republic of China. (2009). *China population statistics yearbook 2009*. Beijing: China Statistics.

National Bureau of Statistics of the People's Republic of China (2010) *2009 Statistical Bulletin*. Retrieved November 16, 2010 from www.stats.gov.cn/tjgb/ndtjgb/qgndtjgb/t20100225_402622945.htm.

Qiao, X. C. (1998a). An analysis of social support insurance system for the rural aged in China. *Demographic Research [Renkou Yanjiu]*, 5, 8–13.

Qiao, X. C. (1998b). An analysis of social support insurance system for the rural aged in China. *Demographic Research (renkou yanjiu), 1998*(5), 8–13.

Samson, M., van Niekerk, I., & MacQuene, K. (2006). *Designing and Implementing Social Transfer Programs*. Cape Town: EPRI Press.

Tang, X. Q. (1998). An analysis of social support system for the rural aged in the primary stage of socialism of China. *Journal of Sociological Research [Shehuixue Yanjiu]*, 2, 101–103.

Tang, J. (2007). *Strategies to population aging in rural China (zhongguo nongcun renkou laolinghua de duice)*. Beijing: China Social Science Documentation Press.

The Central Government of the People's Republic of China (2009) *The state council's guidelines for trials on implementing new rural old age insurance*. Retrieved September 20, 2009 from www.gov.cn/gongbao/content/2009/content_1417926.htm.

Wang, H. (2004a). Survey of elderly support in underdeveloped rural areas. *Agriculture Economy [Nongye Jingji]*, 1, 54–55.

Wang, S. X. (2004b). *Research on old age support in china [Zhongguo Yanglao Baozhang Yanjiu]*. Beijing: Hualing.

Wang, X. M. (2007). *Welfare provision for the aged in rural governance [Xiangcun Zhili zhong de Laoren Fuli]*. Wuhan: Hubei People's Press.

Wang, M., & Xia, C. L. (2001). The current state of burden of family support for the elderly in China: Analysis and discussion. *Chinese Sociology and Anthropology, 34*(1), 49–66.

Xi, C. Q., Lu, Z. G., & Hu, Z. C. (1996). Quest for social support system for the rural aged in China in the 21st century. *Journal of Sociological Research [Shehuixue Yanjiu]*, 1, 105–111.

Xia, H. Y. (2003). Investigation and analysis of old-age care for rural old population in Taicang City. *Market and Demography Analysis [Shichang yu Renkou Fenxi]*, 1, 40–53.

Xinhua News Agency (2006, 27th November) Twelve percent Chinese rural labourers participate in old-age pension. *People's Daily Online*. Retrieved April 21, 2007 from http://english.people.com.cn/200611/27/eng20061127_325683.html.

Xinhua News Agency (2007) Constructing Socialist New Villages. Retrieved April 9, 2009 from http://news.xinhuanet.com/ziliao/2006-02/07/content_4146460.htm.

Xinhua News Net (2009) Unexpected speed of population aging in China Retrieved April 9, 2009 from http://news.xinhuanet.com/politics/2009-10/26/content_12331024.htm.

Xu, X. B. (2001). Family support for old people in rural China. *Social Policy and Administration, 35*(3), 307–320.

Yang, C. Y., & Du, G. Z. (1997). Positive analysis of economic and social conditions of construction of social pension insurance plan for the rural aged. *China Rural Observation [Zhongguo Nongcun Guangcha]*, 5, 55–59.

Yang, S. H., & Wu, Y. X. (2003). Common sense in Chinese rural communities and the status quo of family's provision for the aged. *Exploration and Free Views [Tansuo yu Zhengmin]*, 2, 23–25.

Ye, J. (2005). The analysis of a community care model of support for the aged in rural areas of China. *Journal of China Agriculture University (Social Science Edition) [Zhongguo Nongye Daxue Xuebao]*, 1, 37–40.

Zhang, X. L., & Xu, Y. B. (2003). Construction of a developmental family policy in China. *China Social Sciences [Zhongguo Shehui Kexue]*, 6, 84–96.

Zhang, L. X., Yi, H. M., & Rozelle, S. (2010). Good news, bad news: results from a national representative panel survey on China's NCMS. *IDS Bulletin, 41*(4), 95–106.

Zheng, R., & Wang, D. (2005). A study of the current system for agricultural pension security in Zhejiang Province. *Journal of Hangzhou Normal University (Social Science Edition) [Hangzhou Shifan Xueyuan Xuebao]*, 3, 66–69.

Zhou, S. B. (2001). The choice of rural aged support model. *Jiangxi Journal of Social Science [Jiangxi Shehui Kexue]*, 6, 178–180.

Chapter 6
Social Policy, Family Support, and Rural Elder Care

Guifen Luo

Abstract This chapter will focus on the social protection issues of rural residents including the rural elderly. I begin by reviewing the key features of the system introducing a brief developmental history and a description of the context of the system as characterized by the urban bias. Part one of the chapter describes the social protection schemes for rural residents. It illustrates the fact that the rural residents are excluded from the formal social security system. I shall explain the content, nature, and the problems of the rural pension pilot program and the rural cooperative medical scheme. Part two of the chapter explores the nature of government's social policy in terms of the development of policy goals and its implications for both traditional welfare institutions and rural elder care issue. It presents an interpretation on the interaction of social policy, traditional family support pattern, and rural elderly care.

Keywords Modern social security system • Rural community • Rural social security pilot programs • Social policy • Traditional family support pattern

Who Cares: Modern Social Security System or Traditional Family Support Pattern

Formation of the Urban Bias Social Security Provision

From the very beginning of Chinese social security development, partly due to ideology and partly due to the state's capacity, the policy makers limited their concern to the urban areas. During the 1950s, the Chinese government enacted a series of

G. Luo (✉)
Social Security Research Centre of China, Renmin University of China, Beijing, China
e-mail: guifenluo@hotmail.com

S. Chen and J.L. Powell (eds.), *Aging in China: Implications to Social Policy of a Changing Economic State*, International Perspectives on Aging,
DOI 10.1007/978-1-4419-8351-0_6, © Springer Science+Business Media, LLC 2012

regulations on social insurance which were applied in state-owned enterprises (SOEs), government units, public institutions, and nonprofit organizations. The Labor Insurance Regulations, which were initiated in 1951, were amended in 1953 and 1958 and the welfare benefits were further expanded to employee's families. In addition, a separate pension insurance scheme for Civil Servants was introduced in 1952 (Ministry of Labor and Social Security [MOLSS], 2000a).

The insurance schemes were quite extensive but they were only for the urban residents. The benefits included old-age pension, free medical care, and welfare housing allocation. The administration was left to the work units, which became responsible for the running of the schemes (MOLSS, 2000a). Under these regulations, urban employees, who were mostly employees of state enterprises, were in practice tied to their work units but in return they enjoyed generous benefits: the level of old-age pension benefits ranged 50–70% of the claimant's previous wage, depending on the individual's work history; the retirement age was set at 60 for men and 50 for women (State Council [SC] 1951); income support when disabled; free healthcare; virtually free housing; free child care; and variety of heavily subsidized welfare fringe benefits which the work units provided. The urban collective enterprises could offer some social security benefits that were not as good as in the state sector and the generosity and quality of benefits varied according to the financial condition of the work unit. As there were hardly any private enterprises in China at that time, most of the employed urban labor force was in the state-owned and collective firms and therefore, a great deal of urban employees were actually covered by social insurance schemes. Hence, as early as the 1950s China had set up an extensive social security system for the urban residents.

In contrast to the, comparatively speaking exceptionally positive picture in cities, rural China and its massive rural residents got almost nothing from formal social security and rural people were intentionally neglected when the government planned and made social policy. In the countryside, aside from the so-called disaster relief program which aims to mitigate extreme poverty caused by serious natural disasters, the government introduced a marginal social assistance scheme which was limited to the most destitute people, mainly those who had no means of living and no families from whom they could get support (Ministry of Civil Affairs, 2000). Furthermore, the rural social assistance scheme was and still is largely operated at the grass root level and financed by the villagers themselves. This implies that the state or local governments are not, with the exception of the disaster relieve program, responsible for the provision of social security. In fact, the rural areas did not appear to exist the formal social security system. For the rural residents, the main source of protection came from the family.

Since its establishment in the 1950s, the fractional and urban biased social security provision framework has remained more or less unchanged up to now. However, recently some changes have been taking place. While the urban social security reform has been in process during the last two decades and pre-reform occupational-based welfare has been shifted to a new system, as an anticipated and unintended result, the social safety net for urban residents is gradually shrinking; relatively speaking however, social protection for the rural residents, has not improved.

The long-standing urban policy bias has created a dual social and economic structure in China. First of all the dual structure can be seen in the resident's status. According to the household register system, Chinese are divided into two categories, rural residents and urban residents. Rural residents are excluded from getting formal jobs in cities and receiving relevant social security benefits. Second, the dual structure is highly visible on the level of social and economic development. In 2002 (NBOS, 2003), the income of urban households was 3 times that of rural households; the proportion of telephone subscribers among urban residents approached 3 times that of the rural residents. In 2000, life expectancy at birth for rural residents was 5 years less than that of the urban residents (China Information [CI], 2002); the number of people with higher education per 100 persons among urban residents was 18 times that of rural residents, whereas, the illiterate rate among rural people is twice that of urban people (NBOS, 2001c).[1] As a result, part of the government's development strategy and part of the different prerequisites previously existing in urban and rural areas, uneven development has created huge regional gaps. The gaps not only exist in the economic sphere in terms of income disparities, but they also appear in almost all aspects of social life. To some extent, urban China is a rapidly developing society, and some cities are becoming postmodern, high-tech metropolises; whereas rural community is forcefully maintaining traditional features and all the characteristics of underdeveloped societies. The Chinese fragmented social security system developed under this dual socialeconomic structure and reflects the above-mentioned structural and social inequalities.

Social Security Pilot Programs Conducted in the Rural Community

There are two social security pilot programs that have been introduced and guided by the state since the 1990s: the old-age pension program and the collective medical program. The programs emphasize the role of self-responsibility with the intention of not increasing the state's financial burden. The government's role in rural social security affairs, according to the official definition, is to provide "guidance" and "policy support" (MOLSS, 2000a). Thus, these pilot schemes are mainly financed by the individual farmers themselves and they are organized at the grass root level of villages and towns.

---

[1] In 2002 the annual per capita disposable income of urban households was 7703 Yuan, and the per capita net income of rural households was 2376 Yuan. The number of people with higher education per 1,000 persons among urban residents was 89 and that number was 4.9 among rural residents, while the illiteracy rate was 8.25% among the rural people and 4.04% among the urban people (NBOS, 2001c, pp. 49, 51). Life expectancy at birth for rural residents was 69.55 years and 75.21 years for urban residents (CI 2002).

The Rural Old-Age Pension Program

The old-age pilot program for rural residents was initiated in 1991 with the approval of the State Council. It is a voluntary scheme and is effectively working only in economically developed regions. At the national level, the program covered approximately 12% of the total rural labor force by the year 2001(MOLSS and NBOS, 2002). The scheme is based on individual accounts and a full savings accumulation model where the main financial source is individual contributions. The size of the pension follows the defined contribution principle. The contribution rate is set by the Ministry of Civil Affair and it ranges from 24 to 240 Yuan per year (MOLSS, 2000a). In some cases the pension funds are expected to be supplemented by collectives providing that communities or collectives have the fiscal capability to do so. In some developed regions, additional financing comes primarily from the profits of village and township enterprises.

In principle, the rural pension funds are managed at the county level but in practice, the villages, especially in rich areas, run some of them. Since different localities in rural China are at different stages of development, there are a number of patterns in the rural pensions system that vary greatly when it comes to the qualifying conditions, the contribution rates, the benefit levels, and the management frameworks. For example, in Guangdong province, one of the most developed areas in China, some rural pension plans are financed by the bonuses of the village and township enterprises' stock. The stock share is dispensed to each individual villager and increased according to the individual's age. The procedure favors the older villagers considerably (Zhang, 2001).

There are some uncertainties and the future development of the rural pension program is not guaranteed. These uncertainties are associated with the administrative and policymaking spheres. In respect to the policymaking, there are worries about the potential financial burden on the government. Even though the government is totally free of the financial responsibility and the current scheme is operated entirely as a savings accumulated system, the pension funds run may be affected by inflation and other serious long-term political and economic challenges. Should these risks be realized, the government will have to take the final responsibility for the security of the funds. There are some basic solutions when it comes to dealing with these challenges. Some decision-makers argue that the rural pension plans be transferred totally to the markets and instead of being administered by government agencies, the management and supervision should be carried out by commercial insurance companies. Some others insist that old-age support in rural areas should be the task of the family. Thus, the two alternative options for rural pension policy are marketization and familism, or combinations of these measures. The administrative capability is another open question and it is getting worse. Like in the urban insurance schemes, the rural pension funds have suffered from corrosion and loss of legitimacy due to the misconduct of management agencies and local governments (Zhang, 2001; Yue, 2001). The crucial issue for the healthy operation of the scheme has been of how to ensure the security of pension funds. In fact, as a result of these policy debates and the practical problems to overcome, the rural pension pilot schemes have been limited to previously selected localities. The rural pension pilot schemes have stagnated and the number of new pilot areas has not increased since 1998 (Wang, 2002a).

The Rural Cooperative Medical Scheme

In rural China, there was once broad coverage for cooperative medical schemes. Most villages had their own the so-called "barefoot doctors" (nonprofessional medical personnel that received limited primary medical training) and the farmers could get basic medical service for a small patient fee. The main financing came from the individual village collectives that both financed and managed the program. Although the level of medical service provided by the collective medicine was low both in terms of quality and quantity, it still played an important role in improving the health status of the average farmers (Wang, 2001). After the implementation of the agricultural reform in the early 1980s the rural collective economy collapsed and collective welfare no longer functioned. The collective medical programs disappeared—by the middle of the 1980s, with the exception of a few developed areas, the proportion of villages that sustained the cooperative medical schemes decrease from the previous period high of 90–5% and the majority of rural residents lost their healthcare protection.

Like the old-age pension program, the cooperative medical schemes for farmers are run on a voluntary basis. The scheme is mainly financed by individual contributions and subsidies coming from township enterprises (SC, 1997). The benefits cover 15–90% of the medical service costs in different localities, which usually only covers the minor diseases. Thus there is a huge variation between areas. By the end of 1990s, the cooperative medical scheme covered about 6.5% of the total rural population. In addition, there is about 5% of the rural population that is covered by various other medical insurance schemes. All in all, the medical care schemes cover only about one tenth of the rural population (Wang, 2001, p. 282).

Generally speaking, current rural healthcare is chiefly a fee-for-service system. Grass-roots doctors, who are semiprofessionals, conduct most medical activities and the service price is decided by the market. Farmers have to cover the medical service expenditure themselves. When prices of medical services increase rapidly and continuously, the number of farmers who cannot afford medical services increases correspondingly. From 1990 to 1999, the cost of clinic treatment increased five-fold, the cost of the hospitalization increased six-fold, while during the same period the income of the rural labor only doubled. Due to lack of finances, about 40% of the rural patients could not receive treatment and about 65% of the patients could not get the surgery or hospitalization they needed at the end of the 1990s (Wang, 2001, pp. 291–292). For the majority of rural residents, the absence of medical care is the most severe social risk they face nowadays.

Since 1993, as a response to the worsening situation, the Chinese government has tried to reestablish the rural collective medical program. But so far, the results have not been especially convincing. In fact, the failed attempts at reestablishing the rural collective medical program does not come as any surprise. There are several reasons for the failure.

First of all, healthcare resources are distributed extremely unequally between urban and rural China. Since the 1950s, 20% of urban residents consumed 80% of the total national healthcare resources. The most recent official statistics show that inequality continues to exist. The number of rural hospital beds accounts for only 20% of all beds and the professional medical personnel in rural areas is only one quarter of the total national medical professionals (NBOS, 2003). As a result of the lack of public health resources in rural areas, the price of medical services is extremely high, and in addition the cooperative programs in most localities usually do not cover these expensive costs. Therefore, in a situation where there is an extreme shortage of basic healthcare resources, the collective medical program no longer meets farmers' medical needs in the face of new risks of disease and the high demand for medicine. The second reason for the failure is attached to the framework of the program: the state is not willing to take any responsibility for financing the rural collective medical scheme. It is actually a mutual-aid program with lower contribution rates and consequently with lower benefits levels. The result is that for the richer rural residents low collective healthcare means nothing and they prefer to buy commercial health insurance polices to cover the costs incurred from serious illness. However, for most rural residents, the low contribution rate is welcomed while the cost incurred from serious illness is excessive in comparison to the low benefits. Third, precisely as with the pension scheme, there are many administrative problems. The rural collective medical program is run at the grass-roots level and managed by the leaders of the village; there are not enough channels for average villagers to supervise how their contributions are used which inevitably weakens the legitimacy of the program. Moreover, there is an unequal distribution of benefits and a misuse of funds at the microlevel. Taking these factors into consideration, it is no wonder that the majority of villagers declined to join the collective medical program (Wang, 2001).

Finding a scheme that functions seems to be the key to solving rural resident' healthcare problems. In this respect, the government must make much more effort to change the diminishing trend in investment in the countryside and to try to reorganize the rural medical service system. To some extent, the policy makers are aware that rebuilding the rural healthcare system is an urgent issue. In 2001, the central government formulated a reform plan, which aims to establish a rural medical service network (Wang, 2002a). According to the plan, the rural medical service reform would focus on the control of infectious disease, the prevention of noninfectious chronic illness and the provision of basic treatment services for rural residents.

There are some positive changes in the social security situation of the rural residents. The latest development has been the initiation of minimum living standards programs in a few developed localities. The first broad-range program was carried out in Zhejiang province in 2001. The program is a universal social assistance scheme implemented within the provincial areas. The local governments finance the program entirely and the expenditures are shared by government budgets at different levels (Zhejiang, 2001). According to the provincial regulations, in 2001, the average monthly cash benefit was 180 Yuan for urban resident and 95 Yuan for rural villagers.[2]

As a whole, the state has not played the role of welfare provider in rural areas at all. In fact the government has never promised guaranteed social security for the massive number of rural residents. The two social security pilot programs for rural residents are in fact individual saving accounts rather than fully fledged public social security arrangements. In fact, due to the nonexistence of financial foundations, and the lack of trust in the schemes among the farmers, it is too optimistic to expect these schemes to play a key role in the affairs of rural residents social security.

In sum, the Chinese social security system does not provide any protection or benefits to rural residents, it is essentially a medium for the social exclusion of rural residents. In addition to other rural–urban inequalities, social security has created a new kind of social inequality between rural and urban residents. Due to lack of formal social protection, the traditional measures, principally families and individual labor capacity, have remained the primary source of security for rural residents including rural elderly.

Interaction of Social Policy, Traditional Family Support Pattern, and Rural Elderly Care

From the perspective of industrial functionalism the development of the modern social security system is determined by the process of modernization or structural differentiation (Wilensky, 2002). For instance, Flora and Heidenheimer (1981, p. 8) interpret the welfare state "as a general phenomenon of modernization, as a product of the increasing differentiation and the growing size of societies on the one hand and of processes of social and political mobilization on the other." According to Flora and Heidenheimer (1981, p. 8), the growth of welfare state "is an answer to the growing needs and demands for socioeconomic security in the context of an increasing division of labor, the expansion of markets, and the loss of security function by families and other communities." Meanwhile, Wilensky (2002, p. 211) insists further that "the welfare state is one of the great structural uniformities of modern society." Based on an examination of the development of social security in the 22 richest countries, including 19 rich democracies and three communist regimes, Wilensky argues that although they vary greatly in civil liberties and civil rights, they vary little in their general strategy for constructing a floor below which no one sinks. The richer these countries became, the more likely they were to broaden the coverage of both population and risks.

---

[2] The local average wage was 1090 Yuan per month in 2000 (NBOS 2001b).

China's case appears not to fit so well into this industrial logic, and the socialist state has never accepted the ideas of the welfare state nor followed the welfare state's practice on the process of industrialization. Instead of the welfare state China has developed an "economic state." According to Chen (1996), the Chinese state is characterized by its economic function. Compared to welfare states which are deeply committed to providing welfare for their residents, Chinese governments concentrate in practice intensely on economic administration and ignore residents' social welfare needs in principle. However, the idea of the "economic state" does not mean that all social security issues are always ignored by the Chinese government, "Actually, once they make up the major economic loss, or if they sense that the social problems will threaten their economic ambition or even political ruling, it is not so hard for them to sharply increase social spending by relocating the emphasis of funds appropriation, though they may still not be interested in the welfare state doctrine" (Chen, 1996, 271). The function of social security in maintaining social order in the context of socioeconomic transformation has been a prosperous field in social policy study. Richard M. Titmuss (1974, p. 48) points out in his notable model of social policy, the residual welfare model, one objective of social security is to function as a means of social control and maintain law and order. Frances Piven and Cloward (1993) analyze the history of public welfare system in the prevention of social disorder in the United States, and argue the role of social policy as a guarantor of social peace. In the Chinese case, it is widely recognized among Chinese policy makers and researchers that social policy can serve as a peace-maker, especially in rapid social transformation processes (Song and Gao, 2001; Zheng 2002). As a consequence, in practice, the primary function of social policy has been to maintain social stabilities (Guan, 2001a, 2001b), and the development of a social security system is to a great extent in response to immediate political and economic pressures (Saunders and Shang, 2001).

If we trace the shape of China's social security system which emerged with the process of state industrialization, we find that the objective of the system always served both the political control and economic development strategy exerted by the government. Although welfare programs have changed greatly during different periods, the general goals of social policy keep the same trajectory in several developed stages; that is, institutionalized social security systems simply provide very limited protection to residents at the level that, in Baldwin's terms (1990), is both the necessary economical minimum and the lowest level as determined by politics. In other words the basic objective of social policy is to maintain social stability and guarantee fulfillment of state's industrial strategy. As in other East Asian countries, social policy in China is constantly secondary to economic policy (Tang and Ngan, 2001). As a part of the state's overall transition, the development of a social security system has been guided by a logic which has had to suit the requirements of state's industrialization strategy or market economy (Luo, 1993). In contrast, improving social justice and providing basic social protection to all residents have never become direct objectives of social policies.

From the beginning of the 1950s, institutional social security was designed to serve the state's rapid industrialization strategy and was heavily biased toward state industrial sectors and only involved groups which were politically and economically important. Therefore, occupation-based labor insurance benefits—the core programs of social security system—became the privilege of industrial workers and state employees in urban areas. This was the norm until recently. In the 1980s, along with the overall economic transformation, the objectives of the social security reforms were adjusted to fit in with the market reform of the state industrial sectors. The primary intention of urban welfare scheme rearrangement was to liberate state enterprises so they could pursue purely economic objectives (Zhou, 2000). Moreover, welfare programs were redesigned to be dovetailed with the aim of improving labor productivity (White, 1998, p. 179). Since the 1990s, as a result of the explosion of social conflict caused by increasing social gaps between well-off and the worse-off groups, the Chinese government realized that proper social policies were necessary to maintain social stability. Therefore, the party-state appears to be more aware of social needs than ever before; however, it is not in a position to concentrate unremittingly on the "unproductive issues"—that is, the welfare of its residents. The general goal of social policy is to concentrate on maintaining political control and economic growth rather than improving residents' well-being or social justice (Guan, 2001a, 2001b; Sun 2002). As a consequence, institutional social security benefits are still socially regressive in the sense that "to him that hath shall be given" that is, the industrial labor and urban residents which constitute a small part of the society "get." The other much larger sectors of the society—rural populations and migrant workers—are excluded from the security system (White, 1998, p. 187).

If we examine the objectives of China's social policy from a developmental perspective, which links economic and social policies within the framework of a planned development processes, it might be at least a reasonable option for the government to maintain a balance between economic development and social protection, even though this may not be the best way to deal with the welfare issues affecting all residents. The developmentalist perspective regards economic development as a desirable and essential element in social welfare (Wang, 2002b; Tang and Ngan, 2001); according to developmentalism, there should be a creation of organizational arrangements at the national level that harmonize economic and social policies within a comprehensive commitment to sustainable and people-centered development (Midgley, 1999). In this sense, some analysts argue that China's social policy programs, in particular the existing social insurance policies, have several advantages: they facilitate economic development and foster social integration. In addition, all things considered, an economically and socially viable social security system will support the economic development of the country in the twenty-first century (Tang and Ngan, 2001). However, this judgment is only from the perspective of the party-state goals and the beneficiaries protected by these schemes. If we keep in mind the fact that the overwhelming majority of the population lack basic social protection, it is hardly possible to draw the same conclusion.

Meanwhile, in the domain of Western social policy studies there are other perspectives attempting to understand the interrelationship between the "economic miracle" and the underdeveloped social protection which have emerged in East Asia. Among them, for instance, there are some that explain a model in which achieving sustained economic growth without proper social welfare is a kind of "social dumping." The social dumping policies refer to the model which intentionally forestalls the development of welfare state schemes in order to keep social benefits low, or to retrench existing schemes during rapid economic development (Esping-Andersen, 1996, p. 261; Hort and Kuhnle, 2000). White and Goodman (1998, p. 18) point out, when assessing the cost of the social protection programs from the perspective of the governments measured in monetary terms, East Asian welfare systems appear to be cheaper than Western equivalents. Some studies demonstrate that some East and Southeast Asian countries have introduced social security programs at a lower level of "modernization" than Western European countries (Hort and Kuhnle, 2000). This argument implies that there is no reality, considering the social dumping policies in those countries. However, in China's case, given the fact that the state is still reluctant to expand institutional social security to the largest sectors of society, which are perceived as politically and economically unimportant after the industrialization initiated half century ago, and the "economic miracle" has still come around in recent decades, the Chinese model stands out as an excellent example for having a "social dumping" policy.

Reality suggests that, from whatever perspective we try to explain social policy development in China, it is the survival of the traditional family networks in rural areas that makes it possible for the party-state to leave most of the social protection issues of the rural households off the agenda, and concentrate welfare resources disproportionately on the urban state sectors. The present study (Chap. 6) illustrates how the institution of traditional family support adapted to the new situation in order to cope with the challenge of decreasing capacity. However, it would be misguided to assume that the existing rural family support system is robust enough to resist social dumping policies indefinitely. Existing family support in rural China, in particular the pattern practiced among migrating-mother households, reflects the complex responses both from the traditional institutions and the rural female workers to the paradox of the polices designed and implemented by the party-state during the industrialization process.

As for the survival of traditional family support: "the inherent nature of traditional family is not considered by the dogmatic communists as congruent with the proletarian idea of selflessness and their ultimate goal of communist society" (Chen, 1996, p. 31). In addition, some basic functions of the family were systemically externalized outside the household by party-state policies, yet the provision of welfare was left totally in the hands of the family. The fact that there are no alternative agencies providing social security in rural society forces the rural households to try to internalize the costs of social security by pooling resources across generations within the family. However, the problem is that the households have lacked the financial strength needed to take the burden. On the other hand, the social dumping policies, which arise as a result of both the ideological and economic objectives the party-state pursues, have meant an intentional delay in the process of the demolition of rural family support networks. The new pattern practiced by migrating-mother households is an example of the survival of the traditional system in the wake of state failure.

When it comes to the issue of the dynamic interaction between the state social policies, elderly care, and rural women's nonagricultural labor force participation, it is widely recognized that when the working woman has to work in modern labor markets, she can no longer walk out the back door, and the modern tension between work and family emerges (Wilensky, 2002, p. 10). In the case of Chinese rural female workers, there are indeed conflicts between women's labor market participation and the care duty for family members, in particular problems concerning dependent elderly. However, the evidence also proves that massive rural–urban migration by female laborers does not necessarily make the situation worse; under these circumstances, female workers' nonagricultural employment injects new elements into the traditional family support system and permits it to continue to be an option in the new situation. With a shaped national goal of industrialization and a social dumping security system having been implemented over the past five decades, China's economic efficiency-oriented social policy has reached a crossroads. These welfare arrangements performed reasonably well until recently in terms of the government's objectives. However, if we look at the real welfare situation at the rural family and elderly level, it turns out that those policies are de facto social dumping policies and lead to crises in rural family support. Under China's policy making framework, however serious the crises, or however strong those structural pressures may have been, neither of the two were transformed into statutory social protection autonomously (cf. Kangas, 2000). It seems that an overall reshuffle of the objectives of social policy and, more profoundly, institutional changes are needed to maintain the sustainable function of family support. The crucial question here is to what extent the policy changes will be manageable.

In reference to reidentifying the objectives of social policy, if we consider that precious ideological and institutional heritage retain an influence despite the fact that "the continued hegemony of the Chinese Communist Party may well require political leaders to control potentially fractious mass constituencies by means of welfare benefits" (White, 1998, p. 194), it is hardly feasible to expect the policymakers to fundamentally be able to shift the basic principle of policy strategy rapidly. In addition, other constraining factors also constrict the trajectory of policy formation. Among them the lack of ability of the government to manage the system and the social trust issue; in other words, the government's moral deficit is the crucial problem when it comes to shaping an effective social policy, and also in the task of leading the reformed social security system soundly.

In the Western welfare state discourse, much attention has been given to the state structure and the state's administrative capacity to carry out reforms (Orloff and Skocpol, 1984; Skocpol, 1992; Immergut, 1992; Evans et al. 1985). A central argument in these state-centered studies is that in explaining cross-national differences in social policies we must not only look at the political power relations but rather analyze institutional and administrative aspects of the state. For example, Orloff and Skocpol (1984) in their comparison of the United States and England argue that due to the lack of efficient state bureaucracy, the planned social policy reforms were undermined in the States, whereas the more centralized bureaucracy in the UK was able to carry through the corresponding reforms. In Western countries, particularly the Nordic countries universal welfare is the norm, there are strong, welfare-oriented governments which are motivated by the idea that one of the important functions of the state is to redistribute income and provide protection for the citizens. Thus, in small, homogeneous Nordic countries it is significantly easier for the central governments to carry out the reforms they wish to (Kangas and Palme, 2005).

This administrative issue is more than true for China and it effectively blocks the implementation of any nationwide social policy program. In China, a strong government and a great number of officials are undeniably present. However, it is a control-oriented framework and the state's activities are more focused on social control than the provision of services and welfare to citizens (Wang [1948]1981; Chen, 1994; China Encyclopedia 1992; Wu, 2003). In the sphere of social security, particularly in rural China, neither the local government framework nor the skills of official's live up to the administrative qualifications needed for successful implementation, not to mention long-term planning and anticipation of the future of social policy programs. The current government administration in rural china is simply not designed to deliver welfare benefits and social services. Besides this general question of government incapacity, there are huge technical problems, such as the absence of an efficient taxation system, lack of management and supervision of social security schemes, effective labor market information systems, and so on. Any major changes in the social security program in rural China will be impossible due to problems caused by these elements. Given the enormous size of the rural population and the vastness of the countryside, creating an efficient system will be a complex and long-term task.

An additional problem is connected to social trust. In China it is a major issue that social insurance funds have been misused. Due to loopholes in fund management, there have been numerous examples of pension funds being misused or diverted by local governments for other purposes, resulting in huge losses to the funds and deficits in financing (Chow and Xu, 2001). In that sense, the party-state administers a corrupt and inefficient bureaucracy, not surprisingly this situation has created an atmosphere of suspicion and undermines social trust. The lack of trust, for its part, means that it is difficult to collect taxes and raise funds necessary for the running of social security schemes.

It is for these reasons that, the radical change in the social security system in order to include the entire rural population is institutionally ruled out in the short- or medium term. And so, what kind welfare programs are affordable and accessible for both the government and rural residents, in particular the elderly, in the foreseeable future? The "family support" policy seems to be the most effective national initiative. This policy strategy should include comprehensive and integrative measures to ensure the family support system is sustentative in the future. Addressing the problem of elderly care in rural China, a more basic need will do some kind of social security equalization between the rural and urban residents. In every respect, the Chinese social security system can be regarded as an arrangement that creates and maintains societal inequalities. It will be a long-term task to set up an equal and universal social security system in China.

In sum, if the capacity crisis of traditional family support in rural China is set to continue with the expectation that China's social policy goal will be adjusted somewhat in line with the idea that "citizens aren't assumed to be too involved with things going wrong" (Giddens, 1998, p. 163), the effective social policy will be implemented in order to provide basic support.

My argument is that, in the Chinese's case, the communist party-state has been playing an overpowering role in the policymaking process. In the field of social security, the party-state's will and capability are two crucial elements in the formulation and reformation of welfare schemes. The analysis indicates that the following aspects of policymaking enable us to extract at least part of the explanation. First, there is an issue of the balance between social stability and being totally inactive. Maintaining social peace is the overwhelming goal of the Chinese party-state. The government will seemingly never be willing to take measures to improve rural residents' welfare situation unless the dissatisfaction of the rural population threatens the political stability. Second, the unwillingness to act on the issue of care of the rural population partly results from the narrow understanding of modern social security's function and purpose. It has in reality simply been accepted in the Chinese political system that rural residents have no political channels through which to strive for their welfare rights and interests in the policymaking process. Third, due to the incapacity of the state to take the responsibility for the welfare of all Chinese citizens, it will be impossible to fundamentally change the current social security system in the near future. The administrative incapacity of the government has effectively blocked any attempt to set up a universal social security system.

Moreover, it is the state's polices that have shed doubt on the continual viability of the family support pattern which practiced in rural community. The Chinese party-state's social policy resembles the East Asian model in which welfare programs are overwhelmingly introduced by those in power rather than as a result of popular demand (White and Goodman, 1998, pp. 14–15). As a result of the enthusiasm of those in power pursuing industrialization at any cost, China has ended up with a social dumping social policy strategy. The goals of these policies are focused on economic growth and political stability, deliberately ignoring rural residents', in particular the rural elderly, urgent need for basic social security. In actual fact, it is those social dumping policies which exhaust the capacity of the traditional welfare institution and make rural family support systems increasingly fragile. As long as social dumping policies are maintained, crises in the rural family support system will exist.

Considering that China's government is bound to encourage an "economic state" at least in the foreseeable future (Chen, 1996, p. 271), it is impossible to expect ground-breaking policy changes in the short- and medium term. The potential adjustment on the objective of social policy should be "the reinforcement of family support." In addition, if social security is recognized in its fullest sense, it is dependent in the final instance not only on governments but also on families and communities, both formal and informal (Jones, 1993, p. 213). Particularly in a country in which the traditional informal systems have not yet been destroyed, a policy reinforcing family support could be a reasonable solution.

References

Baldwin, P. (1990). *The politics of social solidarity: Class bases of the European welfare state 1875–1975*. Cambridge: Cambridge University Press.

Chen, S. (1996). *Social policy of the economic state and community care in china culture*. Aldershot: Ashgate.

Chow, N., & Xu, Y. (2001). *Socialist welfare in a market economy*. Aldershot: Ashgate.

Esping-Andersen, G. (1996). After the golden age? Welfare state dilemmas in a global economy. In G. Esping-Andersen (Ed.), *Welfare states in the transition: National adaptations in global economies*. London: Sage.

Evans, P., Rueschemeyer, D., & Skocpol, T. (Eds.). (1985). *Bring the state back in*. Cambridge: Cambridge University Press.

Flora, P., & Heidenheimer, A. J. (1981). Introduction. In P. Flora & A. J. Heidenheimer (Eds.), *The development of welfare states in Europe and America*. New Brunswick: Transaction books.

Giddens, A. (1998). In A. Giddens & C. Pierson (Eds.), *Conversation with Anthony Giddens, making sense of modernity*. Cambridge: Polity Press.

Guan, X. (2001a). Globalization, inequality and social policy: China on the threshold of entry into the world trade organization. *Social Policy & Administration, 35*(3), 242–257.

Hort, S. E. O., & Kuhnle, S. (2000). The coming of East and South-East Asian welfare states. *Journal of European Social Policy, 10*(2), 162–184.

Immergut, E. (1992). *Political construction of interests: National health insurance politics in Switzerland, France and Sweden, 1930–1970*. New York: Cambridge University Press.

Jones, C. (1993). The pacific challenge: Confucian welfare states. In C. Jones (Ed.), *New perspective on the welfare state in Europe*. London: Routledge.

Kangas, O. (2000), From workmen's compensation to working women's insurance: Institutional development of work accident insurance in OECD countries, ZeS-Arbeitspapier Nr. 10/2000, Universität Bremen.

Kangas, O., & Palme, J. (2005). Social policy in a developmental perspective: The Nordic experience. In O. Kangas & J. Palme (Eds.), *Social policy and economic in the Nordic countries*. London: Palgrave.

Midgley, J. (1999). Growth, redistribution and welfare: Toward social investment. *Social Service Review, 73*, 3–21.

Orloff, A., & Skocpol, T. (1984). Why not equal protection? Explaining the politics of public social spending in Britain, 1900–1911, and the United States, 1880s–1920s. *American Sociological Review, 49*(6), 726–750.

Piven, F., & Cloward, R. (1993). *Regulating the poor: The functions of public welfare*. New York: Vintage Books.

Saunders, P., & Shang, X. (2001). Social security reform in China's transition to a market economy. *Social Policy and Administration, 35*(3), 274–289.

Skocpol, T. (1992). *Protection soldiers and mothers: The political origins of social policy in the United States*. Cambridge: Harvard University Press.

Tang, K.-L., & Ngan, R. (2001). China: Developmentalism and social security. *International Journal of Social Welfare, 10*, 253–259.

Titmuss, R. (1974). In B. Abel-Smith & K. Titmuss (Eds.), *Social policy: An introduction*. London: George Allen & Unwin.

White, G. (1998). Social security reforms in China: Towards an East Asian model? In R. Goodman, G. White, & H. Kwon (Eds.), *The East Asian welfare model*. London: Routledge.

White, G., & Goodman, R. (1998). Welfare Orientalism and the search for an East Asian welfare model. In R. Goodman, G. White, & H. Kwon (Eds.), *The East Asian welfare model*. London: Routledge.

Wilensky, H. L. (2002). *Rich democracies: Political economy, public policy, and performance*. Berkeley: University of California Press.

Works in Chinese

Chen, Zaogang. (1994). Zhengfu zhineng lilun xuyao gengxin' (Renovation of the government function). In Zhireng Zhang & Gongmin Xu (Eds.), *Xingzheng Tizhi Gaige yu Zhuanbian Zhengfu Zhineng (Administration system reform and change of the government's role)* (pp. 224–324). Beijing: Shehui Kexue Wenxian Chubanshe (Social Science Documentary Press).

China Information (2002). Zhongguo Xinxibao Wangluoban (China Information). http://www.zgxxb.com.cn. Retrieved 21 June 2002

China Encyclopedia (1992). *Zhongguo Da Bai ke quan shu Zhengzhi xue Juan (Big encyclopedia of china: Political science)* (Pp480). Beijing, Shanghai: Zhongguo Da Bai ke quan shu Chubanshe (China Encyclopedia Press).

Guan, X. (2001b). 'Quanqiuhua beijingzhong de zhongguo shehui zhengce' (China's social policy in the context of globalization). In J. Tang (Ed.), *Shehui Zhengce: Guoji Jingyan yu Guonei Shijian (Social policy: International experiences and practices in China)* (pp. 19–26). Beijing: Huaxia Chubanshe (Huaxia Press).

Luo, G. (1993). 'Fengxian yu anquan: Shichang jingji zhongde shehui baozhang zhidu' (Risks and security: Social security system in market economy). In Y. Li (Ed.), *Shichang Jingji Xinzhixu (New order of the market economy)* (pp. 185–289). Beijing: Zhongguo Renmin Daxue Chubanshe (Renmin University of China Press).

Ministry of Civil Affairs. (2000). *Zhongguo Shehui Fuli (China's social welfare handbook).* Beijing: Min Zheng Bu Bian (Ministry of Civil Affairs).

Ministry of Labor and Social Security. (2000). *Zhongguo de Shehui Baoxian (China's social insurance handbook).* Beijing: Laodong he Shehui Baozhang Bu Bian (Ministry of Labor and Social Security).

MOLSS & NBOS. (2002). '*2001 Du Laodong he Shehui Baozhang Sheyi Fazhan Tongji Gongbao'(Communique 2001 on labor and social security).* Laodong he Shehui Baozhang Bu & Guojia Tongjiju (MOLSS & NBOS). http://www.stats.gov.cn/tjgb/.

NBOS. (2001a). *2001 Nian Zhongguo Tongji Nianjian (Statistics book 2001).* Beijing: Zhongguo Tongji Chubanshe (China Statistics Press).

NBOS. (2001b). *Zhongguo Renkou Yongji Nianjian 2001(China population statistics yearbook 2001).* Beijing: Zhongguo Tongji Chubanshe (China Statistics Press).

NBOS. (2003). *2002 Nian Guomin Jingjin he Shehui Fazhan Tongji Gongbao (Statistical Communique 2002).* Beijing. http://www.stats.gov.cn/tjgb/. 28 March 2003.

Song, X., & Gao, S. (2001). 'Zhongguo shehui baozhang tizhi gaige: Zhuyao jinzhan, yanjun xingshi yu zhengce jianyi' (Social security system reform in China: Achievement, challenge and policy advice). In Xiaowu Song (Ed.), *Zhongguo Shehui Baozhang Tizhi Gaige yu Fazhan (Report on the reform and development of China's social security system)* (pp. 3–40). Beijing: Zhongguo Renmin Daxue Chubanshe (Renmin University of China Press).

State Council. (1951). 'Zhonghua renmin gongheguo laodong baoxian tiaoli' (Labor insurance regulations). In Xiaowu Song (Ed.), *Zhongguo Shehui Baozhang Tizhi Gaige yu Fazhan (Report on the reform and development of China's social security system)* (pp. 315–321). Beijing: Zhongguo Renmin Daxue Chubanshe (Renmin University of China Press).

State Council. (1997). 'Guanyu fazhan he wanshan nongcun hezuo yiliao de ruogan yijian' (Regulations on rural cooperative medical system). In Xiaowu Song (Ed.), *Zhongguo Shehui Baozhang Tizhi Gaige yu Fazhan (Report on the reform and development of China's social security system)* (pp. 401–404). Beijing: Zhongguo Renmin Daxue Chubanshe (Renmin University of China Press).

Sun Liping. (2002). 'Ziyuan chongxin jiju beijingxia de dicing shehui xingcheng' (The formation of the bottom of the social structure under the background of resources redistribution). *Zhanlue yu Guanli (Strategy and Administration)*, No.1 (2002). Beijing.

Wang Y. N. (Yanan) ([1948]1981), *Zhongguo Guanliao Zhengzhi Yanjiu (The study of china bureaucratic system).* Beijing: Zhongguo Shehui Kexue Chubanshe (China Social Science Press).

Wang, Y. Z. (2001). 'Zhongguo nongcun yiliao baozhang zhidu baogao' (The development of health care system in rural China). In Jiagui Chen, Zheng Lv, & Yanzhong Wang (Eds.), *Zhongguo Shehui Baozhang Fazhan Baobao 1997–2001 (China social security system development report 1997–2001)* (pp. 268–311). Beijing: Shehui Kexue Wenxia Chubanshe (Social Science Documentary Press).

Wang, Y. Z. (2002a). 'Zhongguo nongcun shehui baozhang zhidu bianqian yu pinggu' (Rural social security system transformation and evaluation). In Gongcheng Zheng (Ed.), *Zhongguo Shehui Baozhang Zhidu Bianqian yu Pinggu (Social security system transformation in China)* (pp. 239–274). Beijing: Zhongguo Renmin Daxue Chubanshe (Renmin University of China Press).

Wang Xiaoqiang. (2002b). 'Chengshihua yu jingji zengzhang' (Urbanization and economy growth). Beijing: *Weilai yu Xuanze Canyue Wengao (Views: Future and Choice)*, No. 5.

Wu Jinglian. (2003). 'Jianshe yige gongkai, touming he ke wenze de fuwuxing zhengfu' (Shaping an open, transparent, responsible and service-oriented government). *Caijing Zazhi(Financial Journal)*, 20 June 2003 Beijing.

Zhou Xiaochuan. (2000). 'Shehui baozhang yu qiye yingli nongli' (Social security and the profit of the enterprises). *Shehui Jingji Tizhi Bijiao (Comparative Study on Social and Economic System)* No. 6. Beijing.

Yue, S. (2001), 'Woguo renkou laolinghua qushi jiqi duize' (Aging trend in china and the policy response). *Zhongguo Renmin Daxue Fuyin Baokan Ziliao: Shehui Baozhang Zhidu (china Renda social sciences information center: Social security system)*, No. 5. Pp 13–17. Beijing.

Zhang, Jinchang. (2001). 'Zhongguo nongcun yanglao baozhang zhidu baogao' (Challenges of family support for the old in rural China). In Jiagui Chen, Zheng Lv, & Yanzhong Wang (Eds.), *Zhongguo Shehui Baozhang Fazhan Baobao 1997–2001(China social security system development report 1997–2001)* (pp. 234–267). Beijing: Shehui Kexue Wenxia Chubanshe (Social Science Documentary Press).

Zhejiang. (2001). *'Zhejiang nongmin shouci naru baozhang fawei'* (*The initiation of regulations on minimum living standard in zhejiang rural areas*) (No. 12, 61pp). Beijing: Zhongguo Renmin Daxue Fuyin Baokan Ziliao: Shehui Baozhang Zhidu (China Renda Social Sciences Information Center: Social Security System).

Zheng Gongcheng. (2002). 'Zhongguo shehui baozhang: Zhidu bianqian, pinggu yu fazhan' (Reform of social security system in China). In Zheng Gongcheng (Ed.), Zhongguo Shehui Baozhang Zhidu Bianqian yu Pinggu (Social security system transformation in China) pp. 1–76. Beijing: Zhongguo Renmin Daxue Chubanshe (Renmin University of China Press).

Chapter 7
China's Family Support System: Impact of Rural–Urban Female Labor Migration

Guifen Luo

Abstract Based on interviews with rural–urban migrants in Anhui and Sichuan provinces of China, this chapter focuses on the coping strategies adopted by Chinese rural–urban migrant families to deal with the tensions caused by changes in generational care chains. By illustrating how and why the traditional family support system managed to survive and function after women's emigration, a new pattern of generational contract was identified in response to the lack of support from the formal social security system. The dependent children and the elderly were left in the countryside by the migrating mothers. The able grandparents took care of the grandchildren while the young migrating couples reciprocated by giving their parents financial support, other material help and promises of better support in the future. The chapter demonstrates that this new pattern of informal intergenerational support renewed the capability of family support. It is not a simple replica of the old fashioned preindustrial welfare nexus; at a more general level, the new generational contract can be seen as a model of interaction between tradition and modernity.

Keywords China • Rural–urban female migrant workers • Traditional family support system • New informal generational support pattern

Introduction

Since the late 1970s, China has been undergoing a gradual transformation from a centrally planned command economy to a market-based system; this great transformation brings rapid economic growth and social changes. One significant aspect

G. Luo(✉)
Social Security Research Center of China, Renmin University of China, Beijing, China
e-mail: guifenluo@hotmail.com

S. Chen and J.L. Powell (eds.), *Aging in China: Implications to Social Policy*
of a Changing Economic State, International Perspectives on Aging,
DOI 10.1007/978-1-4419-8351-0_7, © Springer Science+Business Media, LLC 2012

is that there has been an increasing number of rural–urban migrants, including massive numbers of rural women moving from their home villages to distant urban areas to find jobs. By 2003, the number of rural–urban migrant laborers was as high as 114 million, which accounted for over 20% of the total 500 million rural laborers (NBOS, 2004).[1] At the beginning of the new millennium, about 50% of all migrant workers in China were estimated to be women (UNRISD, 2005, 83). Rural–urban migrants shape the largest population movement during peacetime in China and perhaps the largest movement in the recent world history (Robert, 2000).

The dual role of rural women as breadwinners and main caregivers within the family raises interesting questions about the impact of industrialization upon agrarian societies. Women's labor market participation is without exception linked to numerous issues of family responsibilities. In the case of China's rural–urban migration, both social science research and common sense tell us that there are conflicts between capitalist demands for the free movement of labor and traditional family care practices based on Confucian values. Under the current Chinese social security system, the family is supposed to be the main source of social support for rural residents and women play a vital role in this social security arrangement. When more and more rural women move to distant urban areas to get paid employment, the traditional chains of care responsibilities are challenged and changed. This chapter will show that this modern vs. traditional confrontation leads to a number of practical solutions where new circumstances are met with a mixture of reshaped traditional care arrangements and new kinds of generational care contracts. The practices offer some potential solutions for China's rural social security crisis. Therefore, exploring how rural women and their households deal with these conflicts, and what kinds of new social bonds are being created during this process are very important both theoretically and for policy purposes. The knowledge drawn from the chapter will hopefully contribute both to a better understanding of China's social security system, and to on-going discussions on the welfare state from a comparative perspective.

The family has been functioning as an overwhelming support system in many preindustrial societies. If it is true that in most Western countries in addition to the family the church has traditionally also been a welfare provider (Kersbergen, 1995), we could say that in China the family and kin system have for centuries acted as almost the only source of social support. Familism is characterized as the core of the Confucian ethic (Fei, 1947/1985; Liang, 1990; Tu 1998) and a metaphor for Chinese social structure (Feng, 1948/1985; Jin, 1986; Tu, 1996). Family has played a profound role in the formation of the Asian form of capitalism, in power politics, in creating social stability, and in moral education in China (Tu, 1996). Classical Confucian works have provided a strong normative basis for family relationships and intergenerational contracts prevailing in China. These relationships were described as the well-known *Wu Lun*, namely, the five basic relationships: ruler–subject, parent–child, husband–wife, siblings, and friends (Analects of Confucian, 1990, 30/chapter 12.11).

---

[1] Here the term "rural–urban migrant" refers those who were migrated from their home villages more than one month during the year 2003. The number of the migrants was 90 million in 2001 (NBOS, 2004).

Since an individual was located in a well-structured network of human relationships, in *Wu Lun* systems, everybody was assigned duties and obligations, including mutual welfare responsibilities. With respect to narrow kin obligations and generational support patterns, the primary principle is the so-called *Fu Ci Zi Xiao* ethic, that is, father-goodness and children-filial piety. These values and normative principles concerning children's support for the elderly and intergenerational reciprocity have been sustained by Confucian ethics and legitimized by the Chinese authorities via both the formal laws and informal conventions (Xiao, 2001, 174; Zheng, 1997). Filial piety is considered by many people as one of the main moral and ethical principles guiding behavior. The traditional notions of social welfare maintain their influence during the social transformation of the country after 1949 (Chan & Tsui, 1997); the family currently plays the crucial role in welfare provision in rural areas in particular. During the 1950s, the government introduced a Stalinist welfare model in urban society and ran generous social insurance programs for industrial employees (Chen, 1996, 131–133; Hussain, 2000; White, 1998, 178). By contrast, the rural areas do not appear to exist in the formal social security system. For the rural residents, the main source of protection comes from the family.

In social policy research, the important role of the family and familism in the social security system in contemporary China has been emphasized by most, if not all, studies on this subject. This more or less established and accepted distinction has been captured in the notion of the "East Asian welfare model" (Goodman & Peng, 1996, 193; White & Goodman, 1998), or the "Confucian welfare system" (Jones, 1993), characterized by a strong reliance on the family as the locus of social welfare and service delivery. However, there are contested views as to trajectories of social change in this welfare model. Both classical "industrial functionalism" and postmodernism predict changes in the model. In his statement on the convergence theory, Wilensky (2002, 5) draws strong conclusions about the net effect of industrialization on the family itself and its support capacity in modern society: "… the massive structural changes associated with industrialization have brought major changes in family size, composition, functions, and lifestyles." At the same time he declares that "… it also reduces the family motivation and resources to care for aging parents and to meet the risks of invalidism, sickness, job injuries, and other shocks." Against this background, fundamental changes in the traditional Chinese welfare arrangements are to be expected in the wake of industrialization. With this assumption in mind, I explore what has happened to the traditional family support systems in the process of de-agrarianization in rural China, in particular in the context of female migration.

Under the current circumstance in rural China, supporting elderly parents is considered to be the unquestioned duty of adult children, especially sons. It is expressed by the old saying "having sons for old age support." Taking into consideration that rural women usually move to their husbands' families when they get married, the sons and daughters-in-law have traditionally been given the duty of giving support to the elderly parents and parents-in-law in rural areas. Despite the fact that women were subordinated to men in the Confucian "ideal" family (Tu, 2001), under the traditional welfare model women have played a key role in fulfilling welfare responsibilities within the family. However, the deepening and ever accelerating marketization in China over the past 2 decades has pushed more and more rural men to seek nonagricultural jobs in distant cities. This rural–urban migration is changing the profile of the family support system. The traditional family-based welfare arrangement is facing increasing challenges. On the one hand, from the women worker's point of view, there are severe tensions between the desire for the free movement to cities and the demands of traditional family responsibilities. On the other hand, the dependent family members, both the children and the elderly who are left behind in the countryside, are excluded from the formal social protection system and are exposed to the risk of lack of care due to the departure of the main care provider. One link in the chain of care is missing and needless to say this causes problems and tensions that must be somehow solved.

This chapter focuses on the coping strategies adopted by the migrant families to deal with the tensions caused by changes in generational care chains. To be more specific, I try to answer the following questions: How are rural family support systems functioning under the circumstances of female labor migration? What has happened to the traditional generational contract in this process of rapid economic and social change? What kind of welfare arrangement have been developed among the households from which women are migrating in order to meet the care deficits caused by their absence and the new demands brought forth by market forces?

In this chapter, the concept of the "rural migrant worker" refers to a group in the Chinese labor force who move between rural and urban areas, who seek and find temporary jobs in nonagricultural sectors in the urban areas, and who when out of work return to their home villages and engage in farming again. According to China's *Hukou* household registration system, this group of laborers maintains permanent rural resident status whenever and wherever they work. In the Chinese literature, both in scientific analyses and in policy documents, a rural–urban migrant without a Hukou is considered to be a "floating rural laborer" (*Nongcun liudong laodongli* or *Nongmin gong*) instead of a "migrant laborer." In this chapter, I use the term "migrants" in line with the mainstream understanding of this term in migration studies.

The data used in the chapter were collected during 1999 and 2000 in Anhui and Sichuan provinces[2] of China as part of a research project on rural–urban migrant workers in China—"Study on Out-Migrants and Return Migrants" (SOMRM). The project was conducted by the Research Center for Rural Economy, Ministry of

---

[2] Province Profile (as of 2001):

Sichuan province is located in the southwest of China; population of 86 million; it has the fourth largest population of the 31 provincial regions in China; 9,000 ha. cultivated land which accounts for approximately 7% of total national area; 5,250 Yuan/person (about 530 Euro/person) per capital GDP—the highest is 37,400 Yuan/person (about 3,740€/person) in Shanghai and the lowest is 4,200 Yuan/person (about 420€/person) in Gansu province; composition of gross domestic product: primary industry 22%, secondary industry 40%, tertiary industry 38% (national average of composition of gross domestic product: primary industry 15%, secondary industry 51%, tertiary industry 34%).

Anhui province: located in the south-east of China; population of 63 million; it has the eighth largest population of the 31 provincial regions in China; 6,000 ha. of cultivated land, which account for about 5% of total national area; 5,200 Yuan/person (520€/person) per capital GDP; composition of gross domestic product: primary industry 23%, secondary industry 43%, tertiary industry 34%.

Source: NBS (National Bureau of Statistics of China) (2002).

Agriculture, from 1997 to 2001. The author of the chapter was a member of the research team.[3] The data for the project was collected using case interviews and sample surveys. The analysis of this chapter relies primarily on the case interview material.

The interviewing strategy was designed to collect distinct information about out-migration and return-migration experiences at the individual and village level (Bai & He, 2002). Three hundred and forty-four individual informants and 12 village focus groups were selected from four counties located in Anhui and Sichuan. All in all, among the 344 individual interviewees, there were 129 female respondents. In terms of counties, villages, and individuals the cases were selected nonrandomly.[4] Interviews were conducted in 1999 and they were organized on three levels: individual rural workers, village leaders, and county government officials. A semi-standardized interview technique was applied in the all interviews: questions were typically asked of each interviewee in a systematic and consistent order; however, the interviewers were expected and allowed to probe far beyond the answers to the standardized questions (Berg, 1995, 33). At the individual level, the predetermined questions concentrated on interviewee individual and family experiences in the out-migration and return-migration process. A broad range of aspects of migration experiences were involved, mainly including the reasons for out-migration and migrating home, opinions on working and living conditions in cities and the gains, difficulties, problems, and

---

[3] The author of the paper, as a member of the research team, was responsible for the study of the female workers of the project, and with other members of the research team conducted interviews in Anhui province. Preliminary results were published in China in 2001 and 2002, in Finland in 2005.

[4] According to Stanley Lieberson (1992, 115), the choice of cases for study is itself critical and requires great thought about the appropriate procedure for choosing them. A case study approach was applied in an attempt to follow Lieberson's advice. In order to guarantee the representativeness of the data, the scheme of case selection designed by the project team abides by the following rules (He & Bai, 2002):

(1) Cases apportionment: first we selected two representative counties separately from Anhui and Sichuan province, and then we selected three representative villages in each given county.
(2) The criteria of selection for case counties: one requirement was that the number of out-migrant workers accounted for at least 20% of county's total labor force in 1998. There should be a long history of out-migration records compared with other counties and the county should be a typical agricultural county (indicated by gross value of agricultural output) and at the median or average level in terms of economic development (indicated by per capita disposable income) and natural resources (situation about farm land, irrigation, etc.) within the province.
(3) The criteria of selection for case villages: the proportion of out-migrant workers in a village should exceed the average proportion in the given county. The village should also display a relatively long out-migration record. The three villages selected represent different stages of economic development measured as per capita disposable income, and the villages are poor, rich, and middle-income localities. There were 12 sample villages in total.
(4) Within the village, individual interviewees were selected on the basis of a quota sampling procedure according to their migration experiences. Twenty-five interviewees in each village including at least 15 with returnees were conducted. Among the interviewees there were five female returnees. In addition there were five who had never migrated and five relatives to those who were current migrants were interviewed. There were at least seven female respondents in total in each village. In principle, the interviewees belonged to the "labor force," i.e., the villagers were between 16 and 60 years of age for male, and 16 and 55 years of age for female.

changes related to migration. The information on each case usually formed a distinct story, which the respondent relayed to the interviewer. Most of the individual interviews were conducted at the respondents' home.

Information from village and county government officials was collected using focus group interviews. At the village level, focus group interviewing was conducted in each of the 12 villages, and interviews usually consisted of three or two village leaders depending on the village in question. The information collected from the county government officials mainly covered data on the local economic development situation, local policies relevant to rural labor mobility, and so forth.

The chapter is organized as follows: The information presented in the next section will serve as the background in which the story of changes of family support is taking place. Following that, the practices and capacity crisis of family support in rural China which take place in the circumstances of changing socioeconomic background will be illustrated, concentrating in particular on the generational support patterns developed by the female migrating-mother households, and analyzing how this new pattern deals with the capacity crisis of the traditional system. Then, the characteristics of the newly emerging generational contract will be identified, and I shall try to offer some explanations as to why the pattern works. The final section will draw conclusion.

Poverty of Rural–Urban Migrant Families in Cities

Although most researchers have argued that China's social security system was already established in the early 1950s when the first nationwide unified formal social insurance program was introduced (Chow & Xu, 2001; Kong, 2001; Saunders & Shang, 2001; Wong, 1998; Zheng, 1997), until now, there is no nationwide program covering all residents in China. In fact the Chinese social security system is a fragmented system characterized by inequity and incomplete coverage (Saunders & Shang, 2001). Social security provisions have been segmented along the lines of a rural–urban divide, and all schemes are confined to either the urban or rural residents (Hussain, 2000). On the other hand, from the rural resident's point of view, the social security system is an arrangement that is biased toward urban citizens and almost all main social security measures are designed to protect solely the urban labor force and residents. For the rural residents, including a great number of migrant workers, the main source of security comes from their own labor capacity and their families. The formal social security provisions existing in rural China are sparse. Aside from the so-called disaster relief program which aims to mitigate extreme poverty caused by serious natural disasters, the government introduced a marginal social assistance scheme which was limited to the most destitute people, mainly those who had no means of living and no family from whom they could get support (MOCA, 2000). Furthermore, the rural social assistance scheme was and still is largely operating at the grassroots level and financed by the villagers themselves. In fact, the voluminous rural population is not covered by any formal statutory social security system and is excluded from social protection.

During the era of the centrally planned economy, the Chinese government rigidly controlled labor mobility and migration, especially the residence changes from rural to urban areas and job transfers from farming to nonagricultural work. It was almost impossible to migrate from rural to urban areas, except with permission rarely given by the authorities. This control has functioned through the household registration system (Hukou system), which is a unique institutional arrangement that strictly segregates rural and urban areas (Cai, 2003). The Hukou system was set up in the mid-1950s. It functioned as a domestic passport preventing rural residents from entering cities; and rural laborers were excluded from working in nonagricultural sectors.[5] Under this system, an urban Hukou membership was required in order to stay in cities and get employment. The urban Hukou status included a series of social entitlements (Song, Huang, & Liu, 2002) like food quotas, jobs assigned by government, as well as associated social security benefits such as old-age pension, free housing, and free health care. Hukou membership also guaranteed privileged access to urban public services such as education.

Since the economic reforms started in the late 1970s, the system of the centrally planned economy has changed in many respects. Up to the mid-1980s the rigid controls on rural–urban migration were gradually eased. As a consequence of the market-oriented reforms in labor policy in particular, food and housing provision in urban areas, employment, housing conditions, and food supply regulations in cities were gradually changed. Hence, more and more rural laborers are able to move to cities to find employment. This movement led to the so-called phenomena of "the surge of a floating rural labor force" (*Mingong Chao*) (Cai, 2000; Du & Bai, 1997, 2). Until 2000 the central government tried to introduce some policy changes and made attempts to abolish some local discriminatory regulations (Song et al., 2002). However, pervasive legal restrictions for rural–urban migrants still exist. Under current policies, rural–urban migrants have access to some occupations in urban areas, but nevertheless they are just like "foreign laborers" (*Wailai mingong*) working in the cities, and only in very few cases can the rural resident status be changed to urban status. All in all, this means that the majority of rural migrants remain excluded from the urban social security system (Luo, 2000; Song et al., 2002). Despite the fact that most of the rural–urban migrants are no longer self-employed farmers as they have become employed in nonfarming sectors, they are still treated as rural residents in terms of their social identity and continue to be marginalized as secondary workers in the urban labor market. Policymakers still assume that migrant workers have some means of social security because, at least in theory, they own farmland in their home villages. Hence, the majority of migrants are still not covered by mandatory urban social security programs. As a result, migrant workers and their families suffer social security poverty in cities; they do not have social rights—to even basic health care, old-age pension, work injury insurance, or educational rights for their children in cities.

---

[5] The primary purpose of the system was to control rural residents' move to cities. For city people there were no formal barriers to move to countryside.

When we take a closer look at a special group, the rural female workers and their families, the situation regarding social security becomes even worse. Due to the fact that migrants are not covered by mandatory social insurance programs, and employers are not required by law to make any social security contributions, the female migrant workers are left on their own. Moreover, the local city governments regard the migrant workers simply as laborers rather than residents, and therefore they are socially excluded from the surrounding urban society. It is usually impossible for migrant women to live together with their families when in cites. As a consequence, the dependent family members, both the children and the elderly, have to be left behind in the countryside. The case studies we have conducted for this research show that in order to earn the necessary income in the urban labor market, female migrant workers suffer various job security and social security problems. Concerning family care issues, for example, as migrants and their children do not qualify for public services or welfare benefits provided to urban residents, children of rural migrants cannot join urban children's care centers and public schools, unless the parents pay an enormous fee. Usually the fees for child care and education are too expensive for migrants to afford. The majority of our respondents, including male interviewees, complained that they were not financially able to afford such large amounts of money. As a result, most female migrants leave their children in the home villages in the care of the grandparents. In the case that the female migrant workers do take their children with them to the city, instead of sending them to care centers mothers have to take the children who are below school age with them to work and keep an eye on them while they work. It is not unusual to see this kind of scene at rural migrants' tiny business booths in cities. When the children reach school age, the typical process is for the mother and the children go back to the village.

The situation of female migrant workers and family support issue are somewhat contradictory: on the one hand, they are faced with all the risks which accompany *an industrializing* society; on the other hand, their social security is based on traditional safety nets and the crucial issue is whether or how long this situation can be sustained.

The Changing Capacity of the Rural Family Support System

It is widely assumed that the Chinese traditional family support system is a multitiered institutional framework of security (Hou, 1991; Zheng, 1997), which is composed of three lines (Leisering, Gong, & Hussain, 2002, 75). The first line is composed of immediate family support usually provided by the family members who live in the same household or in close vicinity and share a common income. The second line is formed by the extended family, which pools various social risks that go beyond the first security line. The last and more expanded risk pooling institution constitutes the kinship network.

In contemporary rural China, family support systems remain the primary welfare production unit; however, the socioeconomic transformation has nullified two security lines—the extended family and the larger kin network. This change has a strong impact on the capacity of the family support system. In present-day rural China, the capacities of family support are shaped mainly by two factors. The first is the economic resources a family has at its disposal. The main resource is the farmland which a household possesses, factually or nominally, and its products (Chen & Han, 2002; Wen, 1999; Zhao & Wen, 1998). The second factor is the family structure (Wang, 2000; Xu, 2001); both reflect long-term socialeconomic transformation processes on the one hand, and are directly associated with social policy on the other.

First of all, the role that land previously played, as a reliable economic foundation of family support, has shrunk (RTMA, 1995; Tao, 2002). This is linked to the processes of de-agrarianization and marketization. Chinese agriculture has developed in such a way that the income from farming is contributing less and less to farmers' income. It has been a long-term trend that agricultural income takes up a decreasing portion of farmers' total income (Bai & He, 2002, 153; Du, 1997, 179) as one would expect and as has happened historically in many countries. Given the fact that subsistence agriculture barely provides sufficient products and income to guarantee the welfare of all family members, in order to get by, rural households have to find new financial resources to compensate for the decrease in income from agriculture. Nonagricultural activities are usually alternative ways of securing the capacity of family support.

Second, it was commonly assumed that the traditional Chinese family support was directly connected to the extended family and kin systems (cf. Hou, 1991; Zheng, 1997). However, there seems to be a paradox between the existing rural household structure and the support pattern. While the rural family maintains its functions as the primary welfare provider, the traditional household structure has been eroded since 1949—a consequence of the socioeconomic transformations that rural society has been undergoing. Among other socialist restructuring movements that have been carried out in the country, the setting up of cooperatives in the 1950s and the later introduction of communal property eroded the traditional societal fabric and the economic foundation of kinship (Qin, 1999; Xiang, 2001). Although there are different estimates as to the extent to which the extended family de facto existed in traditional China, what is clear is that there are very few instances of such families in contemporary rural China. At the beginning of the second millennium, the size of an average household was 3.7 persons in rural areas and 3.1 in urban areas (CPSY, 2001). Nowadays, in rural areas as well, the common residential pattern is a nuclear family, and only a few households contain extended families with three generations living under the same roof (Chen, 2000; Wang, 2000; Xu, 2001).

In terms of welfare arrangements, in the present context the family is responsible for welfare provisioning in contrast to the much broader responsibility of the kin network in earlier times. Regarding the systems of support for the elderly, according to customary practices the elders were supposed to live in their own households as long as they were able to care for themselves; usually the elderly couples took care of each other, and stayed in the labor force as long as their physical condition allowed them to do so (Chen, 1996, 2000; Liu, 1996; Wang, 2000; Wang & Xia, 1994; Xu, 2001). In these cases children usually provided some complementary support, including financial aid and food supplies and according to the traditional rule these duties were imposed on the son's family. When the elderly parents became weaker and unable to care for themselves, usually after the death of a spouse, the widow or widower moved in with his/her son's family. The elderly person became dependent on his/her children and, de facto, it was the wife of the son who was responsible for the care of the dependent elderly parent-in-law.

Generally, with regard to the changing capacity of family support, the burning topic is the consequences faced as a result of children's reduced capacity to provide the elderly with the support and care they need. It is argued (Chen & Silverstein, 2000; Liu, 1999; Tao, 2002) that rural–urban labor mobility aggravates the crisis of rural family support. The majority of rural–urban migrants and would be support providers are young; while the young are migrating, a great deal of dependent elderly people are left in the countryside. Female migration particularly is supposed to destroy the fabric of traditional family support.

However, the present chapter indicates that the situation is not that simple and not that gloomy. The interview data show that families are capable of inventing new care arrangements. Among the migrating-mother households—the husbands/fathers of children are also gone to cities to work in these cases—a new and commonly practiced family support strategy, which could be called the "new mutual generational support pattern," is evolving. According to the interview data and stories, the real support issue for young migrant couples, particularly migrant mothers with dependent children, is the care needed for the children that they want to leave behind, rather than the care for the elderly. The reason for this is clear: the parents of most young migrants are not that old and they can take care of themselves; they still stay in the labor force. The outlook for children is more precarious. As rural–urban migrants and their children are not entitled to urban public welfare services, the migrant children are not accepted into urban public childcare facilities and public schools. Therefore, most migrant mothers (migrant couples with children) have to leave their children in the countryside and ask the grandparents to take care of them. As mentioned previously, most rural elderly people currently maintain and live in their own households independently, are responsible for managing their household tasks by themselves (Xu, 2001), and are engaged in agricultural activities as well as other kinds of work. Therefore, although the mother-in law is traditionally available to provide day care for the children, taking care of the grandchildren is not an unavoidable duty for them. In other words, as a consequence of having to look after their own households they are in a position where they can refuse to take responsibility for other people's (i.e., their children's) households. Of course, caregiving within the household involves practical obligations as well as giving love and affection (Ngan & Wong, 1995). For the purpose of this chapter, I shall avoid the issue of familial emotion, and discuss the issue of responsibility. For grandparents especially the grandmothers—the main care providers for the children—taking care of the children is likely to be an extra task rather than a responsibility; therefore, if a young mother (or young couple if you like) wants to go to the city, she has to be successful in persuading the elderly person left behind to take on the task of looking after their grandchild. Among the households interviewed, the common practice is for young couples to send extra money to their parents to compensate for the childcare work. The "extra money" is a small amount of cash which the migrant couples give to their parents in addition to the money sent to cover the child or children's living expenses; it also contributes toward the shared support imposed on the brother's (son's) family.

From the elderly person's point of view, the extra cash can be seen as a kind of "wage" for the care of the grandchild. They can also expect to receive gifts on top of the regular cash payment when their migrant children visit the home village or when they return from the city, usually in the form of food and clothing. These gifts have a very important symbolic function and they indicate good familial relationships, especially between daughters and parents-in-law. Maintaining good relationships between generations is crucial in guaranteeing that the elderly get proper support and care when they become weaker and more dependent. The aim of these gifts can be interpreted through Marcel Mauss's (1990 [originally 1925]) argument that there is no such thing as a free gift. There are always elements of demand for reciprocity. "Social contracts take place in the form of presents; in theory these are voluntary, in reality they are given and reciprocated obligatorily" (Mauss, 1990, 3). Thus, there is a great degree of reciprocity that gift-giving symbolically indicates and fortifies. The gift giving visualizes social contracts and obligations. The exchange of various gifts is a thread that knits society together, enchants common norms, shared identities, reciprocal solidarity, and economic ties.

The following story of an elderly woman who was taking care of her grandchildren represents a common pattern related to the previously discussed issues, and therefore, I recite it at length:

Huaqi is 46 years old and lives in a village located in Sichuan province with her 49 year-old husband and 5-year-old grandson who is the child of her oldest son. She works as the female leader of her village. Huaqi has two sons, both of whom are married, and both of the sons and the daughters-in-law are working in Chengdu, the capital of the province. The 25-year-old son and his wife have been in running a cloth business since 1993, and their 5-year-old son is left in the village to be cared for by Huaqi. The younger 23-year-old son and his wife are tailors and they have been working together with their uncle since 1994. They have a son who is less than 1 year-old; as the baby is too young to be left in the home village, the parents take care of him. When the baby grows up a little, they will leave him to be looked after by the grandparents as well. Huaqi herself got married in 1972. Two years after the marriage, Huaqi and her husband moved away from her mother-in-law's household and had some grain for food and a room to live in. At that time they were very poor, often even lacking in food. Huaqi's father-in-law died several years ago and her mother-in-law lives with one of her brother in-law's families. Huaqi and her husband give their old mother 200 kg rice per year as their token of support.

Huaqi had been raising pigs for over 10 years before they moved into a new house in 1998; as the work was too dirty for life in the new house, she gave up it. Her husband does not farm either and manages rental affairs of their other houses and sometimes helps her to look after the grandson. Her household income consists mainly of their sons' remittances, the rent payment from the two other houses they own (240 Yuan, approximately 25€ per month), and her wage (100 Yuan, approximately 10€ per month). She and her husband have been taking care of one of the grandsons, and next year the other grandson (the son of her younger son) will be left with them to be taken care of. Huaqi considers that "our sons and daughters-in law have quite strong sense of filial piety, so I am pleased to take care of the children for them." "Both sons' families send remittances to us (Huaqi and her husband), the flat we inhabit now, which costs 80,000 Yuan (about 8,000€) was bought by our elder son." Although her economic situation is just average in the village, she is quite satisfied with her current life, and feels sure that she will be looked after by the sons' families when she and her husband get older and weaker and cannot manage on their own (cases of interviews F1-202, The data set of "A Study on Out-Migrants and Return Migrants," 2000, Beijing).

In fact, this household has totally moved away from the agricultural sector yet they are still deemed as rural residents under China's household registration system. The number of such cases is ever increasing in many rural areas. In this transfer process, the younger generations migrate to cities and get long-term jobs and the elderly remain in the countryside even though they are no longer engaged in agricultural activities.

The above story is quite a harmonious one. However, there may be strong tensions involved in these kinds of chains of care. In fact, it is a widespread phenomenon that familial conflicts occur within households (Wang, 2000; Wang & Xia, 1994; Xu, 2001). The interview material collected from both village leaders and individual respondents shows that conflicts usually revolve around disagreements over how the support and care responsibilities for the elderly parents should be divided between the sons' families. About 20% of the women interviewed claimed that they experienced this kind of problem. The material from focus group interviews with the village leaders indicated that: "the son's household tends to take on a smaller share of their financial responsibilities compared to other brothers for their parents because of financial pressures in their own households." In some cases, where "disputes on providing support to the parents become very acute, young couples even attempt to avoid providing support for their elderly parents" (village interview F3, F1, Z1, Z2, W1, & S2, The data set of "A Study on Out-Migrants and Return Migrants," 2000, Beijing). Not surprisingly, in these cases the support and care for the elderly are usually not guaranteed due to each of their son's household tending to shift the responsibility to the other. However, disputes over how support should be shared out among the adult children, and the consequent decline in support for elderly parents is only one phenomenon of the rural family support crisis. The essential underlying issue is that the position and role of both the older generation and younger generation in the family support system have changed, and this structural change results from the transformation in land ownership. The story behind this transformation in contemporary rural China is a topical issue. However, it is not the aim of this chapter to give a detailed analysis of the relevant factors and processes.[6] The discussion here is narrowed down to the welfare consequence of these changes, in particular, the impact of the transformation of farmland ownership on the capacity of the family to give support.

---

[6] For a discussion on collectivization in rural areas during the period of before reform see Cheng (1999), and Yang (2004).

For a discussion on rural policy reform see Reisch (1992), and Zhou (1992).

Under the pre-socialist (pre-1949) private land ownership system, which was practiced in traditional rural China, the older people, as the heads of the family, controlled the family property—mainly the right to own land—and had the power to determine the rules of inheritance. Different members of the family as a welfare institution provided support to other family members (Xu, 2001; Yang, 1965, 65; Zhang, 2001; Zhao & Wen, 1998). Under this welfare arrangement, the old parents controlled the financial resources, namely the farmland and the farm products, and they were the decision-makers when it came to delegating welfare responsibilities and distributing benefits within the family. In this contract-like situation where the young family members depended on their parents or other senior kin members in terms of the financial resources, they in turn provided care for the old people of their kin. The elderly had strong power resources (decisions about inheritance, for example) at their disposal. Based in this mutual support arrangement, the old generation provided financial support and the younger generation the care work.

However, the socioeconomic foundation of this family support system was undermined with the creation of the new socialist state and the Communist party when they introduced Stalinist collectivization in the rural areas after 1949 (Duan, 2001; Jiang, 2002). Some researchers praise this form of collectivization and its achievements highly (Guan, 2001; Patnaik, 2003). For instance, Utsa Patnaik (2003, 39) argued that "much of China's good growth, reduction of rural poverty and excellent performance on the human development indicators can be traced back to the initial egalitarian land reform and its consolidation through the decentralized units like cooperatives and the later commune system up to 1980." However, from the perspective of social security outcomes, farmland collectivization can be seen as the cataclysm for the erosion of the traditional family support system. Collectivization made it impossible for family support systems to maintain their capacity. While the farmland collectivization transformed the traditional non-alienable family ownership rights into collective property controlled both by village leaders and governments, it also changed the roles of both the elderly people and their adult children under the traditional welfare arrangement. Collectivization deprived the old people of their power and control over economic resources in the family support institution. This turned the relationship between the giver and receiver within the family support institution upside down. Parents lost their financial resources that fortified their power over other family members, and hence their ability to claim care and support in old age. They thereby became dependent on the benevolence of their children, mainly their sons' family. It was no longer the young who relied on the old, but rather the old who relied on the young for a living. Therefore, the collectivization of the socioeconomic structure shifted family support from a two-way welfare provision system to a one-way support arrangement. The altered structure of the traditional family support system, on top of widespread poverty among the rural households, diminished both the willingness and the capacity of young adults to provide support for their elderly parents. The "classical intergenerational contract implied that the young would care for the aged in exchange for the transfer of wealth" (Esping-Andersen, 1999, 41); when the aged are deprived of the wealth and have nothing to transfer to the young in exchange for the care, family support becomes uncertain.

The situation also appears to have changed in the case of rural women workers' migration. According to the generational care pattern followed in migrating-mother households, old people in particular the grandmothers are restored to the role of welfare givers by caring for their grandchildren and by receiving monetary resources from their own children for providing child care. This setup serves as an alternative to the traditional land ownership arrangement which has vanished. On the other hand, these practices also encourage the younger generation to improve the support they give to the elderly. The opinion of the interviewees may illustrate the cases which are in the process of changing (case W1-301, 45-year-old male villager): "migrants have a great deal of filial piety and do respect their parents." The reasons are complex; some examples include "increased income and being affected by urban resident's manner of treating old people." But among them, the most important incentive for filial piety is the fact that "young migrants fear that their parents will not help them to take care of the children if they do not treat them as well as possible. In particular when there are several young couples working away from the home, they must try their best to show more filial piety to their parents in order to persuade the older people to help them to take care of their dependent children." In this display of filial piety, gift-giving plays an important role in the Maussian sense. One could therefore argue that it is through care-giving that the older generation especially the grandmother seems to have regained some of its power resources.

Here, the priceless child care that older parents are in the position to provide for their children secures financial resources derived from the children (Chen & Silverstein, 2000). For older people especially older women it seems unfair that after fostering their own children, they now face another round of nursing and caring. However, the opinion of a 46-year-old female villager represents rural elderly's attitude toward the issue: "most older people want their children to migrate to cities to find job," in order to "make some money and make the household get out the poverty. I hope my children will going out when they finishing their school, I'd like to take care of my grandchildren someday when I have" (The interviewee has two children, 19-year-old daughter and 16-year-old son; both were in school when interview was conducted.) (case F1-302, The data set of "A Study on Out-Migrants and Return Migrants," 2000, Beijing). In fact, given the exhausted support capacity of rural households, this new pattern of exchange appears as the only possible means by which the two parties, the welfare receivers and givers within the family, can improve the well-being of the entire group. The elderly are able to get some extra money to live on, and young people are given the opportunity to try and adapt their livelihood strategies to a capitalist economy that is producing industrial and service jobs mostly in urban and distant places.

So far, I have demonstrated under which principles the new generational contract makes it possible for the family support system to renew its capability: rural female worker's migration presents a possibility for the family support pattern to reshape the relationship between the elderly and their adult children, and therefore updates the mutual welfare provision arrangement in the new socioeconomic situation.

As to what extent this new pattern works efficiently de facto, the responses of the village leaders and the farmers interviewed on the questions of the impact of female migration on family support in their villages will be illustrated. On the subject of family support in female migrant households, the responses were surprisingly positive. The common remarks were that: "among the migrant households, particularly in female migrant cases, familial conflicts that arose from discussion about family support had drastically declined." This was mainly due to "the improved financial situation among the female migrant households." As a consequence of improved parent–child relationships and the harmonious life of the family, "support for the elderly improved noticeably" (village interview F3, F1, Z1, Z2, W1, & S2, The data set of "A Study on Out-Migrants and Return Migrants," 2000, Beijing). Thus, for instance, when talking about family support among villagers, the village leaders of Daisi village in Sichuan province said: "compared to nonmigrant households, young couples of migrant families displayed much more filial piety." The most visible evidence of this appears in two aspects. One is that "there are fewer quarrels among sons' households over how to share support responsibilities for their old parents and the informal agreement about the support ratio are fulfilled well. Previously they quarreled about the support because they were too poor." The second one is that "the relationship between daughters-in-law and mothers-in-law appears to have improved greatly in female migrant households. Besides providing the agreed financial support to the elderly without dispute, now many young migrant women often buy additional gifts for their mothers-in-law when they return to the village. The migrant families get more earnings, so they have the capacity to provide better support for the elderly." (Village interview V-F1; the data set of "A Study on Out-Migrants and Return Migrants," 2000, Beijing).

A response to this topic given by a female worker who had returned home may illustrate the point from the perspective of migrant children: "People have hearts; no matter how far from their parents they live and work, they always manage to support them. If somebody gets higher earnings when working outside the village, they can give much more to the parents. Staying in the village could mean better care for the older people when they need it, but on the other hand, if that were the case they would have less money and suffer from poverty" (case Z1-103, 44-year-old female returned; the data set of "A Study on Out-Migrants and Return Migrants," 2000, Beijing). Thus, there is a trade-off between the old values and the intensification of care and economic resources.

The messages given above suggest that economic poverty is the main factor restraining farmers' households from performing family support duties. This leads us to another important issue, which is acknowledged and to some extent emphasized by most researchers (for example, Adamchak, 2001; Xu, 1997; Yue, 2001), namely, the impact of the so-called "one-child policy" and the changing family structure regarding family support. Some studies argue that the "one child one family" policy as well as the changes in the household structure impairs the families' capability to provide support and care for the elderly (Liu, 1999; Zhang, 2001). Moreover, they argue that the migration of rural young adults is the cause of the weakening of rural family support. However there are counterarguments as well. Some studies demonstrate that the rule of one child per family was never implemented in the countryside—not even in principle. Only two provinces (besides the four municipalities) carry out the "one child one family" policy in rural areas among the 31 provincial regions (Guo et al., 2003). Throughout China, approximately 35% of the total population enforce the "one-child policy," and 36% of couples got formal permission from the authorities for the birth of their second child, and 1.3% of couples for the third child (Zeng, 2004). Therefore, the so-called "one-child policy" may not correspond to reality, and the change in age and household structure in rural China has not been as strong determining factors for rural family support crisis as they are often deemed to be, at least they will not be in the forthcoming 2 or 3 decades (Chen, 2000).

As to the impact of rural women's migration on family support, given the fact that the rural elderly over 60 years of age usually have more than two children, and that not all children migrate to distant cities, the caregiver shortage is not a serious problem for those migrant families. Moreover, the majority of rural–urban migrants are young (the sample survey of the research project showed the average age of the migrants is 28 years), and their parents are usually in their early 50s and in good health. This condition means that the elderly are able to carry out the tasks of "family care" (Zhao, 2002). Hence the crucial reason for the support crisis of rural family comes from both the lack of economic resources within the households, and from the financial dependency of elderly people upon their children rather than their dependency for personal care (Chen, 2000; Zhao, 1997). As a matter of fact, as most rural elderly people have no monetary pension benefits, financial problems constitute the real source of crisis. As income from off-farm jobs has become indispensable for rural households in maintaining their support capacity, the family strategy is such that the young migrate to cities in order to gain nonagricultural earnings. The old live in villages and take care of their grandchildren while receiving financial support from their migrating children in return. It is as a result of this strategy that the financial capacity of the family support system has been renewed, and the traditional system has managed to maintain its function, despite the fact that the size of the household and the manner of residence have changed significantly.

Transformation of the Generational Contract

The generational support pattern adopted by migrating-mother households is not simply the "preindustrial welfare nexus" in nature, although it certainly reflects a two-way generational interchange of welfare only within the families (Esping-Andersen, 1999, 53–54). In actual fact, as an updated version of the traditional measure, the Chinese case presents certain aspects of the dynamic relationship between the preindustrial institutions and the demands of modern society. In the following section some explanations as to why this preindustrial welfare arrangement has been updated in the case of female workers' (mothers) migration are offered.

Here, the concept of a "new social space" is used to refer to the construction of welfare nexus in the new pattern of family support. The idea of a "social space" was used by researchers in migration studies (Biao, 1999, 215–250; Rosaldo, 1988). Their studies found that migrants usually create their own system in the process of settlement; this system is known as a "new social space." In this social space migrants have their own lifestyles, values, rules of behavior, and networks, and the new practices are distinct from their original and destination societies. This means that the new social space is not integrated very well into the established structures (Biao, 1999, 216, 240). In the Chinese case, due to the lack of institutional support, the rural–urban migrants have created their own formal and informal systems in order to protect themselves (Skeldon & Hugo, 1999, 340). The updated generational support pattern practiced in migrating-mother households can, in terms of welfare provision, be regarded as one aspect of the new social space in the world of migration. Being excluded from the formal social security system has meant that migrants have to find new ways of dealing with the problem of support for dependent children and elderly parents. The traditional measure of generational interchange of welfare opened up possibilities for them to create their own social space for carrying out the family responsibilities. This issue is similar to the situation that has occurred in some Western welfare states when private welfare programs have been introduced at a time when the public sector had neither the economic nor political capacity to respond to the demands of social security. In such instances, these demands will be met in the private sector (Kangas & Palme, 1993). In rural China, the situation is a little different. There are no public sector benefits and the private sector is not functioning well. Therefore, only traditional measures are available for migrating-mother households. However, if we look at these practices more closely, we can see that there are some aspects that are distinct from the typically traditional path. It is in this sense that I apply the idea of a new social space to describe changes in the institutional aspects of this updated support pattern, as well as the interaction between current and traditional practices.

The first noticeable feature is that the practices of migrating-mother households in this new social space alter the role and status of rural women in the family support system, resulting in a change in the balance of power in the intergenerational family relations for support. Under the traditional family support system, major decisions on support arrangements were made by adult sons or other senior male members (Ngan & Wong, 1995). Women were submissive and used to serving their husband as well as their family, and caring for the family members was regarded as their real duty. So, although women, especially when they became someone's wife and consequently daughter-in law, were obliged to be the primary caregivers and had no opportunity to voice an opinion on these matters. Women's lives were simply "a labor of cultural obligations" and they received little appreciation for their caregiving work they did (Ngan & Wong). New practices enable young rural women, even those who have independent children, to go out and take on off-farm employment. Their earnings from nonagricultural activities shift their position from the less respected caregivers to an important financial supporter of the families. Noting that the crisis of rural family support mainly results from the lack of economic resources, the changing role of female migrants not only promotes the elevated capacity of the system but also implies changes in the traditional institutions in terms of power structure. The interview data indicate that in most migrant households, women are considered to have a right to take part in decision-making in issues relating to support for the elderly and children and do not only carry out their traditional obligations. In reality, wives' attitudes toward the care of their parents-in law are often the most crucial and in most cases affect the destiny of independent elderly family members. To some extent, young women now are both primary caregivers and primary decision-makers in the new social space.

The same implication can be indicted from the perspective of grandmothers who are taking care of the grandchildren. The common understanding of China's rural–urban migration model is that the Chinese case, to a certain extent, can be explained by developmentalist theory which assumes that migration decisions are a part of family strategies to raise income and insure against risks (Cai, 2000; Guan & Guo, 1997; Huang, 1996). Migration decisions are made by households instead of individuals in order to benefit whole families' particular household economic gains (Bai & He, 2002; Du, 1997; RTMA, 1995; Zhou, 2001). The new pattern of generational contract take place in the context of the household migration model, and it is an important issue in family decision-making. The messages obtained from the interviews show that the grandmothers participated in family decision-making in issues relating to migration as well as children care, "after counsel with my families, say parents, husband and mother-in law, we decided to leave for the city" is a common response concerning women's migration decision (cases of interviews W1-105, F3-112, F3-111, Z1-105, W3-114, The data set of "A Study on Out-Migrants and Return Migrants," 2000, Beijing). The message implies that the grandmothers are the participator in household migration decision. In this context the family support arrangement is unable to achieve without grandmother's promise to take care of the grandchildren. The grandmothers as decision-makers play a strong role and thereby negotiate their power under the household migration decision model. In addition, according to Confucian ethics, ideally for women, playing a strong role in domestic issues was assumed the active fulfillment of life; therefore, contributing to the collective good was deemed as a means of practicing women's power rather than weakness (Tu, 2001, 202).

Second, the new pattern increases the extent of welfare exchange between the adult children and their elderly parents via family support. In the traditional model, family members worked in the same household-based agricultural unit and the issues of the generational contract simply involved welfare arrangements between the generations within the family. This implied that income redistribution between the old and young generations occurred in a relatively closed space of kinship. The new support nexus hurdles the boundaries of the traditional family economy sphere. From the side of younger generation, this care pattern frees younger workers, especially female workers with dependent children, from the subsistence-based agricultural sector enabling them to get better paid off-farm jobs in urban areas miles away from their home villages. The migration allows rural younger generations to share in the opportunities and benefits offered by urban economic development, and these added financial sources guarantee a great capacity for family support.

From the point of view of the elderly especially the grandmothers—who are taking care of the children, this arrangement enables the rural elderly who are unable to be engaged in competitive urban labor markets to get their share of benefits in the development toward market-based economy. As rural residents are not eligible for public social support, the social security system makes no contribution to the masses of rural elderly. When they lost their earning capacity by moving to urban areas as migrant laborers, they became deprived of access to the benefits of the economic development, both in terms of direct income and in terms of pensions. Under the new support pattern developed by the migrating-mother households, the older parents can share a much bigger portion of the income which their migrant adult children gain from nonagricultural employment. A response to this topic given by a 57-year-old female villager who was taking care of her 7-year-old grandson may illustrate the point from the perspective of elderly parents, in particular the grandmothers: "the households have the migrants no longer suffer from poverty," because "the young people, the sons, the daughters-in-law and the daughters are working in the cities, they can get better wages than they work in countryside" (case W3-201, The data set of "A Study on Out-Migrants and Return Migrants," 2000, Beijing). Here, the remittance and other material support the migrant children provide the elderly parents with not only represent the traditional vertical reciprocity, but also a financial input of resources from outside the family. In this sense, in the "new social space" the welfare exchange between the migrant children and their parents in the countryside is not only redistribution between generations but also between the wealthy urban and the poor rural areas. In this sense, the model seems to work as long as there are huge disparities between urban and rural areas; when they are evened up the model dissolves.

Third, there are two interesting alternatives observed in migrating-mother households regarding the care arrangement for their dependent children. One is that in some cases the grandfather is also engaged in care activities, usually if the grandmother is unable to carry out the task due to bad health, for example, or if she is deceased. The other one is that some young migrant couples leave their children to the wife's parents instead of husband's parents as was the norm in the traditional order. The main reasons for this are the absence of husband's parents—deceased or in bad health—or there are several sons who have migrated and the grandparents are too busy to care for all of the grandchildren. Although there are only a few cases among the interviewees, this option was also out of question under the traditional order, especially in the countryside. These practices suggest that changes have not only taken place in the role of women but also in the role of elderly parents. The traditional form of family support was characterized by two rules: the first rule was that the care-giving always was the responsibility of women, and the second that the bringing up of children was an affair for the husbands' family, and the wives' parents were never involved. In traditional thinking and practice, the daughters, once married, were considered to be a part of their husband's households. Daughters' children were not regarded as the offspring of her original family, and it was seldom that grandchildren were taken care of by the grandparents on the wife's side. However, these principles and practices seem to fade away under the current practices of migrating-mothers households. Even though they apparently offer an alternative rather than being the primary option for young migrating couples, the practices reflect some fundamental changes in the core of traditional values and behavior.

Those changes also have a profound effect on the resources of family support; in particular, they expand the source of caregivers. All of these aspects have profound implications in understanding changing values and behavior of rural residents in the de-agrarianization process.

In response to the lack of state protection, migrating-mother households have developed their own "social space" in welfare provision, and the pattern of generational support appears to work well to some extent. However, this is only a partial representation of a picture that has many elements. After all, this pattern is a kind of preindustrial measure and effective only within certain kinds of family strategies. The generational interchange of welfare can work within the migrating households but is not an option for every household. The obvious limitation of the pattern presents itself in the three aspects: the first is that the new pattern cannot be effective for families with weak labor resources; the second is that illiterate grandparents do not meet grandchildren's education needs; the third is the problems of the psychological well-being of migrating mothers and children left in the home village.

In the first instance, the family has traditionally been the main locus of pooling life-course risks of individuals (Esping-Andersen, 1999, 41), and obviously it is not good at dealing with risks caused by the transformation of the whole society. This transformation affects the whole family. If we say that migrating-mother households successfully "internalized some social risks by pooling resources across generations" (Esping-Andersen, 1999, 37), this family self-help strategy creates risk pooling only with certain preconditions. From the point of view of the elderly, the practice does not enable them to avoid dependence on their children. The elderly are not dependent either on their savings accumulated throughout their working life or on the farm products (Leisering et al., 2002, 80), but rather on the income of their migrating children. This means that the support level and the guarantee for the elderly are dependent on the earning power of their migrating children. Only the elderly parents who have migrant children with greater earning powers or several migrant children are guaranteed to receive support, and thereby lower their risk of falling into old age gloomily. When migrating children, especially daughters-in law, fail to find relatively well-paid jobs in urban areas, get ill, or have work injuries, the new pattern does not work and therefore the "pension" of the elderly parents is reduced and their future becomes more uncertain. On the other hand, when it comes to the younger generation, in particular women with dependent children, the precondition for them to go out is that they have at least one parent who is in good health and willing to look after their grandchildren. In other words migrating women's capacity to exert their earning power is decided by the labor power or care-giving capacity of the elderly. According to our interview data, it is rarely the elderly who refuse to take care of their grandchildren; the usual grandparent-related factors which prevent female workers from migrating are the absence of parents, and in some cases the old parents themselves are in need of care. In these cases, female workers' earning power through participation on urban labor markets is diminished, and consequently the opportunities to strengthen the financial capacity of the family are limited. This implies that the practices are to some extent still based on the traditional generational contract, and therefore the pattern can only be effective in cases where households have, according to Esping-Andersen's (2002, 29) terminology, strong labor resources, and obviously not in "work poor families."

In the second case, illiterate grandparents who do not meet the criteria of modern society cannot provide their grandchildren with education and cultural capital due to their own lack of these resources. In most cases, among the migrating-mother households which practice this support pattern, the grandparents are only responsible for taking care of the grandchildren's daily needs, such as feeding and clothing them while the matter of the children's education is not a part of this informal care contract. The grandparents simply play the role of caregiver not "tutor." "The lack of parental help and supervision with children's studies" becomes "a serious problem for migrant households" (W9-07, W2-114, F1-108, F3-111,W1-105, S2-101,S2-202, S3-116, The data set of "A Study on Out-Migrants and Return Migrants," 2000, Beijing). The story recounted by a 36-year-old female returnee who migrated with her husband is a representative both of the situations mentioned above: "I was pleased that I went out, because I was able to earn money outside the countryside. Why did I leave? Simply to earn money! My household needed money badly, needed money to buy farming tools, fertilizer, and farm chemicals; it needed money for children's school fees, for elder parents-in-law support and for many other things." During the period of the migration, "we left our 6-year-old son with the grandparents in the home village. But one year later, we found that it was difficult for the grandparents to educate and discipline our son." Besides this, "the grandparents themselves were in need of care, so I had to returned to village." Her husband is still working in a construction team in Xinjiang when interview was conducted (case of interviews Z1-105, The data set of "A Study on Out-Migrants and Return Migrants," 2000, Beijing).With regard to the "quality" of bringing up the children, the common strategy of the households is for the informal contract on grandparents who are fostering to only cover preschool children. This strongly suggests that in the family intergenerational care pattern, love and affection are not sufficient; social, cultural, and educational capital is also needed. In order to guarantee the children a proper upbringing, migrating mothers usually quit their jobs and return to their home villages when the children reach school age. Thus the new-space arrangements described above are applicable only to the preschool children.

Moreover, there are potential problems regarding the psychological well-being caused by this pattern. With regard to the impact of the practice of migrating-mothers leaving their dependent children behind with the grandparents on female workers, some researchers argue that due to the female migrants being less likely to be constrained to household work in urban areas than in their rural homes, they have greater opportunities to use the characteristics that contribute to labor productivity in the urban labor force (Zhu, 2002). Others even assume that those female workers have the best of both worlds: they can enjoy the earnings and freedom that are often denied to women in traditional Chinese families, without the double burden that bogs down so many other working women (Pieke, 1999, 7). This is probably true in terms of the efficiency of commodified labor. However, if we consider the issue from the perspective of the humanistic psychological well-being of the children and migrant mothers, the argument could be just the contrary. In fact, our interviewees frequently expressed their anxiety about the negative impact on the psychological well-being of the children who are left behind with the grandparents while their parents are working in urban areas: "lack of discipline and the possibility of the influence of bad manners for instance, aside from the education problems due to absence of the parents' supervision" (W9-07, W2-114, F1-108, F3-111,W1-105, S2-101,S2-202, S3-116, The data set of "A Study on Out-Migrants and Return Migrants," 2000, Beijing). As to the psychological well-being of the migrant mothers who have to leave the children in the home village, almost all the cases show that they suffered rather than gained pleasure from these arrangements. A 28-year-old female returnee's experience is not an unusual case: "I and my husband worked in 5–6 factories one after another in Guangdong province from 1994 to 1998, while the grandparents look after our 2-year-old daughter in the home village." During their migration, "I and my husband were only able to visit our home village and see our daughter and parents once, and could only stay with them for 2 weeks. I was badly homesick." She said: "I liked that well paid job in cities, I was also quite satisfied with that position, but I had to give up the job and return to the village," because "my daughter had reached the age where she needed education and the illiterate grandparents did not have the ability to provide any." Her husband is still working in Guangdong when interview was conducted (case of interviews F3-111, The data set of "A Study on Out-Migrants and Return Migrants," 2000, Beijing). These situations suggest that the preindustrial pattern which exchanges welfare within the household only functions under certain conditions, and suffers as a consequence of problems in industrial society.

Conclusion

The chapter focuses on the coping strategies adopted by the migratory families to deal with the tensions caused by changes in generational care chains. To be more specific, the chapter explores how and why the traditional family support system managed to survive and function under the circumstances of female workers' migration. The chapter illustrates that in comparison to their urban counterparts, rural–urban migrant workers are more exposed to social security poverty. The female migrants and their families have no civil rights in cities, and they are totally excluded from the surrounding urban society. One of the extreme problems the female migrants face during their stay in cities is how to carry out the family care responsibility. Being excluded from the formal social security system has meant that migrants have to find new ways of dealing with the problem of support for dependent children and elderly parents. The updated traditional measure of generational interchange of welfare opened up possibilities for them to create their own social space for carrying out the family responsibilities.

The chapter also examines the dynamic interaction between traditional welfare institutions and the requirements of the industrial society in the context of rural women's migration. Family support has been the basic and most fundamental institutional welfare arrangement in China, especially in rural areas. Since 1949, the societal-economic foundation of this traditional system has gradually been eroded by the party-state's economic and social policies. In the meantime, farmland collectivization implemented in rural China since the 1950s has put a strain on the financial resources intended for family support on one hand, and changed the nature of mutual support which is crucial to maintaining the function of welfare transfer between generations within the households, on the other hand. All of those changes have directly resulted in a continuing decrease in the capacity of family support and as a consequence crises are emerging.

The massive surge of rural–urban female migration also contributes to changes in family support and the generational contract within the rural households. Contrary to the common assumption, this chapter illustrates that the problems do not turn out to be worse when rural women are involved in nonfarm employment in distant factories or cities. There is a concern that the process of industrialization has weakened traditional norms of filial piety, and that labor mobility reduced both the willingness and the capacity of younger generations to support their elderly parents as argued by Wilensky (2002) and other Chinese scholars. No evidence was found to support this argument in my research. The present study indicates that the real crisis of rural family support lies in the lack of financial resources, and the main welfare arrangement problems for female (mother) migrant households are caused by childcare issues. Responding to the lack of support from the formal social security system, the migrating-mother households developed their own strategy in order to cope with the situation. The dependent children and the elderly, who have at least partially maintained their capacity to work and take care of their grandchildren, are left in the countryside—the grandparents in charge of the grandchildren. The young migrating couples reciprocate by providing financial and other material help, and make promises that there will be improved support in the future. This new generational support pattern renewed the badly weakened capability of family support on two dimensions: one is that women's off-farm earnings meant an increase in the financial resources for family support; the other is that by migrating, the working mother helps to restore the mutual support principle which had been wrecked by state policies.

Furthermore, the informal generational support contract practiced among migrating-mother households is not simply the old fashioned preindustrial welfare nexus; it reflects an update in the nature of a traditional institution in the new societal-economic situation. Among the changes in rural residents' values and behavior in respect of the new family support pattern, two aspects are important. One is that the role and status of rural women (daughters-in-law and grandmothers) in the new pattern is shifting from previously being that of "family laborers" without decision-making rights, to being both the primary caregiver and the primary decision-maker. Another updated aspect is that the new generational support contract increased the extent of welfare exchange between the generations, and enabled the elderly who live in the countryside to indirectly share the benefits of economic development. While recognizing the functions of this informal arrangement, it must be acknowledge that this traditional measure has long had its weaknesses. In the case of the care pattern adopted by migrating-mother households, it simply improves the situation of resource-strong, especially human resource-strong, families.

At a more general level, the case of China calls into question industrialization logic of welfare state development. Contrary to the idea that economic growth and industrialization will lead to welfare state expansion, this chapter confirms that in the case of China, the industrial and economic growth over the last couple of decades have not been reflected in the welfare state policies.

Acknowledgments This chapter draws on my PhD thesis on the situation of migrant rural women in China. Support for my study was provided by a grant for a research position in the Department of Social Policy at the University of Turku, Finland from July 2000 to December 2005. The thesis is a subproject in the research program "Challenges of Modernization and Globalization for Chinese Social Policy" financed by the Academy of Finland. I am exceedingly grateful to my financiers. A debt of gratitude is owed to numerous people in Finland, in particular I would like to name Professor Olli Kangas, the leader of the research project, Ismo Söderling, Director at Population Research Institute for helpful comments and/or encouragement on earlier drafts of the thesis. I have greatly benefited from the teamwork in the research group which conducted the fieldwork in China that forms the empirical basis of this chapter. In addition, I would like to thank Shahra Razavi and anonymous reviewer for valuable comments and suggestions regarding an earlier version of this chapter.

References

Adamchak, D. J. (2001). The effects of age structure on the labor force and retirement in China. *The Social Science Journal, 38*, 1–11.

Berg, B. L. (1995). *Qualitative research methods for the social science*. Boston: Allyn and Bacon.

Biao, X. (1999). 'Zhejiang Village' in Beijing: Creating a visible non-state space through migration and marketized traditional networks. In F. Pieke & H. Mallee (Eds.), *Internal and international migration: Chinese perspectives*. Richmond: Curzon Press.

Cai, F. (2003). *Migration and socio-economic insecurity: Patterns, processes and polices*. Geneva: International Labour Office. ISBN 92-2-113590-X.

Chan, R. K., & Tsui, M. (1997). Notions of the welfare state in China revisited. *International Social Work, 40*, 177–189 (0020–8720;1997/04).

Chen, S. (1996). *Social policy of the economic state and community care in China culture*. Aldershot: Ashgate.

Chen, X., & Silverstein, M. (2000). Intergenerational social support and the psychological well-being of older parents in China. *Research on Aging, 22*(1), 43–66.

Chow, N., & Xu, Y. (2001). *Socialist welfare in a market economy*. England: Ashgate.

Esping-Andersen, G. (1999). *Social foundations of post-industrial economics*. Oxford: Oxford University Press.

Esping-Andersen, G. (2002). A child-centred social investment strategy. In G. Esping-Andersen, D. Gallie, A. Hemerijck, & J. Myles (Eds.), *Why we need a new welfare state*. Oxford: Oxford University Press.

Goodman, R., & Peng, I. (1996). The East Asian welfare states: Peripatetic learning, adaptive change, and nation-building. In G. Esping-Andersen (Ed.), *Welfare states in the transition: National adaptations in global economies*. London: Sage.

Hussain, A. (2000). *The Chinese social security system*. Final report: PRC social security program for the Asian Development Bank, Beijing.

Jones, C. (1993). The pacific challenge: confucian welfare states. In C. Jones (Ed.), *New perspective on the welfare state in Europe*. London: Routledge.

Kangas, O., & Palme, J. (1993). Statism eroded? Labour market benefits and the challenges to the Scandinavian welfare states. In E. J. Hansen, S. Ringen, H. Uusitalo, & R. Erikson (Eds.), *Welfare trends in Scandinavia*. New York: M.E. Sharpe.

Kersbergen, K. V. (1995). *Social capitalism: A study of Christian democracy and the welfare state*. London: Routledge.

Leisering, L., Gong, S., & Hussain, A. (2002). *People's Republic of China, old-age pensions for the rural areas: From land reform to globalization, Asian Development Bank 2002*, Publication Stock No. 090802, Philippines.

Lieberson, S. (1992). Small N's and big conclusions: An examination of the reasoning in comparative studies based on a small number of cases. In C. C. Ragin & H. S. Becker (Eds.), *What is a case? Exploring the foundations of social inquiry*. New York: Cambridge University Press.

Mauss, M. (1990). *The gift: The form and reason for exchange in archaic societies*. London: Routledge.

Ngan, R., & Wong, W. (1995). Injustice in family care of Chinese elderly in Hong Kong. *Journal of Aging & Social Policy, 7*(2), 77–94.

Patnaik, U. (2003). Global capitalism, deflation and agrarian crisis in developing countries. In S. Razavi (Ed.), *Agrarian change, gender and land rights*. Oxford: United Nations Research Institute for Social Development, Blackwell.

Pieke, F. (1999). Chinese migrations compared. In F. Pieke & H. Malle (Eds.), *Internal and international migration: Chinese perspectives*. Richmond: Curzon Press.

Reisch, E. (1992). Land reform policy in China: Political guidelines. In E. B. Vermeer (Ed.), *From peasant to entrepreneur: Growth and change in rural China*. Pudoc: Wageningen.

Robert, K. (2000). Chinese labour migration: Insights from Mexican undocumented migration to the United States. In L. West & Z. Yaokui (Eds.), *Rural labour flows in China*. Berkeley: Institute of East Asian Studies, University of California.

Rosaldo, R. (1988). Ideology, place, and people without culture. *Cultural Anthropology, 3*(1), 77–87.

Saunders, P., & Shang, X. (2001). Social security reform in China's transition to a market economy. *Social Policy & Administration, 35*(3), 274–289.

Skeldon, R., & Hugo, G. (1999). Conclusion: Of exceptionalisms and generalities. In F. Pieke & H. Mallee (Eds.), *Internal and international migration: Chinese perspectives*. Richmond: Curzon Press.

Tu, W. (1996). Confucian traditions in East Asian modernity. In W. Tu (Ed.), *Confucian traditions in East Asian modernity: Moral education and economic culture in Japan and the four mini-dragons*. USA: American Academy of Arts and Sciences.

Tu, W. (1998). An alternative vision of modernity: From a Confucian perspective. *Harvard China Review*, Magazine Online summer 1998, Culture & Society, Vol. 1(1).

UNRISD (United Nations Research Institute For Social Development). (2005). *Gender equality striving for justice in an unequal world, policy report on gender and development: 10 Years after Beijing*. UNRISD Publications. Accessed on Dec 1, 2010 from http://www.unrisd.org/80256B3C005BCCF9/.

White, G. (1998). Social security reforms in China: Towards an East Asian model? In R. Goodman, G. White, & H. Kwon (Eds.), *The East Asian welfare model*. London: Routledge.

White, G., & Goodman, R. (1998). Welfare orientalism and the search for an East Asian welfare model. In R. Goodman, G. White, & H. Kwon (Eds.), *The East Asian Welfare Model*. London: Routledge.

Wilensky, H. L. (2002). *Rich democracies: Political economy, public policy, and performance*. Berkeley: University of California Press.

Wong, L. (1998). *Marginalization and social welfare in China*. London: Routledge.

Xu, Y. (2001). Family support for old people in rural China. *Social Policy & Administration, 35*(3), 307–320.

Yang, M. C. (1965). *A Chinese village: Taitou, Shantung Province*. New York: Columbia University Press. (Original work published 1945).

Zhou, F. (1992). Stability first! Chinese rural policy issues, 1987–1990. In E. B. Vermeer (Ed.), *From peasant to entrepreneur: Growth and change in rural China*. Pudoc: Wageningen.

Zhu, N. (2002). The impacts of income gaps on migration decisions in China. *China Economic Review, 13*, 213–230.

Works in Chinese

Analects of Confucian. (1990). *Lun Yu, in Sishu Wujing, Song Yuan Ren Zhu (in The Chinese Classics, Song dynasty version [960–1279 A.D]*). Beijing: Zhonghua Shu Ju.

Bai, N., & He, Y. (2002). 'Huixiang, haishi waichu? Anhui Sichuan liangsheng waichu laodongli huiliu yanjiu' (Return to Native Place or Immigrating to Big cities for Working? A study on the phenomena of farmer-labour's returning home from the cities). *Shehuixue Yanjiu (Sociological Research), 2002*(3), 64–78.

Cai, F. (2000). *Zhongguo Liudong Renkou Wenti (Study on China's floating population)*. Zhengzhou: Henan Renmin Chubanshe (Henan Renmin Press).

Chen, C. (2000). Jingji duli caishi nongcun laonianrn wannian xingfu de shouyao tiaojian (Economic independence and the well-being of rural old people). *Renkou Yanjiu (Population Study), 2*.

Cheng, S. (1999). *Rural development in China: Theory and practice*. Beijing: Renmin University of China Press.

Chen, X., & Han, J. (2002). Nongmin tudi sheyongquan liuzhuan xu jiji wentuo (Issues of farm land transform). *Renmin Ribao (People's Daily)*, September 9, 2002, p. 9, Beijing.

CPSY. (2001). *Zhongguo Renkou Tongji Nianjian 2001 (China population statistics yearbook 2001)*. Beijing: Zhongguo Tongji Chubanshe (China Statistics Press).

Du, Y. (1997). Jiben Qushi he Zhengce Tailun (Trend and policy). In Y. Du & N. Bai (Eds.), *Zouchu Xiangcun (Going out the countryside)*. Beijing: Jingji Kexue Chubanshe (Economics Press).

Du, Y., & Bai, N. (Eds.). (1997). *Zouchu Xiangcun (Going out the countryside)*. Beijing: Jingji Kexue Chubanshe (Economics Press).

Duan, Q. (2001). Zhongguo nongcun shehui baozhang de zhidu bianqian (Institutional transform of social security in rural China). *Ningxia Shehui Kexue (Ningxia Social Science), 1*, 22–30.

Fei, X. (1985). Xiangtu Zhongguo (From the soil). Beijing: Shenghuo, Dushu, Xinzhi, Sanlian Shudian (Sanlian Press). (Original work published 1947).

Feng, Y. (1985). Zhongguo Zhexueshi Jiashi (A history of Chinese philosophy). Beijing: Beijing Daxue Chubanshe (Beijing University Press). (Original work published 1948).

Guan, X. (2001). Quanqiuhua beijingzhong de zhongguo shehui zhengce (China's social policy in the context of globalization). In J. Tang (Ed.), *Shehui Zhengce: Guoji Jingyan yu Guonei Shijian (Social policy: International experiences and practices in China)* (pp. 19–26). Beijing: Huaxia Chubanshe (Huaxia Press).

Guan, Z., & Guo, J. (1997). Cong buliudong dao liudong: hongguan beijinghe jiating juece (From non-mobility to migration: Policy background and the family decision). In Y. Du & N. Bai (Eds.), *Zouchu Xiangcun (Going out the countryside)* (pp. 27–83). Beijing: Jingji Kexue Chubanshe (Economics Press).

Guo, Z., Zhang, E., Gu, B., & Wang, F. (2003). Cong zhengce shengyulv kan zhongguo shengyu zhengce de duoyangxing (Fertility and family policy in China). *Renkou Yanjiu (Population Study), 9*.

He, Y., & Bai, N. (2002). Jincheng haishi huixiang: zhongguo nongmin ruhe xuanze chengshihua (Where rural labour move: Going out to city or return back to countryside). In N. Bai & H. Song (Eds.), *Huixiang, Haishi Jingcheng? Zhongguo Nongcun Waichu laodongli Huiliu Yanjiu (March forward to the city or return back to the country: A study of the returned rural labours)* (pp. 3–13). Beijing: Zhongguo Caizheng Jingji Chubanshe (China Financial and Economic Press).

Hou, W. (1991). *Shehui Baozhang Lilun yu Shijian (Social security theory and practice)*. Beijing: Zhongguo Laodong Chubanshe (China Labor Press).

Huang, P. (1996). Xunqiu shengcun de chongdong: cong weiguan juaodu han zhongguo nongmin feinonghuan huodong de genyuan (Looking for survival: The reasons of Chinese farmer's non-agricultural activities from the micro perspective). *Ershiyi Shiji (Twenty-one Century), 12*. Hongkong.

Jiang, C. (2002). Noncun tudi yu nongmin de shehui baozhang (Farm land and social security in rural China). *Jinji Shehui Tizhi Bijiao (Comparative Study on Social and Economic System)*, *1*, 59–55.

Jin, Y. J. (Yaoji) (1986). Zhongguo Lishi Chuantong yu Xiandaihua (Chinese tradition and modernization: The selected works of Jin Yaoji). Taiwan: Youshi Wenhua Zhuanye Gongsi (Youshi Press).

Kong, J. (2001). Yanglao baoxian zhidu gaige de fengxian yu fangfan (Reform of the old-age insurance system: Possible risks and preventive measures). In X. Song (Ed.), *Zhongguo Shehui Baozhang Tizhi Gaige yu Fazhan (Report on the reform and development of China's social security system)* (pp. 43–58). Beijing: Zhongguo Renmin Daxue Chubanshe (Renmin University of China Press).

Liang, S. (1990). *Zhongguo Wenhua Yaoyi, Liang Shuming Quanji Disanjuan (Essence of Chinese culture, collected edition of Liang Shuming, Vol. 3)* (p. 85). Jinan: Shandong Renmin Chubanshe (Shandong Renmin Press).

Liu, C. (1996). Fazhan yi Ziwo baozhang weizhu de nongcun shehui yanglao baoxian (Self-reliance of the elderly and rural pension programme). *Renkou Yanjiu (Population Study)*, *6*.

Liu, G. (1999). Woguo nongcun jiating yanglao zunzai de jichu yu zhuanbian de tiaojian (Changing base of family support in rural China). *Renkou Yanjiu (Population Study)*, *3*.

Luo, G. (2000). Zhongguo shiye baoxian zhidu de gaige shijian ji pingjia (Unemployment insurance reform practice in China). In W. Yang & G. Luo (Eds.), *Shiye Baoxia (Unemployment insurance)*. Beijing: Zhongguo Renmin Daxue Chubanshe (Renmin University of China Press).

MOCA (Ministry of Civil Affairs). (2000). *Zhongguo Shehui Fuli (China's social welfare handbook)*. Beijing: Minzheng Bu Bian (Ministry of Civil Affairs).

NBS (National Bureau of Statistics of China). (2002). *China Statistical Yearbook 2002*. Retrieved from http://www.stats.gov.cn/english/statisticaldata/yearlydata/.

NBOS. (2004). *2003 Nian Nongcun Waichu Wugong Laodongli 110 millions (The employment of rural mobile laborers 2003)*. Nongdiao Zongdui (Rural Survey Team of NBOS). http:/www.sannong.gov.cn/fxyc/ldlzy/200405270846.htm.

Qin, H. (1999). Chuantong zhongguo shehui de zairenshi (The revaluation of traditional China). *Zhanlue yu Guanli (Strategy and Administration)*, *6*, Beijing.

RTMA (Research Team, Ministry of Agriculture). (1995). 1994: nongcun Laodongli kuaquyu liudong de shizheng miaoshu (Description of rural labour force mobility in 1994). *Zhanlue yu Guanli (Strategy and Administration)*, *6*.

Song, H., Huang, H., & Liu G. (2002). Nongcun laodongli liudong de zhengce wenti (Policies on rural labour force mobility). In N. Bai & H. Song (Eds.), *Huixiang, Haishi Jingcheng? Zhongguo Nongcun Waichu laodongli Huiliu Yanjiu (March forward to the city or return back to the country: A study of the returned rural labours)* (pp. 163–188). Beijing: Zhongguo Caizheng Jingji Chubanshe (China Financial and Economic Press).

Tao, Y. (2002). Ergyuan jingji jiegou xia de zhongguo nongmin shehui baozhang zhidu toushi (Social security of rural residents in the context of dual social structure). *Zaijing Yanjiu (Financial Research)*, *11*.

Tu, W. (2001). *Dongya Jiazhi yu Doyuan Xiandaixing (Confucian values and pluralism modernity)*. Beijing: Zhongguo Shehui Kexue Cgubanshe (China Social Science Press).

Wang, M., & Xia, Z. (1994). Beijing zhongqingnian jiating yanglao xianzhuang fenxi (Research on family support in Beijing). *Renkou Yanjiu (The Population Study)*, *4*.

Wang, Y. C. (Yicai). (2000) Jiating yanglao, tudi baozhang yu shehui baoxian xiangjiehe shi jiejue nongcun yanglao de biran xuanze (The combine of family support, land security and social insurance in the rural area). *Renkou Yangjiu (The Population Study)*, *5*.

Wen, T. (1999). Bange shiji de nongcun zhidu bianqian (Institutional transformation in rural China in the latest half of century). *Zhanlue yu Guanli (Strategy and Administration)*, *6*. Beijing.

Xiang, J. (2001). '*Jiazu de biaqian yu cunzhi de zhuanxing: guanyu jiazu zai woguo xiangcun zhili zhong de zuoyong de yixiang hongguan kaocha' (Changing network of the kin and the transformation of social control at village level), Zhongguo Nongcun Yanjiu 2001 nian Juan (China Rural Research 2001)*. Wuhan: Zhongguo Huazhong Shifan daxue Zhongguo Nongcun Wenti Yanjiu Zhongxin (China Rural Research Centre of Huazhong Normal University).

Xiao, Q. (2001). *Xiao yu Zhongguo Wenhua (Filial piety and Chinese culture)*. Beijing: Renmin Chubanshe (Renmin Press).

Xu, Q. (1997). Nongcun de Jiating yanglao nongzou duoyuan (How long it will go: Family support in rural China). *Renkou Yanjiu (The Population Study)*, 6.

Yang, X. (2004). *Bainian Zhongguo Jingjishi Biji: Cong Wanqing Dao 1949 (Economic history in China: From the late Qing dynasty to 1949)*. Retrieved Dec 1, 2010 from http://www.sinoliberal.com/scholar/china%20history%20of%20modern%20economy%200.htm.

Yue, S. (2001). Woguo renkou laolinghua qushi jiqi duize (Aging trend in China and the policy response). *Zhongguo Renmin Daxue Fuyin Baokan Ziliao: Shehui Baozhang Zhidu (China Renda Social Sciences Information Center: Social Security System)*, 5, 13–17. Beijing.

Zeng, Y. (2004). Shengyu zhengce xu Pingwen Guodu (The possibility of family policy change). *Zhongguo Renkou Xinxi Wang (CHINA POPIN)*. Retrieved Dec 1, 2010 from http://www.cpirc.org.cn/yjwx_detail.asp?id.

Zhang, Q. (2001). Teda chengshi zanzhu renkou xianzhuang de yanjiu: yi Beijing shi weili (Research on situation of temporary residents in Chinese metropolises: the example of Beijing). In L. Ke, B. Gransow, & H. Li (Eds.), *Dushi Li de Cunmin: Zhongguo dachengshi de liudong renkou (Villagers in the city: Rural migrants in Chinese metropolises)* (pp. 165–173). Beijing: Zhongyang Bianyi Chubanshe (Zhongyang Translation and Edit Press).

Zhao, C. (1997). Shuchudi: waichu jiuye dui nonghu ji shequ de yingxiang (Impact of the rural-urban migration on the farmer household and the rural community). In Y. Du & N. Bai (Eds.), *Zouchu Xiangcun (Going out the countryside)* (pp. 127–158). Beijing: Jingji Kexue Chubanshe (Economics Press).

Zhao, C. (2002). Waichu, huiliu yu xiangcun shehui bianqian (Out-migration, return, and rural social change). In N. Bai & H. Song (Eds.), *Huixiang, Haishi Jingcheng? Zhongguo Nongcun Waichu laodongli Huiliu Yanjiu (March forward to the city or return back to the country: A study of the returned rural labours)* (pp. 77–100). Beijing: Zhongguo Caizheng Jingji Chubanshe (China Financial and Economic Press).

Zhao, Y., & Wen, G. (1998). *Zhongguo nongcun shehui baozhang yu tudi zhidu (Social security in rural China and the farm land system)*. Hongkong Zhongwen Daxue Zhongguo Yanjiu Fuwu Zhongxin Wang (The Chinese University of Hongkong). Retrieved Dec 1, 2010 from http://www.usc.cuhk.edu.hk/wk_wzdetails.asp?id=1869.

Zheng, G. (1997). *Lun Zhongguo Tese de Shehui Baozhang Daolu (The characteristic Chinese road of social security development)*. Wuhan: Wuhan Daxue Chubanshe (Wuhan University Press).

Zhou, D. (2001). Yongheng de zhongbai: zhongguo nongcun laodongli de liudong (Everlasting pendulum: the floating of rural labouers in China). In L. Ke, B. Gransow, & H. Li (Eds.), *Dushi Li de Cunmin: Zhongguo dachengshi de liudong renkou (Villagers in the city: Rural migrants in Chinese metropolises)* (pp. 304–326). Beijing: Zhongyang Bianyi Chubanshe (Zhongyang Translation and Edit Press).

Chapter 8
The Utility of Enhancing Filial Piety for Elder Care in China

Jacky Chau-kiu Cheung and Alex Yui-huen Kwan

Abstract One way to judge the utility of enhancing filial piety in society is to assess its contribution to the utility of individual filial piety and of family elder care, as opposed to state elder care. Filial piety in society refers to filial piety as an aggregate social norm. The utility of individual filial piety and of family elder care refer to the expectations and desire for them. A survey of 1,219 older Chinese in six cities in China provided data for assessing the utility of enhancing filial piety. The results generally show that the social norm of filial piety does not consistently strengthen the utility of family elder care. Importantly, the effects of the social norm vary substantially among the six cities. These findings imply that enhancing filial piety in Chinese society would not raise the preference for family elder care over state elder care among the older population. That is, the social norm of filial piety would not sustain individual filial piety or family elder care in ways favorable to older Chinese.

The utility of the social norm of filial piety for elder care is a vital but uncharted issue. Such utility means the usefulness, efficacy, or quality of elder care (Kosloski & Montgomery, 1993). Indicators of utility are an individual's expectations, desire, or preferences (Johansson-Stenman, Carlsson, & Daruvala, 2002). For example, when an older person expects to be cared for by his or her offspring, he or she believes in the utility of such care. This utility is valuable in view of the desire to raise the quality of elder care (Greengross, 1997). In light of the concern for quality, the utility of the elder care provided by offspring or family then appears to be at issue. On the one hand, family or filial elder care may be efficient because of insider knowledge or reduced transaction costs (Wolf, 1999). On the other hand, family or filial elder care may be ineffective due to a lack of professional skill (Montgomery, 1999).

J.C.-k. Cheung (✉) • A. Yui-huen Kwan
Department of Applied Social Studies, City University of Hong Kong,
Kowloon Tong, Kowloon, Hong Kong
e-mail: ssjacky@cityu.edu.hk

S. Chen and J.L. Powell (eds.), *Aging in China: Implications to Social Policy of a Changing Economic State*, International Perspectives on Aging,
DOI 10.1007/978-1-4419-8351-0_8, © Springer Science+Business Media, LLC 2012

Specifically, the utility of filial or grandfilial elder care is believed to be contingent on filial or grandfilial piety. This is the case when filial piety is able to sustain an elder's life satisfaction (Huang, 2011). However, the direct contribution of filial piety to the utility of filial or grandfilial elder care is not evident. Therefore, the extent to which filial piety as a social norm contributes to the utility of elder care requires empirical investigation. Such an investigation is crucial to address the utility of promoting filial piety as a social norm to optimize the utility of elder care. As the promotion of filial piety is a relevant concern in China, the present empirical investigation conducted in six cities in China is pertinent to the country's need for increased elder care (Zimmer & Kwong, 2003). The objective of the study is thus to assess the utility of enhancing filial piety as a social norm to optimize family elder care. This assessment relies on the utilitarian logic that the utility of filial piety transpires in the contribution of filial piety to the desire for filial piety. In this connection, utilitarian logic holds that something is good when it leads to a desirable outcome (Kane, 1991). Accordingly, this assessment is relevant to the justification for enhancing filial piety.

In the present study, the social norm of filial piety is a focal factor in relation to the utility of filial or grandfilial elder care. Here, filial piety refers to its modern and generalized manifestation as being caring, respectful, welcoming, pleasant, obedient, and financially supportive (Liu, Ng, Weatherall, & Loong, 2000; Ng, 2002). This conceptualization retains the affectionate and amiable emphasis of China's Confucian heritage and minimizes the impact of the country' authoritarian and feudalist history. According to Confucian teachings, filial piety is a virtue that serves as a preparation for benevolence and righteousness (Lin, 1992). Most important is that filial piety needs to be ethical and reasonable, and not violate the virtues of human nature, including reciprocity (Li, 1997). The latter justifies filial piety as a return to childbirth and rearing. Nonetheless, while filial piety tends to be the root of personal and interpersonal virtue, the contribution of filial piety as a social norm to societal well-being is less clear. The term social norm refers to the common practice of a custom such as filial piety in society. In one Confucian doctrine, filial piety is generalized to the support and care of elders other than one's parents. Clearly, grandfilial piety, as provided by a grandchild to his or her grandparent, is a generalization of filial piety (Even-Zohar & Sharlin, 2009). Moreover, norm theory also justifies the contribution of the social norm of filial piety to personal filial piety and elder care (Stuifbergen, Van Delden, & Dykstra, 2008). Simply put, norm theory holds that social norms facilitate or constrain personal action because of the general need for conformity, learning, imitation, and social comparison (Hooyman, 1990). The normative contribution of filial piety to societal well-being is a rationale for introducing a social policy to promote filial piety as a social norm. This rationale rests on the premise that the social norm of filial piety is beneficial to the population as a whole, for instance, by raising the quality of life in society (Yao, 2001). Another premise is that such a policy would reduce public expenditure on elder care (Ng, Loong, Liu, & Weatherall, 2000). However, counterarguments against filial piety and its promotion refer to the modern or even postmodern nature of society

(Croll, 2006; Inglehart, 1986). From this perspective, filial piety and family elder care are regarded as unreliable, ineffective, humiliating, and outdated (Brewer, 2001; Izuhara, 2002; Silverstein, Bengtson, & Litwak, 2003). The lack of empirical evidence for and against such arguments necessitates the present investigation. At any rate, the social norm of filial piety warrants investigation because of its possible socio-normative influence on the individual. This investigation conducts a test of the socio-normative vs. individualistic bases for elder care.

The investigation of the utility of the social norm of filial piety for filial, grand-filial, or family elder care is mostly important because these are all possible and feasible (Engelhardt, 2007; Nagata, Cheng, & Tsai-Chae, 2010; Whyte & Qin, 2003). In this sense, the utility of elder care refers to an elder's expectation, desire, or preference for filial, grandfilial, or family elder care or piety, as opposed to elder care by the state. Essentially, the former involves the state as a facilitator, whereas the latter requires the state as a care provider. The state's facilitatory role includes the promotion of filial piety and respect for elders, whereas its provider role involves the provision of financial, material, and emotional support to elders. The differentiation between family and state elder care is crucial in policymaking, as the two forms of elder care involve different strategies and expenditures. In one sense, the two forms are competing or contradictory, in view of the possibility that they will crowd each other out (Cheung, Kwan, & Ng, 2006; Kunemund & Rein, 1999). Therefore, the two forms of elder care should be substitutable, as is expected by policymakers. For example, the substitution of state elder care by family elder care has been used as a way of tackling the fiscal crisis (Papadimitriou, 2007), most notably in Europe (Boersch-Supan, 2007). This crisis purportedly stems from the aging of the population. It also raises the issue of generational equity concerning the aging population's overdependence on the state (Williamson, McNamara, & Howling, 2003), which would favor shifting the elder care role from the state to the family. Family elder care, in theory, also offers the merits of reducing the moral, political, demographic, responsibility, and bureaucratic hazards of state elder care (Engelhardt, 2007). The moral hazard refers to the overuse of state elder care resources, as families relinquish their responsibility and care for the elderly. This has the effect of draining state resources. The political hazard is that politicians promise too much in state elder care to earn public support. This would jeopardize the efficiency and integrity of the state. The demographic hazard is that a shrinking younger population is expected to support an increasingly older population. This demographic change will undercut the sustainability of state elder care. The responsibility hazard means that state elder care is poorer in quality than the care provided by the family because the state does not have a naturally obligated responsibility for elder care. This would worsen the overall quality of elder care. The bureaucratic hazard is that state elder care is neither easily accessible nor timely, due to red tape and lengthy administrative procedures. This also again would vitiate elder care. The possible hazards in state elder care thus favor family elder care as a substitute. Promoting family elder care, in turn, is dependent on promoting the social norm of social piety, according to norm theory. The present empirical investigation aims to substantiate this supposition.

Normative and Other Influences on the Utility
of Family Elder Care

The crucial issue is whether the social norm of filial piety should be enhanced as a means of upgrading the utility of family elder care. This means that the contribution of the social norm to the utility of elder care is paramount for empirical investigation. The value of the contribution of filial piety rests on norm theory, which posits that conformity to such a social norm is salutary, helpful, and beneficial in promoting successful aging and various aspects of the quality and practice of life (Baltes & Baltes, 1990; Maehara & Takemura, 2007). As such, individuals tend to seek to conform to the social norm of filial piety (Friedman, Hechter, & Kreager, 2008; Kosberg, 2005). Hence, personal filial piety should naturally follow societal filial piety. In addition to conformity, the social norm is likely to guarantee the quality of normative practice, as the norm provides guidance for the personal practice of elder care. Accordingly, the norm, similar to professionalization, may set the quality standard of family elder care, which in turn would raise the utility of family elder care (Kane, 1999). This special case of normative support reflects the cultural-fit mechanism of norm theory (Beugre & Offodile, 2001; Cheung & Chan, 2005; Cheung & Yeung, 2011; Dorfman, 2004). Here, cultural or social norms function to safeguard the quality of personal practice. This means that personal practice is more effective when it is congruent with the cultural or social norms associated with that practice. In this vein, family, filial, or grandfilial elder care is likely to be desirable and efficacious in the presence of the social norm of filial piety. The desirability or utility of filial piety may stem from the social support received from society (Wang, Laidlaw, Power, & Shen, 2010). This social support, in turn, is likely to arise from the endorsement and practice of the social norm (Franks, Pierce, & Dwyer, 2003). The provision of social support, therefore, may enable the social norm of filial piety to enhance the utility of family elder care. Conversely, the desirability or utility of state elder care is higher when the personal expectations of filial piety are lower (Daatland & Herlofson, 2003). In this connection, the social norm of filial piety is likely to underlie individual personal expectations of filial piety (Cheng & Chan, 2006; Wang et al., 2010). This means that the social norm of filial piety may dampen the social preference for state elder care.

Therefore, the focal hypothesis of this investigation is that the social norm of filial piety sustains the utility of family elder care. Indicators of this utility are elders' expectations of filial and grandfilial piety, and desire for family elder care, as opposed to their desire and preference for state elder care. The contribution of the social norm of filial piety would, in turn, indicate its utility, in that it is worthwhile enhancing as a social norm. This implication rests on the utilitarian logic of policymaking, which emphasizes the maximization of utility for the whole of society (Greengross, 1997). Accordingly, the social norm of filial piety would be valuable, if it was to safeguard the utility of family elder care, as opposed to state elder care. In this case, the social norm of filial piety would maximize utility by minimizing the public cost of state elder care.

Nevertheless, the influence of the social norm of filial piety on individual preferences for family, as opposed to state elder care is not yet certain. In contrast with the foregoing theory and evidence, is the finding that filial piety enhances the preference for state elder care (Killian & Ganong, 2002; Ward, 2001). One possibility for this positive influence on the preference for state elder care is the association between filial piety and respect of elders in general. That is, respect for, or sympathy with elders escalates the preference for state elder care (Huddy, Jones, & Chard, 2001). Meanwhile, the support for elders, in general, is a generalization of filial piety (Chen, 2003; Whyte & Qin, 2003). The contending arguments concerning the contribution of filial piety to the desirability or utility of elder care provide impetus for the present empirical investigation.

Apart from the normative influence of filial piety expected from norm theory, other influences on the utility of elder care are likely to comply with the theory of self-interest. Self-interest theory maintains that expected material or economic gains are motivators of action and preference (Blekesaune & Quadagno, 2003). The theory primarily emphasizes gains that appeal to self-interest, by giving priority to the self. As such, elder care that is beneficial or caters to self-interest is likely to be preferable. Self-interest in family or state elder care conceivably hinges on an individual's age, and the fact that they live apart from family, and have a low income, as these indicate the relative need for and benefits of elder care (Fortinsky, Fenoter, & Judge, 2004; Guan, Zhan, & Liu, 2007; Lum & Lightfoot, 2005).

Variations Within China

This study surveyed older residents in six Chinese cities located in different areas of China. The cities, Beijing, Shanghai, Nanjing, Guangzhou, Xiamen, and Xian, have clearly different socio-demographic and economic indicators (Cheung & Kwan, 2009). For instance, Shanghai has the largest population, Guangzhou has the highest income and gross domestic product per capita, Xiamen has the smallest population, and Xian has the lowest gross domestic product and built area per capita. The cities also differ in regard to filial piety and its prevalence as a norm, and these variations allow the impact of filial piety on individuals' personal preferences for elder care to be examined across the cities. Moreover, differences in various background characteristics suggest the possibility of variations in the normative influence of filial piety across cities. These variations are likely because Chinese society is not monolithic and the regional differences in China are substantial (Huang, 2008). Moreover, examining the variation in the normative influence is essential to assess the generality of the influence of filial piety across different places in China.

Methods

Individual-level data were obtained from surveys of 1,219 older Chinese (aged 60+) during early 2002. This total composed of 200 respondents from Beijing, 203 from Guangzhou, 201 from Nanjing, 215 from Shanghai, 200 from Xiamen, and 200 from Xian. These cities are scattered across different parts of China, with Beijing located in north, Shanghai in the east, and Xian in the west, and Guangzhou and Xiamen in the south. Because of the household registration system implemented in mainland China, inhabitants cannot freely change their residency across cities. This system thus helped perpetuate the differences among cities and is likely to have fueled the normative influences within, rather than across the cities.

All the surveys relied on a random sampling procedure to locate respondents at their household address. The sampling benefited from support from China's street committees, which are the lowest administrative unit of government in urban China. First, a random sampling procedure selected street committees from each of the mainland cities and the selected committees then provided sampling lists for the second stage of random sampling. As expected, the administrative support guaranteed a response rate of target respondents of above 90% (Shi, 1996), therefore the respondents are representative of community-dwelling older Chinese in each city. For the implementation of the survey, collaborators affiliated with universities in the cities helped arrange for the interviewers to conduct survey interviews with the respondents.

The respondents had an average age of 70.4 years. Among them, 46.2% were female, 74.8% married, 20.2% religious, 9.2% living alone, 58.2% living with children, 34.5% living with grandchildren, and 38.0% living with in-laws.

Measurement

The individual respondents rated items concerning their expectations of filial and grandfilial piety, perceived filial piety practice in society, desire for family elder care, and desire for state elder care on a five-point scale. The scale generated scores ranging from 0 to 100 to indicate intensity, with 0 for "very little," 25 for "rather little," 50 for "average," 75 for "rather a lot," and 100 for "very much." Higher scores meant higher expectations, perceptions, and desires.

Expected filial piety. As an indicator of the utility of family elder care, the expectation of filial piety was a composite of six items concerning the extent to which the respondents currently thought that older people expected their children to be caring, respectful, welcoming, pleasant, obedient, and financially supportive (Liu et al., 2000; Ng, 2002). The reliability (α) of the six-item composite measure was 0.762.

Expected grandfilial piety. As another indicator of the utility of family elder care, the expectation of grandfilial piety was a composite of six items concerning the extent to which the respondents currently thought that older people expected their grandchildren to be caring, respectful, welcoming, pleasant, obedient, and financially supportive (Liu et al., 2000; Ng, 2002). The reliability (α) of the six-item composite measure was 0.789.

Perception of filial piety. As an indicator of the social norm of filial piety, this question was a composite of six items concerning the extent to which the respondents perceived that children in society were generally caring, respectful, welcoming, pleasant, obedient, and financially supportive to their older parents during the preceding 6 months (Liu et al., 2000; Ng, 2002). The reliability (α) of the six-item composite measure was 0.882.

Filial piety in the city. As another indicator of the social norm of filial piety, this item represents the average of the level of filial piety perceived in each city. The perceptual data were obtained from a survey of 3,973 Chinese of various ages. They included 600 inhabitants of Beijing, 888 inhabitants of Shanghai, 600 inhabitants of Nanjing, 685 inhabitants of Guangzhou, 600 inhabitants of Xiamen, and 600 inhabitants of Xian. The aggregate-level reliability was 0.958 (O'Brien, 1990). This high reliability was due to the large number of respondents who provided perceptual data on the practice of filial piety in Chinese society.

Desire for family elder care. As another indicator of the utility of family elder care, this question was a composite of nine items concerning the extent to which the respondents currently endorsed instilling the spirit of filial piety in school, sharing the expenses of elder care among siblings, commending the example of filial piety among various strata in society, extolling the virtues of filial piety in the mass media, supporting caregivers in various sections of the community, supporting caregivers by kin and kith, providing various benefits to caregivers from the state, sharing the responsibility for the daily care of parents among siblings, and promoting respect for elders by the state. The reliability (α) of the nine-item composite measure was 0.892.

Desire for state elder care. As an indicator of the utility of state elder care or an inverse indicator of utility of family elder care, this question was a composite of ten items concerning the extent to which the respondents currently endorsed the state provision of daily care for the of elderly, the state providing community facilities to elders, the state showing concern for elders, and the state subsidizing elders by providing emotional support, accommodation, medical care, social and recreational facilities, and/or material support to elders. The reliability (α) of the ten-item composite measure was 0.941.

Preference for state elder care. As an indicator of the utility of state elder care or an inverse indicator of utility of family elder care, this item was the second factor in factor analysis that contrasted desire for state elder care ($\lambda = 0.305$) with desire for family elder care ($\lambda = -0.305$). The composite reliability was 0.996 (Bollen, 1989).

Acquiescence. As a control variable used in the regression analysis, acquiescence meant that the respondent rated all items consistently highly (Ferrando & Lorenzo-Seva, 2010). Acquiescence was then used as a response set, reflective of carelessness and other biases, and was the mean score of all rating items. A high score meant high ratings of all items and a low score meant low ratings of all items.

Analytic Technique

Linear regression analysis based on a mixed-effect model was used to adjust for random variation among the six cities (Peugh & Enders, 2005). In this analysis, indicators of the utility of family elder care or state elder care were outcome variables and the social norm of filial piety and background characteristics were explanatory variables. The analysis thus adjusted for any bias due to the clustering of cases within the selected cities. In addition to the analysis of data from all the six cities in one model, separate analyses of data from each city served to highlight the findings for each city.

Results

Generally, the desire for elder care was very high among older Chinese ($M>80$, see Table 8.1). The desire for family elder care was generally higher than the desire for state elder care. That is, family elder care was slightly more desirable than state elder care. Despite the apparent variation in the various indicators of the utility of family and state elder care (see Table 8.1), the variations were not statistically significant ($p>0.05$). Hence, the variation among the respondents within each city was relatively sizable, such that the variation among cities became insignificant. In view of the predominance of individual-level variation over city-level variation, the utility of family elder care vs. state elder care at the individual level warrants analysis and explanation.

Table 8.1 Means

Variable	Beijing	Shanghai	Nanjing	Guangzhou	Xiamen	Xian
Expected filial piety	84.2	73.4	65.2	69.8	81.7	67.6
Expected grandfilial piety	86.5	72.2	58.4	69.0	81.6	68.0
Desire for family elder care	87.3	84.6	83.4	88.0	88.2	82.1
Desire for state elder care	76.9	78.0	75.7	83.3	86.2	71.5
Perceived filial piety in society	70.9	58.1	58.5	59.0	64.5	56.1
Filial piety in the city	67.7	59.4	60.8	59.7	64.0	56.1
Acquiescence	77.0	53.6	69.6	62.5	82.5	70.6
Age	70.5	71.3	70.8	68.0	72.5	69.2
Female	47.5	48.4	49.8	44.7	49.5	37.5
Married	80.5	69.8	63.7	89.8	67.5	78.4
Nonreligious	92.0	72.9	95.5	83.2	47.6	87.4
Living alone	6.5	–	11.4	–	6.0	13.0
Living with parents	3.5	–	1.0	–	1.0	3.0
Living with children	57.0	–	58.7	–	60.5	56.5
Living with grandchildren	48.5	–	41.8	–	32.0	15.5
Living with in-laws	48.0	–	38.3	–	26.5	39.0
Living with friends	0.0	–	0.0	–	0.0	0.0
Living with siblings	0.5	0.0	0.5	0.0	1.5	1.5
Income (logged)	6.5	6.8	6.1	7.3	6.5	6.3

Table 8.2 Standardized fixed effects

Predictor	Expected filial piety	Expected grandfilial piety	Desire for family elder care	Desire for state elder care	Preference for state elder care
Perceived filial piety in society	0.044	0.141#	0.072*	0.065**	−0.014
Filial piety in the city	0.034	0.009	0.213**	0.220***	−0.001
Acquiescence	0.409#	0.389#	0.441#	0.429#	−0.046
Age	0.121#	0.099#	−0.013	0.061**	0.117#
Female	0.016	0.031	−0.010	0.000	0.014
Married	−0.019	−0.022	0.028	0.041	0.021
Nonreligious	0.014	−0.044	0.008	0.005	−0.012
Living alone	0.034	−0.018	−0.002	0.031	0.050
Living with parents	0.000	0.021	0.004	0.054	0.079***
Living with children	0.024	0.049	0.010	0.015	0.007
Living with grandchildren	0.064	0.072**	0.000	0.001	−0.007
Living with in-laws	−0.021	−0.111*	0.030	0.022	−0.010
Living with friends	0.264#	0.253*	0.215**	0.184	−0.051
Living with siblings	0.026	0.015	−0.010	−0.012	−0.001
Income	−0.035	−0.001	0.023	−0.018	−0.062**
Variance component					
City	0.003	0.020	0.019	0.054	0.025
Residual	0.821	0.762	0.717	0.678	0.946
R^2	0.176	0.218	0.264	0.268	0.029

#$p<0.001$; *$p<0.01$; **$p<0.05$; ***$p<0.10$

Predicting the Utility of Family Elder Care in All Cities Combined

As a focal predictor, the perceived level of filial piety in society displayed statistically significant positive effects on two indicators of the utility of family elder care, expected grandfilial piety ($\beta=0.141$, see Table 8.2) and the desire for family elder care ($\beta=0.072$). Moreover, the perceived social norm of filial piety aggregated in the city emitted a significant positive effect on utility in terms of the desire for family elder care ($\beta=0.213$). These effects were supportive of the focal hypothesis concerning the contribution of the social norm of filial piety. However, the two social norm indicators also manifested significant positive effects on the desire for state elder care, which was the alternative to family elder care. The effects of the social norm indicators on expected filial piety were not significant. These findings were unfavorable to the focal hypothesis. On balance, the focal hypothesis concerning the contribution of the social norm of filial piety did not receive wholesale support. Filial piety appeared to increase the desire for both family elder care and state elder care.

As a background characteristic, age showed significantly pervasive effects on four of the five indicators of the utility of elder care. In other words, the utility of both family and state elder care increased with age. Moreover, the preference for state elder care over family elder care also increased with age ($\beta=0.117$, see Table 8.2). Another important and consistent background predictor of the utility of family elder care was living with friends. Accordingly, living with friends engendered significant

Table 8.3 Standardized random effects (variations) across cities

Outcome	Perceived filial piety in society	Age
Expected filial piety	0.142	0.050
Expected grandfilial piety	0.083	0.020
Desire for family elder care	0.128	0.095
Desire for state elder care	0.095	0.087
Preference for state elder care	0.137	0.084

Note: The effects are in absolute values, as there was no need to differentiate positive and negative changes

positive effects on expected filial piety, expected grandfilial piety, and the desire for family elder care. Other significant findings included the positive effect of living with grandchildren and the negative effect of living with in-laws on expected grandfilial piety, and the positive effect of living alone and the negative effect of income on the preference for state elder care. None of the findings were attributable to any bias due to acquiescence, as acquiescence already served as a control variable in the analysis.

Differential Effects of the Social Norm Among the Cities

There were considerable, albeit statistically insignificant, variations among cities in terms of the effects of the filial piety perceived in society on various indicators of the utility of elder care (see Table 8.3). The variations in the effects of perceived filial piety were greater than those in the effects of age. These random variations appeared to be substantial, when compared with the fixed effects of filial piety in society and age (compare Table 8.3 with 8.2). In other words, the cities differed in regard to the effect of the perceived social norm of filial piety on the utility of elder care. This finding warranted separate analyses of the six Chinese cities.

The separate analyses of the cities showed that the filial piety perceived in Shanghai had significant positive effects on four of the five utility indicator (see Table 8.4). That is, the social norm of filial piety in Shanghai tended to raise the utility of both family elder care and state elder care. As such, the focal hypothesis did not find clear support in Shanghai. In contrast, the filial piety perceived in Nanjing had significant negative effects on the desire and preference for state elder care. This finding was consistent with the focal hypothesis. Another finding supportive of the hypothesis was the significant negative effect of the filial piety perceived in Beijing on the desire for state elder care. The significant effect of the filial piety perceived in Guangzhou on expected grandfilial piety also agreed with the hypothesis. However, the perceived filial piety in Xian exerted a significant negative effect on expected filial piety and a significant positive effect on the preference for state elder care. This finding contradicted the focal hypothesis. Furthermore, the filial piety perceived in Xiamen did not exhibit a significant effect on the utility of elder care. These findings offer a mixed picture in regard to the hypothesis. In all, they do not consistently support the hypothesis concerning the utility of the social norm of filial piety. Noticeable differences also appeared in the effects of the filial piety perceived in different cities.

Table 8.4 Standardized effects of filial piety in society perceived in each of the cities

City	Expected filial piety	Expected grandfilial piety	Desire for family elder care	Desire for state elder care	Preference for state elder care
Beijing	−0.063	−0.112	−0.076	−0.081[#]	−0.002
Shanghai	0.264*	0.302*	0.319*	0.316*	−0.028
Nanjing	−0.113	−0.016	−0.050	−0.314*	−0.405*
Guangzhou	0.078	0.118**	−0.059	0.004	0.103
Xiamen	−0.077	−0.071	−0.021	−0.031	−0.014
Xian	−0.307*	0.044	−0.065	0.072	0.216***

[#]$p<0.05$; *$p<0.001$; **$p<0.10$; ***$p<0.01$

Discussion

The general conclusion is that the social norm of filial piety did not have sufficiently consistent effects on the utility of family elder care to support the focal hypothesis. Admittedly, some findings were supportive of the hypothesis. These include the general contribution of the perception of filial piety as a social norm to the expected grandfilial piety and the desire for family elder care in the six Chinese cities. Other supportive findings appeared in the separate analyses of the social norm in Beijing, Nanjing, and Guangzhou. Nevertheless, the findings of the analyses were not consistently supportive of the hypotheses. Most findings suggested that, as a social norm, filial piety tended to raise the utility of both family elder care and state elder care. Such findings would neutralize the utility of the social norm, as the norm does not seem to be the unique basis to the utility of family elder care. The findings nevertheless imply that the social norm of filial piety would not reduce the demand for state elder care and thereby save public resources. A worse finding is the contribution of the social norm to the desirability of state elder care, and this contradicts the focal hypothesis. These diverse findings deserve further elaboration.

The positive contribution of the social norm of filial piety to the desire for family elder care, as hypothesized, embodies the assertion that what is common is likely to be good (Hsee, Loewenstein, Blount, & Bazerman, 1999). This influence of prevalence on desirability is plausible through a number of mechanisms, including the provision of social support, guidance, and even indoctrination, and processes of institutionalization, in addition to the mobilization of normative action (Peters, 1999). Accordingly, if filial piety was to become a social norm, many forms of support would emerge to facilitate the practice of filial piety or family elder care. Furthermore, the institutionalization of this support would furnish the formal rules, regulation, and enforcement of filial piety or family elder care. As the social norm would also raise people's willingness to perform family elder care, this willingness or morale would provide a basis for improving the quality of elder care (Emami, Torres, Lipson, & Ekman, 2000). All of these mechanisms would function to enhance the effectiveness, efficiency and, therefore, utility of family elder care. Accordingly, these mechanisms may underlie the influence of the social norm has on the desire for family elder care and the expectation of grandfilial piety across the cities. The mechanisms may also contribute to the impact of the social norm has on the desire for family elder care and the expectations of filial and grandfilial piety in Shanghai and Guangzhou. Conceivably, Shanghai and Guangzhou may have better normative support and institutionalization in regard to family elder care, which would consolidate the contribution of the social norm to family elder care.

In contrast to the above prediction of norm theory is the possibility that the social norm of filial piety weakens elders' expectations of filial piety. One reason for this possibility is that a lesser social norm of filial piety would invoke a higher need for filial piety. This reason is consistent with the influence that living far away from children has on parents' expectations of filial piety (Burr & Mutchler, 1999). The negative effect of the social norm on expectations was found in Xian. In this city, the prevalence of the social norm also slightly reduced the desire for family elder care. These findings suggest that filial piety and family elder care are desirable when they are unavailable. Conversely, the findings may imply that filial piety is satiable, such that it is no longer required when the elder has enjoyed it. This means that the need for, or desirability of filial piety is not enduring and, thus, that the utility of filial piety is short-lived, as least in Xian.

A third possibility is that the social norm of filial piety has no impact on the desire for family elder care. This would mean that the social norm engenders neither support nor satisfaction in regard to family elder care. The simplest argument in support of this possibility is that the elders pay no attention to the social norm or the social norm exerts no social influence on the elders' personal desires. This implies an individualistic scenario in which the elders do not care about the social norm. The individualistic image of older adults has been associated with successful and productive aging, independence, dignity, grand-parenting, and overall quality of life in the Chinese society of Hong Kong (Bengtson, Schmeechle, & Taylor, 1999; Block, 2000; Ho & Chiu, 1994; Moody, 2001; Plath, 2008; Pyke, 1999). Social norms also appear to exert no effect on older individuals' desires for decision making (Funk, 2004). Accordingly, the individualistic older adult lives in his or her own personal world, with no regard to social norms. The older adult then achieves successful and productive aging and other desirable conditions without experiencing any interference from the social environment. Hence, in this context, the desire for family elder care would be exclusively an individual decision (Linden & Horgas, 1997). Alternatively, the social norm of filial piety may not be sufficiently influential to facilitate the support to optimize family elder care and to boost elders' desire for the elder care. This view is consistent with the null finding concerning the impact of the social norm on elders' expectations of filial piety. The null finding specifically indicates that the expectation of filial piety is a matter of personal choice, in accordance with the neoliberal view (Kemp & Denton, 2003). Moreover, the null finding pervaded Beijing, Nanjing, and Xiamen.

By incorporating the view that the desire for family elder care and the desire for state elder care are mutually exclusive, norm theory would lead to the prediction that the social norm of filial piety discourages the desire for state elder care. Simply speaking, if filial piety and family elder care were commonly practiced and accepted, state elder care would be unnecessary and unattractive. This is the thesis of crowding out between family elder care and state elder care (Motel-Klingebiel, Gords, &

Betzin, 2009). The crowding-out effect illustrates the specificity of normative influences, such that the social norm of filial piety would raise the desire for filial piety and discourage desire for other associated norms. Such specificity is common in regard to social influences (Gurung, Taylor, & Seeman, 2003). In support of such normative influences are the negative effects of the social norm found in Nanjing and Beijing. From this perspective, the social norm of filial piety in the two cities would then generate utility by reducing the demand for state elder care. Probably, older residents in the two national/provincial-capital cities are more thoughtful in this regard, due to their greater exposure to state policy and propaganda.

The social norm of filial piety may also elevate the desire for state elder care. This is consistent with the commonly detected crowding-in effect, such that family support and state support are compatible and mutually reinforcing (Kunemund & Rein, 1999). One reason for the crowding-in effect is the ever-growing desire for elder care. In this case, the prevalence of filial piety simply uplifts the desire for more elder care, including from the state. Possibly, the social norm of filial piety demonstrates the need for, and utility of elder care, when awareness of the norm implies having received information about elder care. Another possibility is that the practice and social norm of filial piety are not effective or desirable. Their undesirability would shift the older population's desire towards state elder care. The positive effect of the social norm on the desire for state elder care occurred generally and in Shanghai and Xian especially. In Xian, in particular, the social norm of filial piety had a positive effect on the preference for state elder care and a negative effect on the desire for filial piety and family elder care. This finding shows that the older residents who were aware of the social norm in Xian would prefer to substitute state elder care for family elder care.

The last possibility is that the social norm of filial piety has no impact on the older population's desire for state elder care. This may occur when filial piety has no relation with state elder care. Filial piety and state elder care may be so compartmentalized that neither crowding out nor crowding in occurs (Motel-Klingebiel et al., 2005). In this case, family elder care and state elder care would simply be mutually fungible, rather than mutually exclusive. Findings concerning the null effect of the social norm emerged in Guangzhou and Xiamen. In Xiamen, in particular, the social norm of filial piety exhibited no effect at all on the desire for family or state elder care. Accordingly, the desire for elder care in that city appears to be a personal concern, insusceptible to the influence of the social norm. Notably, Guangzhou and Xiamen are both near the southern coast and may be more receptive to individualistic influences than the inner cities. Older residents in the two cities are then more likely to discount the influence of the social norm of filial piety.

Apart from the social norm of filial piety, age, living arrangements, and income had some significant influences on the desire for family elder care. These background influences generally reflect the role of self-interest. In accordance with self-interest theory, those who have a higher need for, or self-interest in elder care would have a higher corresponding desire for elder care. As such, those who are older, live with friends, have a lower income, or are not living with in-laws, would have a greater need or desire for elder care.

Limitations

The validity of the findings of this study may be compromised due to the cross-sectional survey design and the selection of only community-dwelling older Chinese in the six cities. These findings, therefore, have limited generality, as they do not aptly describe what happen in older Chinese in places other than the urban communities of the six cities. The cross-sectional design carries the risk that the social norm of filial piety is not an early cause of the desire for elder care. More importantly, the mechanisms underlying the effects of the social norm on the desire for elder care are not transparent, simply because the study did not include explicit measures of the mechanisms. Mechanisms, such as social support and cultural fit, derived from the social norm deserve further investigation to account for their normative influence. In addition to extending the measures to the mechanisms of the social norm, further research is needed to provide a more explicit assessment of the utility of elder care. In this respect, desire or expectation may not aptly capture the scope of utility. Pleasure, effectiveness, quality, and perceived utility are additional indicators of utility (Drakopoulos, 1995; Kosloski & Montgomery, 1993). Rather than relying solely on older adults' self-reported data, further research will benefit from the triangulation of data from multiple informants or sources. Family caregivers and other stakeholders would be vital informants on the utility of family and state elder care, and the inclusion of professional assessment and documentary data would also be useful. Importantly, the pairing of data obtained from different sources is needed for crosschecking and refining the data.

The study is also limited in that the philosophy of utilitarianism is referenced as the sole foundation for assessing the value of filial piety. From this perspective, filial piety is seen to be valuable when it is expected or desired and when the social norm of filial piety contributes to the desire for family elder care as opposed to state elder care. This philosophy is limited in scope, as it does not consider other principles of valuation, including deontological ethics, egalitarianism, equity, and social justice. These latter principles would extol the intrinsic value of filial piety, without tackling the consequences of filial piety (Pickard, 2010). As such, filial piety is good or virtuous no matter how much people prefer it and enjoy its benefits. To a certain extent, this means that the value of filial piety is beyond any empirical inquiry, as it remains an ethical judgment. This is a perspective quite different from the scientific approach.

Implications

This study finds that filial piety and family elder care are definitely desirable to older Chinese. They are also slightly more desirable than state elder care. These findings are grounds for facilitating filial piety and family elder care cater to for the self-interest of older adults. However, filial piety does not appear to have created a social norm that consistently boosts the utility of family elder care and weakens the desire for state elder care. In general, the social norm of filial piety shows some effects in raising the desirability of grandfilial piety and family elder care. These effects, however, are inconsistent, in view of the variation detected across the six Chinese cities. Specifically, the utility of the social norm of filial piety is more evident in Nanjing than in the other cities. The general picture is that the social norm either sustains the desirability of both family and state elder care or favors neither family nor state elder care. This inconsistency in the role of the social norm of filial piety means that the utility of promoting filial piety as a social norm is not clearly demonstrated. Instead, filial piety and family elder care tend to be personal concerns, unrelated to the social norm. This means that enhancing filial piety or family elder care is a personal or at best familial striving, which has no consistent relationship with the enhancement of filial piety as a social norm. In other words, enhancing filial piety as a social norm does not help to improve the utility of filial piety or family elder care. Accordingly, older Chinese in the six cities appear to be individualistic, as they are insensitive to the social norm of filial piety. This has some resemblance to the loss of filial obligation in the West (Silverstein et al., 2002). As such, the individual rather than society is seen as the focus of promoting filial piety and family elder care. Creating a norm or ethos of filial piety is not helpful for the promotion of elder care at an individual level.

The finding that the social norm of filial piety does not consistently sustain the utility of family elder care does not mean that filial piety is valueless on other accounts. Filial piety is believed to be intrinsically virtuous. However, what is questionable is the quality and benefit of filial piety with respect to elder care (Izuhara, 2002). The solution to the question would be to improve the quality and benefit of filial piety. The means of achieving this solution, nevertheless, certainly needs rigorous investigation. Besides, the solution may be costly, impractical, and ineffective and, thus, unjustifiable from a utilitarian point of view (Cooke, McNally, Mulligan, Harrison, & Newman, 2001).

At any rate, state elder care is inevitable, particularly when the social norm of filial piety favors state elder care. The preference for state elder care grows with age and with lower income. Those who are older and have a lower income are likely to have a higher need and, therefore, preference for state elder care (Pickard, 2006). Notably, the negative effect of income suggests that state elder care is not a luxury good. Older adults with higher incomes tend to prefer family elder care. Having a well-off older population is, therefore, a condition for reducing the reliance on state elder care. Conversely, state elder care is helpful and justifiable for the needy and those of advanced age.

References

Baltes, P. B., & Baltes, M. M. (1990). Psychological perspectives on successful aging: The model of selective optimization with compensation. In P. B. Baltes & M. M. Baltes (Eds.), *Successful aging: Perspectives from the behavioral sciences* (pp. 1–34). Cambridge, UK: Cambridge University Press.

Bengtson, V. L., Schmeechle, M., & Taylor, B. (1999). Using theories to build bridges in social gerontology. In *Memorial symposium: Quality of life, the 6th annual Asia/Oceania regional congress of gerontology* (pp. 19–24). Korea: Seoul.

Beugre, C. D., & Offodile, O. F. (2001). Managing for organizational effectiveness in Sub-Saharan Africa: A culture-fit model. *International Journal of Human Resource Management, 52,* 535–550.

Blekesaune, M., & Quadagno, J. (2003). Public attitudes toward welfare state policies: A comparative analysis of 24 nations. *European Sociological Review, 19,* 415–429.

Block, C. E. (2000). Dyadic and gender differences in perceptions of the grandparent-grandchild relationship. *International Journal of Aging & Human Development, 51,* 85–104.

Boersch-Supan, A. (2007). European welfare state regimes and their generosity toward the elderly. In D. B. Papadimitriou (Ed.), *Government spending on the elderly* (pp. 23–47). Houndmills: Palgrave.

Bollen, K. A. (1989). *Structural equations with latent variables.* New York: Wiley.

Brewer, L. (2001). Gender socialization and the cultural construction of elder caregivers. *Journal of Aging Studies, 15,* 217–235.

Burr, J. A., & Mutchler, J. E. (1999). Race and ethnic variation in norms of filial responsibility among older persons. *Journal of Marriage & the Family, 61,* 674–687.

Chen, J. (2003). The effect of parental investment on old-age support in urban China. In M. K. Whyte (Ed.), *China's revolutions and intergenerational relations* (pp. 197–221). Ann Arbor: Center for China Studies.

Cheng, S. T., & Chan, A. C. M. (2006). Filial piety and psychological well-being in well older Chinese. Journal of Gerontology: *Psychological Sciences, 61B(5),* P262–P269.

Cheung, C. K., & Chan, C. F. (2005). Philosophical foundations of eminent Hong Kong Chinese CEOs' leadership. *Journal of Business Ethics, 60,* 47–62.

Cheung, C. K., & Kwan, A. Y. H. (2009). The erosion of filial piety by modernisation in Chinese cities. *Ageing & Society, 29,* 179–198.

Cheung, C. K., Kwan, A. Y. H., & Ng, S. H. (2006). Impacts of filial piety on preference for kinship versus public care. *Journal of Community Psychology, 34,* 617–634.

Cheung, C. K., & Yeung, J. W. K. (2011). Meta-analysis of relationships between religiosity and constructive and destructive behaviors among adolescents. *Children & Youth Services Review, 33,* 376–385.

Cooke, D. D., McNally, L., Mulligan, K. T., Harrison, M. J. G., & Newman, S. P. (2001). Psychosocial interventions for caregivers of people with dementia: A systematic review. *Aging & Mental Health, 5,* 120–135.

Croll, E. J. (2006). The intergenerational contract in the changing Asian family. *Oxford Development Studies, 34,* 473–491.

Daatland, S. O., & Herlofson, K. (2003). Lost solidarity or changed solidarity: A comparative European view of normative family solidarity. *Ageing & Society, 23,* 537–560.

Dorfman, P. W. (2004). International and cross-cultural leadership research. In B. J. Punnett & O. Shenkar (Eds.), *Handbook for international management research* (2nd ed., pp. 265–355). Ann Arbor: University of Michigan Press.

Drakopoulos, S. A. (1995). *Values and economic theory: The case of hedonism.* Aldershot: Avebury.

Emami, A., Torres, S., Lipson, J. G., & Ekman, S.-L. (2000). An ethnographic study of a day care center for Iranian immigrant seniors. *Western Journal of Nursing Research, 22,* 169–188.

Engelhardt, H. T., Jr. (2007). Long-term care: The family, post-modernity, and conflicting moral life worlds. *The Journal of Medicine and Philosophy, 32,* 519–536.

Even-Zohar, A., & Sharlin, S. (2009). Grandchildhood: Adult grandchildren's perception of their role towards their grandparents from an intergenerational perspective. *Journal of Comparative Family Studies, 40,* 167–185.

Ferrando, P. J., & Lorenzo-Seva, U. (2010). Acquiescence as a source of bias and model and personal misfit: A theoretical and empirical analysis. *British Journal of Mathematical and Statistical Psychology, 63,* 427–448.

Fortinsky, R. H., Fenoter, J. R., & Judge, J. O. (2004). Medicare and Medicaid home health and Medicaid waiver services for dually eligible older adults: Risk factors for use and correlates of expenditures. *The Gerontologist, 44,* 739–749.

Franks, M. M., Pierce, L. S., & Dwyer, J. W. (2003). Expected parent-care involvement of adult children. *Journal of Applied Gerontology, 22*, 104–117.

Friedman, D., Hechter, M., & Kreager, D. (2008). A theory of the value of grandchildren. *Rationality and Society, 20*, 31–63.

Funk, L. M. (2004). Who wants to be involved? Decision-making preferences among residents of long-term care facilities. *Canadian Journal on Aging, 23*, 47–58.

Greengross, S. (1997). Working for consumers. In P. P. Mayer, E. J. Dickinson, & M. Sandler (Eds.), *Quality care for elderly people* (pp. 3–15). London: Chapman & Hall.

Guan, X., Zhan, H. J., & Liu, Q. (2007). Institutional and individual autonomy: Investigating predictors of attitudes toward institutional care in China. *International Journal of Aging & Human Development, 64*, 83–107.

Gurung, R. A. R., Taylor, S. E., & Seeman, T. E. (2003). Accounting for changes in social support among married older adults: Insights from the MacArthur studies of successful aging. *Psychology and Aging, 18*, 487–496.

Ho, D. Y., & Chiu, C. (1994). Component ideas of individualism, collectivism, and social organization: An application in the study of Chinese culture. In U. Kim, H. C. Triandis, C. Kagitchibasi, S. Choi, & G. Yoon (Eds.), *Individualism and collectivism: Theory, method, and applications* (pp. 137–156). Thousand Oaks: Sage.

Hooyman, N. R. (1990). Women as caregivers of the elderly: Implications for social welfare policy and practice. In D. E. Biegel & A. Blum (Eds.), *Aging and caregiving: Theory, research, and policy* (pp. 221–241). Newbury Park: Sage.

Hsee, C. K., Loewenstein, G. F., Blount, S., & Bazerman, M. H. (1999). Preference reversals between joint and separate evaluations of options: A review and theoretical analysis. *Psychological Bulletin, 125*, 576–590.

Huang, T. Y. M. (2008). Beyond the governance of global city-regions: Discourses and representations of Hong Kong cross-border identities. *Journal of Geographical Science, 52*, 1–30.

Huang, Y. (2011). Family relations and life satisfaction of older people: A comparative study between two different hukous in China. *Ageing & Society*.

Huddy, L., Jones, J. M., & Chard, R. E. (2001). Compassionate politics: Support for old-age programs among the non-elderly. *Political Psychology, 22*, 443–471.

Inglehart, R. (1986). Intergenerational change in politics and culture: The shift from materialist to postmaterialist value priorities. *Research in Political Sociology, 2*, 81–105.

Izuhara, M. (2002). Care and inheritance: Japanese and English perspectives on the generational contract. *Ageing & Society, 22*, 61–77.

Johansson-Stenman, O., Carlsson, F., & Daruvala, D. (2002). Measuring future grandparents' preferences for equality and relative standing. *The Economic Journal, 112*, 362–383.

Kane, R. A. (1991). Personal autonomy for residents in long-term care: Concepts and issues of measurement. In J. E. Birren, J. E. Lubben, J. C. Rowe, & D. E. Deutchman (Eds.), *The concept and measurement of quality of life in the frail elderly* (pp. 315–334). San Diego: Academic.

Kane, R. A. (1999). Goals of home care: Therapeutic compensatory, either, or both? *Journal of Aging and Health, 11*, 299–321.

Kemp, C. L., & Denton, M. (2003). The allocation of responsibility for later life: Canadian reflections on the roles of individuals, government, employers and families. *Ageing & Society, 23*, 737–760.

Killian, T., & Ganong, L. H. (2002). Ideology, context, and obligations to assist older persons. *Journal of Marriage & the Family, 64*, 1080–1088.

Kosberg, J. I. (2005). Meeting the needs of older men: Challenges for those in helping professions. *Journal of Sociology & Social Welfare, 32*, 9–31.

Kosloski, K., & Montgomery, R. J. V. (1993). Perceptions of respite services as predictors of utilization. *Research on Aging, 15*, 399–413.

Kunemund, H., & Rein, M. (1999). There is more to receiving than needing: Theoretical arguments and empirical explorations of crowding in and crowding out. *Ageing and Society, 19*, 93–121.

Li, C. (1997). Shifting perspectives: Filial morality revisited. *Philosophy East & West, 47,* 211–232.

Lin, A. H. (1992). *Study of filial piety in Confucian thought.* Taipei, Taiwan: Wen Jin.

Linden, M., & Horgas, A. L. (1997). Predicting health care utilization in the very old. *Journal of Aging and Health, 9,* 3–27.

Liu, J. H., Ng, S. H., Weatherall, A., & Loong, C. (2000). Filial piety, acculturation, and intergenerational communication among New Zealand Chinese. *Basic & Applied Social Psychology, 22,* 213–223.

Lum, T. Y., & Lightfoot, E. (2005). The effects of volunteering on the physical and mental health of older people. *Research on Aging, 27,* 31–55.

Maehara, T., & Takemura, A. (2007). The norms of filial piety and grandmother roles as perceived by grandmothers and their grandchildren in Japan and South Korea. *International Journal of Behavioral Development, 31,* 585–593.

Montgomery, R. J. V. (1999). The family role in the context of long-term care. *Journal of Aging and Health, 11,* 383–415.

Moody, H. R. (2001). Productive aging and the ideology of old age. In N. Morrow-Howell, J. Hinterlang, & M. Sherraden (Eds.), *Productive aging: Principles and perspectives* (pp. 175–196). Baltimore: Johns Hopkins University Press.

Motel-Klingebiel, A., Gords, L. R., & Betzin, J. (2009). Welfare states and quality of later life: Distributions and predictors in a comparative perspective. *European Journal of Aging, 6,* 67–78.

Motel-Klingebiel, A., Tesch-Roemer, C., von Kondratowitz, H.-J. (2005). Welfare strategy: Do not crowd out the family: Evidence for mixed responsibility from comparative analyses. *Ageing & Society, 25,* 863–882.

Nagata, D. K., Cheng, W. J. Y., & Tsai-Chae, A. H. (2010). Chinese American grandmothering: A qualitative exploration. *Asian American Journal of Psychology, 1,* 151–161.

Ng, S. H., Loong, C. S. F., Liu, J. H., & Weatherall, A. (2000). Will the young support the old? An individual- and family-level study of filial obligations in two New Zealand cultures. *Asian Journal of Social Psychology, 3,* 163–182.

Ng, S. H. (2002). Will families support their elders? Answers from across cultures. In T. D. Nelson (Ed.), *Stereotyping and prejudice against older persons* (pp. 295–310). Cambridge: MIT Press.

O'Brien, R. M. (1990). Estimating the reliability of aggregate-level variables based on individual-level characteristics. *Sociological Methods & Research, 18,* 473–504.

Papadimitriou, D. B. (2007). Economic perspectives on aging: An overview. In D. B. Papadimitriou (Ed.), *Government spending on the elderly* (pp. 1–19). Houndmills: Palgrave.

Peters, B. G. (1999). *Institutional theory in political science: The new institutionalism.* London: Continuum.

Peugh, J. L., & Enders, C. K. (2005). Using the SPSS mixed procedure to fit cross-sectional and longitudinal multilevel models. *Educational and Psychological Measurement, 65,* 811–835.

Pickard, J. G. (2006). The relationship of religiosity to older adults mental health service use. *Aging & Mental Health, 10,* 290–297.

Pickard, S. (2010). The good career: Moral practices in late modernity. *Sociology, 44,* 471–487.

Plath, D. (2008). Independence in old age: The route to social exclusion? *British Journal of Social Work, 38,* 1353–1369.

Pyke, K. (1999). The micropolitics of care in relationships between aging parents and adult children: Individualism, collectivism, and power. *Journal of Marriage & the Family, 61,* 661–672.

Shi, T. (1996). Survey research in China. *Research in Micropolitics, 5,* 213–250.

Silverstein, M., Bengtson, V. L., & Litwak, E. (2003). Theoretical approaches to problems of families, aging, and social support in the context of modernization. In S. Biggs, A. Lowenstein, & J. Hendricks (Eds.), *The need for theory: Critical approaches to social gerontology* (pp. 181–199). Amityville: Baywood.

Silverstein, M., Conroy, S. J., Wang, H., Giarrusso, R., & Bengtson, V. L. (2002). Reciprocity in parentchild relations over the adult life course. *Journal of Gerontology: Social Sciences. 57B,* S1–S13.

Stuifbergen, M. C., Van Delden, J. J. M., & Dykstra, P. A. (2008). The implications of today's family structures for support giving to older parents. *Ageing & Society, 28*, 418–434.

Wang, D., Laidlaw, K., Power, M. J., & Shen, J. (2010). Older people's belief of filial piety in China: Expectation and non-expectation. *Clinical Gerontologist, 33*, 21–38.

Ward, R. A. (2001). Linkages between family and societal-level intergenerational attitudes. *Research on Aging, 23*, 179–208.

Whyte, M. K., & Qin, L. (2003). Support for aging parents from daughters versus sons. In M. K. Whyte (Ed.), *China's revolutions and intergenerational relations* (pp. 167–195). Ann Arbor: Center for China Studies.

Williamson, J. B., McNamara, T. K., & Howling, S. A. (2003). Generational equity, generational interdependence, and the framing of the debate over social security reform. *Journal of Sociology & Social Welfare, 30*, 3–14.

Wolf, D. A. (1999). The family as provider of long-term care: Efficiency, equity, and externalities. *Journal of Aging and Health, 11*, 360–382.

Yao, Y. (2001). *Study on family support for elderly in China*. Beijing: China Population.

Zimmer, Z., & Kwong, J. (2003). Family size and support of older adults in urban and rural China: Current effects and future implications. *Demography, 40*, 23–44.

Chapter 9
Gendered Social Capital and Health Outcomes Among Older Adults in China

Qingwen Xu and Julie A. Norstrand

Abstract The gender dimension of social capital remains underinvestigated, particularly in the Eastern context. *Goal*: This chapter examines the relationship between social capital and health among a sample of older adults from China, particularly from a gender perspective. *Method*: Sample ($N = 1,854$) of older Chinese (60 years and over), 46.4% female, were obtained from Chinese General Social Survey (CGSS) 2005, a representative sample survey of China's urban and rural households. Principal component analysis generated eight dimensions of social capital on which male and female elders were compared using *t*-tests. Treating physical and emotional health outcomes as continuous variables, multiple regressions tested relationship between social capital and health outcomes for male and female groups separately. *Results*: Significant differences were found on some dimensions of social capital by gender. Regressions indicated different associations between dimensions of social capital with health outcomes by gender. *Implications*: Accounting for gender may be important when developing interventions to maximize social capital in communities of China.

Background

Social capital has grown in popularity over the last couple of decades (Harpham, Grant, & Rodriguez, 2004) and the connection between social capital and health is becoming increasingly established (Yip et al., 2007). However, much of this research has been carried out in Westernized countries. Research on social capital and health

Q. Xu (✉)
Tulane University School of Social Work, New Orleans, LA, USA
e-mail: qxu2@tulane.edu

J.A. Norstrand
Graduate School of Social Work, Boston College, Chestnut Hill, MA, USA

S. Chen and J.L. Powell (eds.), *Aging in China: Implications to Social Policy of a Changing Economic State*, International Perspectives on Aging,
DOI 10.1007/978-1-4419-8351-0_9, © Springer Science+Business Media, LLC 2012

among older Chinese is extremely limited. The need for focus on the older sector of the population is vitally important especially in countries such as China where aging of the population is occurring rapidly both in absolute numbers and at a faster pace. From 2001 to 2010 the rate of increase is expected to be approximately 3.28% each year (Chen, Yu-Cheung, & Chan, 2009); and it is estimated that the older population is projected to grow to 25% of total population by 2050 (Chen, 2006). Alongside the aging of the population, considerable demographic and social changes have been occurring in China such as shrinking family size, improvements in education, increasing migration and transformation of family care (Bloom et al., 2010; Chen et al., 2009; Yip et al., 2007). These changes could have important impacts on the relationship between social capital and health among older adults in China.

Meanwhile, gender issues in lifelong health shows that physical, mental, and social status of men and women in old age is rooted in the gender context throughout their lives (Read & Gorman, 2010). In patriarchal societies such as China, females and males are exposed to different experiences in life on the ground of their gender. For instance, access to social resources and vulnerability to risks are influenced by the cultural values, such as low valuation of girls and women as compared to that of boys and men (Smolin, 2011); discrimination against women and discrimination against the elderly have created accumulative disadvantages to the elderly women (Wong, 2011); and gendered power structure in the society and at home limit women's capacity in forming meaningful ties (Castells, 2007) even while some level of personal connectivity is increasingly available everywhere due to technological progress. Thus, difference in health outcomes, as well as the possessing and utilization of social capital across the gender line, could affect the relationship between social capital and health among older adults in China. However, there is little research that has been done on the role of social capital in relation to health through the lens of gender among older Chinese.

With the growing gap of life expectancy between men and women (Poston & Min, 2008), China's growth of the aging population, and in particular the increase in the number of widows and older single women, has become a compelling challenge. A better understanding of the gender dimension of healthy aging is particularly appealing as this may have critical implications for the life cycle of women and men in today's China. Furthermore, it may help the new aging society cope with the growing concern of how to care for the vast elderly population, and guide China's policy initiatives which recently have centered around activities to do with social capital such as productive participation (Morrow-Howell & Lou, 2009). Ultimately an important question that this paper aims to answer is: how the relationship between individual level social capital with physical and emotional health among older Chinese varies by the male–female dimension. The knowledge generated here would help develop appropriated community interventions, maximize the benefits social capital brought to elders' life, and contribute to elders' health and well-being in China.

Dimensions of social capital. Despite its many definitions and wide range of measures (Stephens, 2008), social capital emphasizes social relationships between groups of people (Onyx & Bullen, 2000). Typically, social capital is operationalized and assessed through trust, networks, and norms of reciprocity (Ferlander, 2007). A widely cited definition of social capital in the public health field is that of Putnam (1995): "features of social organization such as networks, norms, and social trust that facilitate coordination and cooperation for mutual benefit" (p. 67).

There continues to be considerable debate regarding how social capital should be examined and assessed. The two most common distinctions have been in terms of cognitive vs. structural social capital, and bonding vs. bridging social capital (Claridge, 2004). Structural social capital involves various forms of social organization, and a variety of networks that contribute to cooperation and cognitive social capital emphasizes norms, values, attitudes, and beliefs (Uphoff, 1999). Alternatively, bonding social capital refers to trusting and cooperative relationships between members of a network who share similar sociodemographically such as age, ethnicity and/or education, and bridging social capital is based on heterogeneous and outward-looking connections that include people across different social groups. Recently, Szreter and Woolcock (2004) have further developed the bonding/bridging conceptualization of social capital by adding a third dimension, linking social capital, which emphasizes the quality of vertical ties that exist between individuals and groups who are explicitly recognized as unequal in terms of power, such as between local governments and citizenry (Ferlander, 2007).

Similar concepts of social capital may exist in the Chinese context. The Mandarin term *Guanxi* denotes interpersonal connections and has been described along the bonding, bridging, and linking dimensions (Bian, 1997, 2001). The conceptualization of "*quanxi* capital" has focused on the "web of extended familial obligations," the relationships that extend beyond familial relationships to ensure "favors exchanges," and "asymmetric transactions" that resources flow from favor givers to favor receivers only as favor givers are recognized as resourceful (Bian, 2001, pp. 275–279). While the three dimensions of "*guanxi* capital" mirror to a certain extent that of the bonding, bridging and linking dimensions, *guanxi* capital nevertheless is not the same as social capital as conceptualized in the Western context. Culturally, *guanxi* capital lies in the skillful mobilization of moral and cultural imperatives, such as obligation and reciprocity "in pursuit of both diffuse social ends and calculated instrumental ends" (Yang, 1989, p. 35). Unlike social capital, *guanxi* capital emphasizes the materialization of interpersonal connections—such personal gains as employment (Bian, 1997), acquiring scare goods and services for family (Riley, 1994), and organizational gains such as business opportunities (Su & Littlefield, 2001). Thus, *guanxi* capital is particularly sensitive to power (Yang, 2002). Conceptualizations of social capital in China seem to ignore the structural/cognitive dimensions. Indeed, empirical studies have suggested that Western-based social capital theory and its conceptualization, based on organizational connections, might not be applied in Chinese society (Xu, Perkins, & Chow, 2010).

Social capital and health. Putnam argued that "in none is the importance of social connectedness so well established as in the case of health and well-being" (Putnam, 2000, p. 326). A substantial body of research has indicated the significant relationship between social capital and physical and mental health conditions, primarily among middle aged adults (Kawachi, Subramanian, & Kim, 2008). Social capital has been hypothesized to impact health through numerous pathways including individual behaviors (such as community participation) and norms (such as trust in community and reciprocity), as well as at the macro level mechanisms (such as historical, political and economic aspects) (Kawachi et al., 2008; Pollack & Knesebeck, 2004). The nature of pathways linking structural/cognitive and bonding/bridging/linking social capital with health remains poorly understood (Kim, Subramanian, & Kawachi, 2006). While demographic factors, including age, gender, marital status, family size, level of education, family income, and socioeconomic status (SES) are important predictors of physical and mental health (Franks, Gold, & Fiscella, 2003), there is evidence that social capital is associated with health outcomes, even after controlling these demographic characteristics. Relationships have been reported between social capital and health outcomes that include increased life-expectancy, and reduced rates of cancer, cardiovascular disease, obesity and diabetes (Kim, Subramanian, & Kawachi, 2008), as well as reduced rates of depression, mental distress, and substance abuse problems (Nyqvist, Gustavsson, & Gustafson, 2006).

Growing research demonstrates that social capital may be particularly meaningful for older adults due to deteriorating health (Pollack & Knesebeck, 2004), yet little empirical research has focused on older adults (Chen et al., 2009; Yip et al., 2007). The existing literature has for the most part demonstrated positive links between indicators of high social capital and good physical and mental health outcomes among older people (Pollack & Knesebeck, 2004). While significant associations have been identified between social capital indicators (e.g., trust, neighborhood cohesion, and reciprocity) and health outcomes among China's rural and urban poor (Yip et al., 2007), research concerning older Chinese has primarily focused on social support and its relations to health outcomes (e.g., Cheng, Lee, Chan, Leung, & Lee, 2009; Liu, Lianga, & Gu, 1995; Lou & Chi, 2001). Only one study assessed the relationship between social capital (i.e., structural networks, and community norms) and elders' quality of life (QoL) and functional capacity (Chen et al., 2009). The authors indicated "social capital had the most significant explanatory power for variation in the older adults' QoL" (Chen et al., p. 119).

Through the lens of gender. The gender dimension of social capital remains underinvestigated (Fox & Gershman, 2000; Molinas, 1998; O'Neill & Gidengil, 2005). In the literature, social capital is often conceptualized gender-blind, paying little attention to gendered intra- and extrahousehold issues of power and hierarchy (Agarwal, 1997; Silvey & Elmhirst, 2003). Indeed, the gender differences in the possession of social capital have been explained by cultural norms, in that women are more likely to have social networks that involve close networks whereas men's networks are more task-oriented (Gautam, Saito, & Kai, 2007). Scholars with feminism perspective have argued that social capital, true to its psychosocial promotion of within- and between-group cohesion, ignores gendered based conflicts (Muntaner & Lynch, 2002; Westermann, Ashby, & Pretty, 2005). And social capital exists within a broader context of gender inequality and can exacerbate women's disadvantages, as women remain excluded from the more powerful networks of trust and reciprocity that exist among men (Inglehart & Norris, 2003; Silvey & Elmhirst, 2003).

Until recently, there has been limited focus on gendered differences of social capital in the public health literature (Chuang & Chuang, 2008). Yet, from empirical studies substantial differences exist in terms of the sizes and the types of social capital to which men and women possess (Aihara, Minai, Kikuchi, Aoyama, & Shimanouchi, 2009). There is a growing body of evidence to suggest that social networks are gender-segregated; that is, men tend to connect with such associations as political parties, unions, and professional organizations that are often male-dominated, large and carrying much social economic and political power, while women tend to join organizations associated with education, the arts, religious institutions, neighborhood organizations, and care-oriented activities (Inglehart & Norris, 2003; Silvey & Elmhirst, 2003; Sun et al., 2007). Studies have also found that women reported significantly greater feeling of being part of the community, and satisfaction with the neighborhood in which they live (Welsh & Berry, 2009). Among aging populations, Aihara et al. (2009) found no gender differences in involvement in community activities among older Japanese; but men exhibited higher *cognition* of social capital, which in their study represented mutual support or reciprocity in one's own community.

Differences in the relationship between social capital and physical and mental health by gender have been found for general adult populations across countries. In terms of physical health, one study found social trust was beneficial for self-rated health of men whereas neighborhood safety and political participation was beneficial for self-rated health of women in Australia (Kavanagh, Bentley, Turrell, Broom, & Subramanian, 2006). A study looking at gendered differences in relationships between social capital and smoking and drinking behavior among Taiwanese adults found that neighborhood-level social trust had a stronger decreasing effect on smoking as well as drinking for women than for men (Chuang & Chuang, 2008). Another study found trust and religious participation improved nutritional health among men but not for women in the United States (Lochner et al., 2005). Looking at mental health, Welsh and Berry (2009) found gendered differences in terms of the relationships between structural and cognitive components of social capital with mental health and satisfaction with life. Specifically social capital indicators predicted mental health among men whereas among women social capital indicators predicted satisfaction with life (Welsh & Berry).

It should be noted however, that differences by gender have not been consistently reported particularly among older people. For instance, Aihara et al. (2009) reported no differences between older men and women in terms of the relationship between social capital and self-rated health, cognitive difficulties and depression. Inconsistent findings in the literature clearly suggest the inadequacy of research concerning older men and women; it might also imply that the different pattern of social capital possessed by older men and women might contribute to different health outcomes. That is, certain types and/or dimensions of social capital might be particularly meaningful depending on gender. This hypothesis triggers the interest in determining gendered differences in possessing social capital and the relationship between social capital and health among older people in the context of China.

Design and Methods

Questions and hypothesis. Given China's long history of patriarch society and the "gender equality" under its communist regime, it is understandable that male and female Chinese elders may have developed different dimensions of social capital, and these dimensions might have played a different role in elders' health trajectory. Thus, in this chapter, we conceptualized social capital using a hybrid approach by combining the bonding/bridging/linking and structural/cognitive dimensions (see Table 9.1 for detailed definition for each social capital dimension). This approach was adopted particularly given the concern that Western-oriented social capital conceptualization may not be appropriately transferred in the context of China; and it is also noted that the cultural *guanxi* capital does not fully grasp Western social capital's structural/cognitive dimensions.

Table 9.1 Social capital assessment items

Social capital types	Definition	Assessment items
Bonding/cognitive	Trust of people and close ties among family members, friends, and neighbors	Ten 5-point Likert questions asking level of trust participants have on people ranging from family members to strangers, and feeling of closeness participants have with family members, friends, and neighbors
Bridging/structural	Horizontal connections through participation in organizational activities	Seven 5-point Likert questions asking the frequency people participate in organizational activities (i.e., sports/exercise groups, social/entertainment, religious, children's interest groups, personal development groups, networking with alumni/people from the same place of origin, and public service)
Bridging/cognitive	Horizontal connections that can generate broader identities and reciprocity	Seven 5-point Likert questions asking the extent people help or expect to be help beyond activities in organizations (i.e., sports/exercise groups, social/entertainment, religious, children's interest groups, personal development groups, networking with alumni/people from the same place of origin, and public service)
Linking	Connection across vertical differentials up and down the social/power scale	Single item question about communist party membership

This chapter will answer: (1) do older male and female Chinese possess different dimensions of social capital? (2) Is social capital associated with health outcomes of elder Chinese after controlling demographic variations? and (3) does the relationship between social capital and health outcomes differ by male and female elders? While acknowledging that the literature is mixed, we hypothesize that dimensions of social capital will differ by gender. Based on the previous literature, we expect to see older women will possess higher levels of bonding/cognitive social capital and older men will possess higher level of bridging/structural social capital. It is difficult to make specific predictions for all the dimensions of social capital looked at here in this study since the literature is both limited and mixed. Furthermore, the cultural transferability of these many dimensions of social capital is hard to predict due to the Chinese context. Nonetheless, we do hypothesize that higher levels of social capital will be associated with better physical and emotional health. Gender differences in the relationships between social capital dimensions and health outcomes are hard to predict; female elders may enjoy social cohesion and closeness among family members and neighbors, thus result in better health outcomes. On the other hand, male elders residents may possess better health as they have more opportunities to access health care resources through higher level of bridging/structural social capital, and be educated with health information and healthy behaviors as a result of their "superior" status in the family and society.

Source of data. This chapter uses data from the Chinese General Social Survey (CGSS) 2005, a representative sample survey of China's urban and rural households aiming to monitor systematically the changing relationship between social structure and quality of life in urban and rural China. The survey was administrated by People's University and Hong Kong University of Science and Technology in China and Hong Kong, respectively. CGSS uses a four-stage stratified sampling scheme with unequal probabilities (HKUST Survey Research Center, 2004). In detail, the sampling units at each stage are: (1) first stage—125 urban districts (including suburban districts) and rural counties (including county-level municipalities) selected, (2) second stage—four townships, town seats and city subdistricts (streets) selected, (3) third stage—two urban neighborhood committees and rural villager committees selected, and (4) ten households selected and then one eligible household member is selected to be the survey respondent.

Sample. Our sample consisted of 1,854 Chinese elderly (60 years and above at the time of survey; 60 years and above is officially considered elderly in China), including 861 (46.4%) female, with a mean age of 67.9 years (sd=6.0). Only a small proportion (1.1%) of the sample was aged 85 years and older. Female and male older adults (see Table 9.2) did not differ in average age. A significantly higher proportion of females lived in urban areas than in the rural ones, although a larger percentage of female elders were alone at the time of survey. In general, older men had higher levels of education, income, and SES.

Table 9.2 Socioeconomic characteristics, social capital, and health outcomes: male vs. female elder people

	Total (N = 1,854)	Female (n = 861)	Male (n = 993)
Demographics			
Age (mean/sd)	67.9/6.0	67.8/5.9	68.0/6.0
Urban resident (%)*	63.8	68.8	59.4
Married (yes %)*	81.0	74.8	86.5
Family size (mean/sd)**	4.0/2.8	3.8/2.5	4.2/2.9
Years of education (mean/sd)*	7.5/5.4	5.9/5.1	8.9/5.2
Family income (mean/sd)**	17,339.2/	14,904.0/	19,457.2/
	34,549.9	17,648.0	44,187.5
SES (%)*			
Bottom	28.0	32.4	24.2
Lower middle	29.9	30.8	29.1
Middle	34.4	30.8	37.6
Upper middle	6.9	5.2	8.3
Top	0.9	0.8	0.9
Social capital	47.2/9.1	47.1/9.2	47.2/9.0
1. Education oriented***	2.8/2.5	2.7/2.1	2.9/2.5
2. Social oriented	4.4/4.1	4.4/4.2	4.3/4.0
3. Origin oriented	1.4/1.2	1.4/1.2	1.4/1.3
4. Closeness of closely related	10.5/2.2	10.4/2.1	10.5/2.2
5. Trust of closely related	16.0/2.5	16.0/2.4	16.0/2.5
6. Trust of not-so-closely related**	8.1/2.1	7.9/2.0	8.2/2.2
7. Religion oriented*	1.3/1.3	1.5/1.6	1.2/1.0
8. Public service oriented**	2.0/1.7	1.8/1.7	2.1/1.8
Health outcomes[a]			
Physical health (mean/sd)*	10.5/4.0	11.1/4.0	9.9/4.0
Emotional health (mean/sd)*	6.8/2.9	7.2/2.9	6.5/2.9

*$p \leq 0.00$, **$p \leq 0.01$, ***$p \leq 0.05$

[a]*Note*: the higher the number is, the worse the health status is reported

Variables. Health outcomes as dependent variables in this chapter include participants' self-assessed physical and emotional health. Self-assessed physical health was created by summing four questions: "Does physical health affect daily activity?," "Does physical health affect daily work at home?," "Do you feel pain?" and "What is your level of energy." Each of these questions was rated on a 5-point Likert scale (the higher, the worse physical health). Physical health variable ranged from 4 to 20 and the reliability test was acceptable (Cronbach's alpha = 0.863). Emotional health was created by summing two questions: "Do emotional issues affect interaction with family and friends?" and "Do emotional issues affect work study and daily activity?" Each of these questions was rated on a 5-point Likert scale (the higher, the worse emotional health). Emotional health ranged from 2 to 10 and reliability was acceptable (Cronbach's alpha = 0.850).

We operationalized social capital through a hybrid approach by combining the bonding/bridging/linking and structural/cognitive dimensions. Table 9.1 provides definitions, and assessment items for these dimensions of social capital. Our assessment items of structural/cognitive and bonding/bridging social capital were similar to those used by others (e.g., Kawachi et al., 2008). We used "years in the Communist Party of China" as a proxy to assess linking social capital (longer years in the Party refers to higher level of linking social capital), because the Party has a monopoly on social and political power in China for decades. While nonmembers do take certain public positions and are increasingly taking leadership in the private sector that possess social and/or political power, Communist Party membership, as well as tenure of the Party, is still the most important factor for upward mobility along the social/ power scale. Party members are power-holding elites and have enjoyed advantages as they link enterprises, villages, and communities with high-level structures.

This chapter also included seven control measures that have been found to be important predictors of physical and mental health: age, urban–rural residence, marital status, family size, level of education, total family income, and SES (Franks et al., 2003; Ziersch, Baum, Darmawan, Kavanagh, & Bentley, 2009). While SES can often be derived from education and income, self-perceived SES is included in the analyses taking into account China's huge geographic disparity in income and educational attainment. That is, self-perceived SES might not correlate with absolute numbers in income and years of education; for instance, widespread low level of income and educational attainment in rural areas might make older people in rural villages perceive themselves with higher level of SES than their urban counterparts.

Data analysis. The questions are addressed through a three-step process. First, given the multiple dimensions and measures used in the literature for social capital, and also considering the uncertainty of using the Western-based social capital measures for Chinese older people, we conducted a principal component analysis (PCA). PCA helped understand the complexity of a conceptual picture by reducing a large host of social capital related variables down to a few unique underlying social capital "dimensions." This has generated eight distinct dimensions of social capital among the Chinese elderly (Table 9.3). Second, male and female elders were then compared along these eight dimensions with a series of *t*-tests were conducted to see if the gender differences were statistically significant. This has revealed three different areas of social capital—those in which male and female elders showed no major differences; those in which female elders fall behind; and those in which male elders showed some deficit. Third, treating physical and emotional health outcome variables as continuous variables, separate multiple regression analyses were used to predict physical and emotional health. Model 1 included control variables only. Model 2, added the social capital dimensions. Multicollinearity and bivariate correlations were examined to ensure that measures such as education, total family income and SES were not highly correlated in the models (i.e., VIF < 2.5 for all measures). All analyses were carried out for male and female elders separately.

Table 9.3 Social capital dimensions

Social capital items	Dimensions							
	1	2	3	4	5	6	7	8
Organizational activities for children's interest	0.863							
Organizational activities for personal development	0.741							
Mutual help beyond organizational activities for children's education	0.859							
Mutual help beyond organizational activities for personal education	0.753							
Organizational activities for sports/exercise		0.817						
Organizational activities for social/entertainment		0.700						
Mutual help beyond organizational activities for sports/exercise		0.814						
Mutual help beyond organizational activities for social/entertainment		0.689						
Organizational activities for alumni/people from the same place of origin			0.808					
Mutual help beyond organizational activities for alumni/SPO people			0.815					
Closeness to friends/family members				0.566				
Familiarity of neighbors				0.801				
Mutual helps among neighbors				0.817				
Trust neighbors					0.767			
Trust people in the community					0.646			
Trust family members					0.791			
Trust alumni					0.548			
Trust not-so-close friends						0.698		
Trust people from the same place of origin (SPO people)						0.711		
Trust strangers						0.777		
Organizational activities for religion							0.963	
Mutual help beyond organizational activities for religion							0.961	
Organizational activities for public interests (volunteers)								0.679
Mutual help beyond organizational activities for public interests								0.692
Community party membership								0.526

Results

Older men and women possessed similar amount of overall social capital in general. Using PCA, eight dimensions of social capital among older people in China are identified (Table 9.3): (1) education oriented, (2) social oriented, (3) origin oriented, (4) closeness of friends and neighbors, (5) trust of closely related people, (6) trust of not-so-closely related people, (7) religion oriented, and (8) public service oriented. From these dimensions, similarities and differences in terms of possession of social capital were found by gender. Specifically, there were no significant difference between men and women on dimensions 2, 3, 4, and 5. However, older men scored significantly lower on religion oriented social capital and significantly higher on education and public service oriented social capital, as well as trust of not-so-closely related people oriented social capital compared to older women.

In terms of health, elder men reported significantly better physical and emotional health than elder women (Table 9.2). Results from regression models indicate that for physical health (Table 9.4), among female elders ($R^2 = 0.077$, $F(15, 544) = 3.039$, $p < 0.00$), after controlling demographic characteristics, religion-based social capital was negatively associated with physical health. Individual's SES and education were significant, with higher SES and longer years of education being associated with better physical health. Among male elders, Model 2 ($R^2 = 0.084$, $F(15, 730) = 4.46$, $p < 0.00$) suggested that social capital generated from the trust of closely related people (family members, friends, and neighbors) was associated with better physical health. Being younger and having higher level of SES were also associated with better physical health.

New patterns of relationships between social capital dimensions and emotional health were observed (Table 9.5). Among female elders, Model 2, which explained 12.3% variance ($R^2 = 0.123$, $F(15, 544) = 5.108$, $p < 0.00$) indicated that, after controlling demographic characteristics, religion based social capital continued to be negatively associated and social capital from the closeness with closely related people (family members, friends, and neighbors) was positively associated with better emotional health. Once more, female elders with higher level of education and SES reported better emotional health. Among male elders, Model 2 ($R^2 = 0.110$, $F(15, 730) = 6.016$, $p < 0.00$) indicated that social oriented social capital, and social capital generated from the trust of closely related people (family members, friends and neighbors) were positively associated with emotional health; however, education oriented social capital was negatively associated with emotional health. Similar to older women, higher level of education and SES were also significantly related to better emotional health for male elders.

Table 9.4 Physical health outcome by gender

Female (n = 861)

	Model 1			Model 2		
	Est.	SE	β	Est	SE	β
Constant	5.635	2.367		10.339	2.918	
Age	0.056	0.031	0.083	0.042	0.031	0.062
Urban	−0.026	0.424	−0.003	−0.040	0.450	−0.005
Married	−0.116	0.424	−0.012	−0.082	0.427	−0.009
Education	−0.087	0.036	−0.116*	−0.078	0.037	−0.103*
Family size	−0.005	0.071	−0.003	0.011	0.071	0.007
Family income	−2.82E−6	0.000	−0.013	−2.74E−6	0.000	−0.012
SES	−0.545	0.178	−0.132**	−0.499	0.181	−0.121**
Social capital						
1.				−0.027	0.078	−0.017
2.				−0.046	0.044	−0.053
3.				0.137	0.150	0.045
4.				−0.136	0.085	−0.070
5.				−0.130	0.073	−0.080
6.				−0.018	0.087	−0.009
7.				0.210	0.101	0.088*
8.				−0.146	0.110	−0.068

Male (n = 993)

	Model 1			Model 2		
	Est.	SE	β	Est.	SE	β
Constant	3.561	1.830		5.280	2.144	
Age	0.054	0.024	0.083*	0.058	0.024	0.089*
Urban	−0.509	0.333	−0.064	−0.478	0.372	−0.060
Married	0.238	0.438	0.020	0.228	0.436	0.019
Education	−0.046	0.032	−0.059	−0.052	0.033	−0.067
Family size	−0.001	0.055	−0.001	−0.002	0.055	−0.002
Family income	1.08E−6	0.000	0.013	8.12E−7	0.000	0.010
SES	−0.825	0.151	−0.203***	−0.793	0.153	−0.196***
Social capital						
1.				0.089	0.063	0.063
2.				−0.073	0.042	−0.078
3.				0.124	0.126	0.044
4.				0.089	0.071	0.051
5.				−0.128	0.063	−0.084*
6.				−0.120	0.068	−0.069
7.				0.171	0.161	0.039
8.				−0.001	0.095	−0.001

Social capital dimensions: *1.* education oriented; *2.* social oriented; *3.* origin oriented; *4.* closeness of closely related; *5.* trust closely related; *6.* trust not-so-closely related; *7.* religion oriented; *8.* public service oriented

*$p \leq 0.05$, **$p \leq 0.01$, ***$p \leq 0.00$

Table 9.5 Emotional health outcome by gender

	Female (n=861)						Male (n=993)					
	Model 1			Model 2			Model 1			Model 2		
	Est.	SE	β	Est.	SE	β	Est.	SE	β	Est.	SE	β
Constant	2.231	1.726		5.961	2.117		3.509	1.341		5.000	1.561	
Age	0.050	0.022	0.100*	0.034	0.022	0.068	0.022	0.017	0.045	0.028	0.017	0.059
Urban	-0.202	0.310	-0.031	-0.364	0.327	-0.055	-0.368	0.244	-0.063	-0.345	0.271	-0.059
Married	-0.074	0.309	-0.010	-0.074	0.309	-0.010	0.184	0.321	0.021	0.160	0.318	0.018
Education	-0.063	0.026	-0.112*	-0.060	0.027	-0.107*	-0.062	0.024	-0.107**	-0.062	0.024	-0.108**
Family size	-0.013	0.052	-0.011	-0.002	0.051	-0.002	-0.065	0.040	-0.059	-0.065	0.040	-0.059
Family income	-9.15E-6	0.000	-0.056	-1.01E-5	0.000	-0.062	2.719E-7	0.000	0.005	6.020E-8	0.000	0.001
SES	-0.587	0.130	-0.192***	-0.587	0.131	-0.192***	-0.617	0.111	-0.206***	-0.580	0.111	-0.194***
Social capital												
1.				0.033	0.057	0.027				0.120	0.046	0.114**
2.				-0.011	0.032	-0.017				-0.071	0.030	-0.104*
3.				0.044	0.109	0.019				0.051	0.092	0.024
4.				-0.178	0.062	-0.123**				0.065	0.051	0.050
5.				-0.102	0.053	-0.085				-0.169	0.046	-0.15***
6.				0.097	0.063	0.066				0.003	0.050	0.002
7.				0.164	0.073	0.093*				0.152	0.117	0.047
8.				-0.091	0.080	-0.057				-0.022	0.069	-0.014

Social capital dimensions: *1.* education oriented; *2.* social oriented; *3.* origin oriented; *4.* closeness of closely related; *5.* trust closely related; *6.* trust not-so-closely related; *7.* religion oriented; *8.* public service oriented

*p≤0.05, **p≤0.01, ***p≤0.00

Conclusion and Implications

Social capital conceptualization in the context of aging China. Findings of this chapter show that, unlike conventional measures of social capital in Western societies, social capital among older people in China has characteristics emphasizing the materialized social connections. In this chapter, participation in certain organizations (bridging/structural social capital) would only become meaningful when certain level of reciprocity (bridging/cognitive social capital) can be established through this connectedness. This result implies, first of all, that Western-oriented formal member organization based social capital (i.e., bridging/structural social capital through participating in organizational activities) might not fit the Chinese context. In post-1949 China, formal member organizations outside the Chinese Communist Party were few until the "emerging civil society" came in the very late 70s (Zhang, 2003). Recent social political reforms have led to the development of various nongovernmental organizations, including membership-based organizations. However, control of the nongovernmental section by the Chinese government still prevails; hence the growth of organizations for the elderly has been very slow. Therefore, bridging/structural social capital might not function quite as conceptualized in the West. Among older Chinese people, their membership in volunteer organizations could often be simply a practice where "young old" takes care of "very old" (Zhang & Goza, 2006). It is hard to expect that this type of social capital would function the same way in Western societies where participation in volunteer organizations often provide meaningful activities such as skill training that have been shown to have social and economic benefits on perceived well-being for older people (Windsor, Anstey, & Rodgers, 2008).

In addition, the Western conceptualization of social capital emphasizes social capital as a social resource; thus trust is often included in assessing social capital in the sense that the people we trust might want to help us, or add value to our personal or professional life; however, people would not "rely" on such trust or informal human connectedness for help or benefit. Results in this chapter showed that only the trust of closely related people (i.e., neighbors, kin, and classmates) was important for both physical and emotional health and only among older men, whereas trust of other people (i.e., persons from same geographic area, friends without true friendship, and strangers) was not significant for either physical or emotional health by either gender. In actual fact, the finding that close ties (i.e., trust of closely related people) were significant, while trust of not-so-closely related was not significant is what would be expected based on previous studies. However, the "gendered" results, that is, the close ties were only meaningful for older men, suggested that trusting connections might still hint to the underline value of materialized social capital. In China, as traditional care of older people is rooted in Confucianism, parental devotion (filial piety) and ancestor worship, care for older family members have been a normative family duty (Wu, Carter, Goins, & Cheng, 2005; Zhan & Montgomery, 2003). Despite the tremendous social economic changes, and the lack of caregivers in families due to China's decades of population control practice (i.e., one-child policy), majority of older people continue to live in the community and family care responsibility for elders is still a cultural and social norm (Xu & Chow, 2011), and often carried out by female family members (Zhan & Montgomery, 2003). Thus, for older men, those closely related people play an essential role in their life and older men probably do rely on them for both care and company that benefit their physical and emotional health.

It should be noted that China is a place with significant widespread poverty in its prereform era (pre-1978) and long-lasting poverty in its rural areas even after reforms, where resources have been limited and resource competition have been high. Given that the Western conceptualization of social capital is nested in an environment where material resources have been adequate for the individual and family's well-being, individuals in China may only benefit from social capital if they are able to access certain material resources essential to their well-being, such as healthcare, education, nutrition, and anti-pollution projects, etc. (Dorsten & Li, 2010). People might not be able to benefit from their connections with family members, friends, and people like them, who are also resource deprived. It is arguable that power related linking social capital may be specially meaningful in China in that people would be able to bring in needed resources through connections with the elite; we nevertheless cannot overestimate the effect of this type of social resource given the China's developing status. While we hope that Western-oriented conceptualization of social capital, particularly the nonmaterialized aspect of human connectedness, might offer special aspect to understand health disparities, results presented in this chapter imply that the cultural *guanxi* capital still appropriately addressed the usefulness of social relationships for Chinese elderly. This could be true only for older people; further research on the conceptualization of social capital in the Chinese context is clearly needed.

Gendered social capital and health. China's record of female status stands out quite favorably in comparison to other developing countries and its Asian neighbors, as a consequence of communist ideology that upholds gender equality (Zhang, Hannum, & Wang, 2008). Nevertheless, in China, women are still significantly disadvantaged by various measures of human and political capital. Nationwide, the human capital of women remains lower than that of men (Hannum, 2005). In terms of political connections and networks in China, which are often signaled by membership in the Chinese Communist Party, only 21.7% party members are female (Xinhua News, June 28, 2010). Results from this chapter further confirm such gendered stratifications in human and political capital.

Contrary to our first hypothesis, older men and woman possessed overall similar amount of social capital. However, when looking at each of the eight dimensions of social capital, older men did score higher on education oriented, trust-of-not so closely related and public service oriented social capital. Older women, on the other hand, scored higher on religion based social capital. Existing literature (e.g., Aihara et al., 2009; Gautam et al., 2007) imply that when the emphasis of social capital measures is on activity oriented tasks, men are more likely to exhibit higher levels of social capital whereas when social networks are emphasized, women on the other hand are more likely to score higher. Thus, the assessment used in this chapter could offer an important point; we'd rather argue that older men and woman have different strengths in establishing network and connecting with others. They could possess the same amount of social capital; although, the pattern and the quality of their social capital may be more imperative than the total amount of it when addressing gender-based disparities. Not surprisingly, the gender disparity in human and political capital may have led to the disparity in possession of education and power-related social capital. It is noted that older woman had more religion-related social capital. Seen as being antithetical to Communism thoughts, religion in China has been subject to registration, supervision, and odious regulation since 1949, yet religion and religious practice rebounded, being sustained by a 1982 central government policy (i.e., Document 19). Few measures of religiosity are available in China, but of those measures available, women are significantly more religious (Stark, 2002) and indeed Chinese older women engage in more religious activities than men according to Brown and Tierney (2009).

However, religion-based social capital was found to be negatively associated with older women's health outcomes. Although studies undertaken around the world repeatedly point to the robustly positive relationship between religiosity and one's well-being, this finding is in accordance with a previous study which found that religiously active elders in China are less likely to report having "good" or "very good" quality of life compared to those who do not participate in religious activities (Brown & Tierney, 2009). In China, unsanctioned religious organizations and practice, while nominally protected, have been under moderate level of control and regulation. Sanctioned religious organizations and practice, on the other hand, have experienced overt displays of intolerance and severe punishment. Without knowing what types of religious membership survey participants had, it is not possible to definitely explain the negative association between the religion-based social capital and older women's health. Participation in sanctioned religious organizations would explain the negative relationship between religious participation and health outcomes. Even if the elders' religious practices were with unsanctioned religious organizations, religious persecution and maltreatment during China's Cultural Revolution that was commensurate with religious activities (FitzGerald, 1967) may continue to affect elders' physical and emotional health. Thus, religious related social capital in China is unlikely to function as a social resource; women's higher score on this type of social capital may not be beneficial to their health.

For older Chinese men, the trust closely related based social capital is the only dimension that is positively associated with both physical and emotional health. This association did not present among older Chinese women. As we argued previously, materialized social connections characterize the social capital among Chinese older people, which is beyond what social capital can offer as a social resource. Owing to gendered distribution of elderly care responsibility, a primary role taken by women in a patriarchy society and Chinese tradition, older Chinese men depend on the closely related people (i.e., family members, friends, neighbors) for care, financial help and companionship, more heavily than what older Chinese women. Hence, the higher level of trust older men have on these closely related people, the better health outcomes they would have. From this perspective, it is noted that gender disparities in heath have been complicated and nuanced. Despite the structured discrimination against women in any patriarchal society, as well as the biological and medical explanations, simply the gendered nature of working and family-care paths, would have created patterned disparities in a constellation of health related resources, relationships, and risks, as well as reduction in mastery and control (Moen & Chermack, 2005). As the role of trusting social connections may not be so important for women's health, it may simply be a result of men's shorter longevity, hence making them reliant on care before women do; this outcome may also imply that women's traditional family responsibilities contribute to women's resilience in their process of aging.

Our interpretation that links the broad social capital factors to elder health is based on the understanding that the great inequality in the procurement of social services and variations in individuals' access to life opportunities and material resources (e.g., health care, education) result in health disparity. In other words, *social capital does not automatically lead to better health*. The positive role played by social capital is subject to the individual characteristics, as well as, the community and macro level environment. In this chapter it is noted that self-perceived SES and years of education (but not income) showed significant associations with elders' physical and emotional health among both male and female elders in China. On the one hand, socioeconomic related measures (e.g., income) might not be able to capture the elements that affect health, particularly in less wealthy countries such as China (Dorsten & Li, 2010). As previously argued, various structural conditions in China, dominance hierarchies embedded in China's health care distribution system, and long-term underinvestment in human or social infrastructure in China's rural areas, may be more powerful than individual level limitations, (e.g., low income). On the other hand, the connection between relative social status in the community and health outcomes underlines the importance of psychosocial determinates. In China's context, a high level of self-perceived SES could help elders identify and achieve personal aspirations, satisfy needs, and cope with the environment. Ultimately, it is the perception of one's psychosocial status which plays a critical role in the evaluation of health outcomes. Additional research investigating the association between SES, social capital and health outcomes, which fully consider levels of economic development and gendered stratification system, is critical.

Limitations. The nature of the relationship between measures of social capital and health can only be described in terms of associations. Owing to cross-sectional nature of the data, no conclusions regarding causality can be made for this chapter. This is a frequent limitation of social capital research; more longitudinal research on social capital and health indictors is needed. Another important limitation, due to limited data availability, is that the measurements of social capital dimensions and health status have not adopted a fuller scale and have not been validated specifically in the context of China and among Chinese older people. Despite these limitations, a major contribution of this chapter is that it represents one of the first studies to examine the relationship between social capital, using the hybrid conceptualization of structural/cognitive and bonding/bridging/linking social capital, with both physical and mental health outcomes among older people in the Chinese context. To the authors' best knowledge, this chapter is the first to use this conceptualization of social capital; clearly, further research is needed to examine this conceptualization. Furthermore, this chapter provides intriguing insight into the nature of the relationship between social capital and health by gender in China.

Implications. The examination of social capital in relation to gender provides an important insight into the role of social environment (as measured by social capital) for men and women separately. The need for understand how women and men function in the environment is important as much literature written by men have not focused adequately on the constructed subjective identity of women (Butler, 2003). Unless women stop trying to measure up to men, and concentrate instead on exploding the kind of power structures that have made women the "second sex," gender-based disparities in health status and social capital would not disappear simply through the improvement of women's statuses in the society. From this perspective, gendered strengths developed across older women's life course (e.g., family-care responsibilities) may help and translate into improved health outcomes through different mechanisms of interacting with the social environment. Contrary to the view that has tended to depict women's existential experiences only as marginalized politically, economically, and socially, from results in this chapter, we argue for a strength-based perspective that emphasizes women's essential role as mothers and caregivers, and the injustices that mar the relationship between caregivers and care receivers.

The concept of social capital has been mobilized within the health field primarily as a possible explanation for findings with respect to inequalities in health, and in particular for the relationship between income inequality and health (Kawachi, Kennedy, Lochner, & Prothrow-Stith, 1997). In a patriarchal society, particular for older women that have been structured to care for others and not enough for themselves, and to settle for second place through their life course, justice for older women in the realm of health and care could possibly be achieved through the celebration of a set of traditional "feminine" norms and values such as community, sharing, emotion, trust, and process, some of which are essential to older people's health and necessarily better than such norms and values of as autonomy, hierarchy, and domination traditionally labeled as "masculine."

This chapter provides preliminary data and implies that social capital may have unique features among older adults and in the Chinese culture. More important, gender difference in the possession of social capital is significant; such differences may be understood in terms of China's traditional patriarchal culture as well as unequal opportunity structures in education, employment, and social service accessibility. In recent decades, China's unique socioeconomic and political structures, as well as policy and practices, have created greater challenges to the health of more vulnerable groups of people (such as older people in rural villages and poor elderly in urban cities). Our findings suggest that it is important that informal social networks be cultivated for older people's well-being. Development of social and community programs that account for gendered differences in terms of interactions with the social environment, and which encourage meaningful and trusting connections among older people, as well as prevention of policies that result in health disparities (e.g., unequal distribution of social welfare services), will be equally influential. Furthermore, as described above, China has initiated policies focused on activities such as productive participation; this is strongly supported by the authors of this chapter. To conclude, implementation of a social capital agenda must account for sociodemographic and environmental contexts of the focus population, in this case, Chinese elders. With the looming growth of the elderly population in China, the need for policies that can assist this sector of the population is compelling.

References

Agarwal, B. (1997). "Bargaining" and gender relations: Within and beyond the household. *Feminist Economics, 3*(1), 1–51.

Aihara, Y., Minai, J., Kikuchi, Y., Aoyama, A., & Shimanouchi, S. (2009). Cognition of social capital in older Japanese men and women. *Journal of Aging and Health, 21*(8), 1083–1097.

Bian, Y. (1997). Bringing strong ties back in: Indirect ties, network bridges, and job searches in China. *American Sociological Review, 62*(3), 366–385.

Bian, Y. (2001). Guanxi capital and social eating in Chinese cities: Theoretical models and empirical analyses. In N. Lin, K. S. Cook, & R. S. Burt (Eds.), *Social capital: Theory and research* (pp. 275–296). New Brunswick: Transaction Publishers.

Bloom, D. E., Canning, D., Hu, L., Liu, Y., Mahal, A., & Yip, W. (2010). The contribution of population health and demographic change to economic growth in China and India. *Journal of Comparative Economics, 38*(1), 17–33.

Brown, P. H., & Tierney, B. (2009). Religion and subjective well-being among the elderly in China. *Journal of Socio-Economics, 38*(2), 310–319.

Butler, J. (2003). Gender trouble, feminist theory and psychoanalytic discourse. In L. M. Alcoff & E. Mendieta (Eds.), *Identities: Race, class, gender and nationalities* (pp. 201–212). Oxford: Blackwell Publishing Ltd.

Castells, M. (2007). Communication, power and counter-power in the network society. *International Journal of Communication, 1*(2007), 238–266.

Chen, H., Yu-Cheung, W., & Chan, K. (2009). Social capital among older Chinese adults: An exploratory study of quality of life and social capital in a Chinese urban community. *International Journal Interdisciplinary Social Sciences, 4*(9), 107–123.

Chen, W. (2006). China's population trend: 2005–2050. *Population Studies (China), 2006*(4), 93–95.

Cheng, S.-T., Lee, C., Chan, A., Leung, E., & Lee, J.-J. (2009). Social network types and subjective well-being in Chinese older adults. *The Journals of Gerontology, Series B, Psychological Sciences and Social Sciences, 64B*(6), 713–722.

Chuang, Y.-C., & Chuang, K.-Y. (2008). Gender differences in relationships between social capital and individual smoking and drinking behavior in Taiwan. *Social Science & Medicine, 67*(8), 1321–1330.

Claridge, T. (2004). *Social capital and natural resource management.* Unpublished Thesis, University of Queensland, Brisbane, Australia.

Dorsten, L. E., & Li, Y. (2010). Modeling the effects of macro-measures on elder health in China: A "fresh sample" approach. *Global Journal of Health Science, 2*(1), 8–19.

Ferlander, A. (2007). The importance of different forms of social capital for health. *Acta Sociologica, 50*(2), 115–128.

FitzGerald, C. P. (1967). Religion and China's cultural revolution. *Pacific Affairs, 40*(1/2), 124–129.

Fox, J., & Gershman, J. (2000). The World Bank and social capital: Lessons from ten rural development projects in the Philippines and Mexico. *Policy Sciences, 33*(3/4), 399–419.

Franks, P., Gold, M. R., & Fiscella, K. (2003). Sociodemographics, self-rated health, and mortality in the US. *Social Science & Medicine, 56*(12), 2505–2514.

Gautam, R., Saito, T., & Kai, I. (2007). Leisure and religious activity participation and mental health: Gender analysis of older adults in Nepal. *BMC Public Health, 7*, 299.

Hannum, E. (2005). Market transition, educational disparities, and family strategies in rural China: New evidence on gender stratification and development. *Demography, 42*(2), 275–299.

Harpham, T., Grant, E., & Rodriguez, C. (2004). Mental health and social capital in Cali, Colombia. *Social Science & Medicine, 58*, 2267–2277.

Hong Kong University of Sciences & Technology. (2004). *Chinese general social survey (China GSS)*. Hong Kong University of Science & Technology Survey Research Center. Retrieved in March 8, 2008, from http://www.ust.hk/~websosc/survey/GSS_e.html.

Inglehart, R., & Norris, P. (2003). *Rising tide: Gender equality and cultural change around the world*. Cambridge: Cambridge University Press.

Kavanagh, A., Bentley, R., Turrell, G., Broom, D., & Subramanian, S. (2006). Does gender modify associations between self-rated health and the social and economic characteristics of local environments? *Journal of Epidemiology and Community Health, 60*(6), 490–495.

Kawachi, I., Kennedy, B. P., Lochner, K., & Prothrow-Stith, D. (1997). Social capital, income inequality and mortality. *American Journal of Public Health, 87*, 1491–1498.

Kawachi, I., Subramanian, S. V., & Kim, D. (Eds.). (2008). *Social capital and health*. New York: Springer.

Kim, D., Subramanian, S. V., & Kawachi, I. (2006). Bonding versus bridging social capital and their associations with self rated health: A multilevel analysis of 40 US communities. *Journal of Epidemiology and Community Health, 60*, 116–122.

Kim, D., Subramanian, S. V., & Kawachi, I. (2008). Social capital and physical health. In I. Kawachi, S. V. Subramanian, & K. Kim (Eds.), *Social capital and health*. New York: Springer.

Liu, X., Lianga, J., & Gu, S. (1995). Flows of social support and health status among older persons in China. *Social Science & Medicine, 41*, 1175–1184.

Lochner, J., Ritchie, C., Roth, D., Baker, P., Bodner, E., & Allman, R. (2005). Social isolation, support, and capital and nutritional risk in an older sample: Ethnic and gender differences. *Social Science & Medicine, 60*(4), 747–761.

Lou, W., & Chi, I. (2001). Health-related quality of life of the elderly in Hong Kong: Impact of social support. In I. Chi, N. L. Chappell, & J. Lubben (Eds.), *Elderly Chinese in pacific rim countries* (pp. 97–113). Hong Kong: University Press.

Moen, P., & Chermack, K. (2005). Gender disparities in health: Strategic selection, careers and cycles of control. *Journal of Gerontology, 60B*, 99–108.

Molinas, J. R. (1998). The impact of Inequality, gender, external assistance and social capital on local level cooperation. *World Development, 26*(3), 413–431.

Morrow-Howell, N., & Lou, L. (2009). *China's rapidly aging population is part of a worldwide trend*. Press Release: George Warren Brown School of Social Work. Washington University in St. Louis. Retrieved June 29, 2010, from http://gwbweb.wustl.edu/newsroom/PressRelease/Pages/aginginChina.aspx.

Muntaner, C., & Lynch, J. (2002). Social capital, class gender and race conflict, and population health: An essay review of bowling alone's implications for social epidemiology. *International Journal of Epidemiology, 31*(1), 261–267.

Nyqvist, F., Gustavsson, J., & Gustafson, Y. (2006). Social capital and health in the oldest old: The Umeå 85+ study. *International Journal of Ageing and Later life, 1*(1), 91–114. doi:10.3384/ijal.1652-8670.061191 DOI:dx.doi.org.

O'Neill, B., & Gidengil, E. (Eds.). (2005). *Gender and social capital*. London: Routledge.

Onyx, J., & Bullen, P. (2000). Measuring social capital in five communities. *The Journal of Applied Behavioral Science, 36*(1), 23–42. doi:10.1177/0021886300361002 DOI:dx.doi.org.

Pollack, C. E., & Knesebeck, V. O. (2004). Social capital and health among the aged: Comparisons between the United States and Germany. *Health & Place, 10*(4), 383–391.

Poston, D. L., Jr., & Min, H. (2008). The effects of sociodemographic factors on the hazard of dying among Chinese oldest old. In Z. Yi, D. L. Poston Jr., D. V. Vlosky, & D. Gu (Eds.), *Health longevity in China: Demographic socioeconomic, and psychological dimensions* (pp. 121–131). New York: Springer.

Putnam, R. (1995). Bowling alone: America's declining social capital. *Journal of Democracy, 6*(1), 65–78. doi:10.1353/jod.1995.0002 DOI:dx.doi.org.

Putnam, R. (2000). *Bowling alone: The collapse and revival of American community*. New York: Simon and Schuster.

Read, J. G., & Gorman, B. (2010). Gender and health inequality. *Annual Review of Sociology, 36*, 371–386.

Riley, N. E. (1994). Interwoven lives: Parents, marriage, and Guanxi in China. *Journal of Marriage and Family, 56*(4), 791–803. doi:10.2307/353592 DOI:dx.doi.org.

Silvey, R., & Elmhirst, R. (2003). Engendering social capital: Women workers and rural–urban networks in Indonesia's crisis. *World Development, 31*(5), 865–879.

Smolin, D. M. (2011). *The missing girls of China: Population, policy, culture, gender, abortion, abandonment, and adoption in East-Asian perspective*. Retrieved in October 30, 2011, http://works.bepress.com/cgi/viewcontent.cgi?article=1008&context=david_smolin.

Stark, R. (2002). Physiology and faith: Addressing the "universal" gender difference in religious commitment. *Journal for the Scientific Study of Religion, 41*(3), 495–507.

Stephens, C. (2008). Social capital in its place: Using social theory to understand social capital and inequalities in health. *Social Science & Medicine, 66*, 1174–1184.

Su, C., & Littlefield, J. E. (2001). Entering Guanxi: A business ethical dilemma in mainland China? *Journal of Business Ethics, 33*(3), 199–210. doi:10.1023/A:1017570803846 DOI:dx.doi.org.

Sun, W., Watanabe, M., Tanimoto, Y., Shibutani, T., Kono, R., Saito, M., Usada, K., & Kono, K. (2007). Factors associated with good self-rated health of non-disabled elderly living alone in Japan: A cross-sectional study. *BMC Public Health, 7*, 297–306. doi:10.1186/1471-2458-7-297.

Szreter, S., & Woolcock, A. (2004). Health by association? Social capital, social theory and the political economy of public health. *International Journal of Epidemiology, 33*, 650–667.

Uphoff, N. (1999). Understanding social capital: Learning from the analysis and experiences of participation. In P. Dasgupta & I. Seregeldin (Eds.), *Social capital: A multifaceted perspective*. Washington: World Bank.

Welsh, J. A., & Berry, H. L. (2009). *Social capital and mental health and well-being*. National Centre for Epidemiology and Population Health, The Australian National University Paper presented at the Biennial HILDA Survey Research Conference 16–17 July 2009. Retrieved in October 30, 2011, from http://melbourneinstitute.com/downloads/hilda/Bibliography/2009_papers/Welsh,%20Jennifer_paper.pdf.

Westermann, O., Ashby, J., & Pretty, J. (2005). Gender and social capital: The importance of gender differences for the maturity and effectiveness of natural resource management groups. *World Development, 33*(11), 1783–1799.

Windsor, T. D., Anstey, K. J., & Rodgers, B. (2008). Volunteering and psychological well-being among young-old adults: How much is too much? *The Gerontologist, 48*(1), 59–70.

Wong, H. (2011). Quality of life of poor people living in remote areas in Hong Kong. *Social Indicators Research, 100*, 435–450.

Wu, B., Carter, M. W., Goins, R. T., & Cheng, C. (2005). Emerging services for community-based long-term care in urban China: A systematic analysis of Shanghai's community-based agencies. *Journal of Aging & Social Policy, 17*(4), 37–60.

Xinhua News. (2010, June 28). *At the end of 2009, the membership of the communist party of China reached to 77.995 millions*. The Central People's Government of the People's Republic of China. Retrieved from http://www.gov.cn/jrzg/2010-06/28/content_1639416.htm.

Xu, Q., & Chow, J. (2011). Exploring the community-based service delivery model: Elderly care in China. *International Social Work, 54*(3), 374–387.

Xu, Q., Perkins, D., & Chow, J. (2010). Community participation, sense of community, and social capital: China's experience. *American Journal of Community Psychology, 45*(3/4), 259–271.

Yang, M. (1989). The gift economy and state power in China. *Comparative Studies in Society and History, 31*, 25–54.

Yang, M. M. (2002). The resilience of Guanxi and its new deployments: A critique of some new Guanxi scholarship. *The China Quarterly, 170*, 459–476. doi:10.1017/S000944390200027X DOI:dx.doi.org.

Yip, W., Subramanian, S. V., Mitchell, A., Wang, J., Lee, D., & Kawachi, I. (2007). Does social capital enhance health and well-being? Evidence from rural China. *Social Science & Medicine, 64*(1), 35–49. doi:10.1016/j.socscimed.2006.08.027 DOI:dx.doi.org.

Zhan, H. J., & Montgomery, R. J. Y. (2003). Gender and elder care in China: The influence of filial piety and structural constraints. *Gender and Society, 17*(2), 209–229.

Zhang, Y. (2003). China's emerging civil society. *The Brookings Institute*. Retrieved in October 30, 2011, http://www.brookings.edu/~/media/Files/rc/papers/2003/08china_ye/ye2003.pdf.

Zhang, Y., & Goza, F. W. (2006). Who will care for the elderly in China? A review of the problems caused by China's one-child policy and their potential solutions. *Journal of Aging Studies, 20*(2), 151–164. doi:10.1016/j.jaging.2005.07.002 DOI:dx.doi.org.

Zhang, Y., Hannum, E., & Wang, M. (2008). Gender-based employment and income differences in urban China: Considering the contributions of marriage and parenthood. *Social Forces, 86*(4), 1529–1560.

Ziersch, A. M., Baum, F., Darmawan, I. G., Kavanagh, A. M., & Bentley, R. J. (2009). Social capital and health in rural and urban communities in South Australia. *Australian and New Zealand Journal of Public Health, 33*(1), 7–16.

Chapter 10
An East–West Approach to Mind–Body Health of Chinese Elderly

Rainbow Tin Hung HO, Phyllis Hau Yan LO, Cecilia Lai Wan CHAN, and Pui Pamela Yu LEUNG

Abstract Mind–body medicine is rooted in the traditional Eastern models of health emphasizing nurturing life, attaining balance, and an integrated view of being. Meanwhile, scientific evidence for the mind–body connection and the development of such therapies in treating disorders have given rise to an assimilation of the Eastern and Western approaches to health and illnesses. In recent decades, there has been an influx of Western medical approaches in China, and Chinese elderly are subjected to an alternative dimension of health that advocates targeted treatment and disease eradication. This chapter discusses how the interplay of Eastern health philosophies and the Western biomedical model can promote the physical and psychological condition of Chinese elderly and to flexibly adapt to their inevitable decline in health statuses.

Keywords Mind–body • Health • Elderly • Chinese • Body–mind–spirit • Eastern medicine

R.T.H. HO (✉) • C.L.W. CHAN
Centre on Behavioral Health, The University of Hong Kong, Pokfulam, Hong Kong

Department of Social Work and Social Administration, University of Hong Kong,
Pokfulam, Hong Kong
e-mail: tinho@hku.hk

P.H.Y. LO
Centre on Behavioral Health, The University of Hong Kong, Pokfulam, Hong Kong

P.P.Y. LEUNG
Department of Social Work and Social Administration, University of Hong Kong,
Pokfulam, Hong Kong

S. Chen and J.L. Powell (eds.), *Aging in China: Implications to Social Policy
of a Changing Economic State*, International Perspectives on Aging,
DOI 10.1007/978-1-4419-8351-0_10, © Springer Science+Business Media, LLC 2012

Introduction

Wishing you good fortune as the eastern seas and life as long as the southern mountains

Rainbow T. H. HO, Phyllis H. Y. LO, Cecilia L. W. Chan, Pamela, P. Y. Leung

Good health and longevity is one of the most revered greetings one can offer to a Chinese person. In the doctrines of the *Huangdi Nei Jing* (The Emperor's Internal Classic), which founded the fundamentals of Chinese medicine dating back from about 2,000 years ago, it is said that a man should live for up to a 100 years. Despite such ancient wisdom and after two millennia of medical advancement, the average life expectancy in China remains to be 72 years for men and 76 years for women but which is already higher than the global norm of about 68 for both sexes (World Health Organization, 2004).

In the *Huangdi Nei Jing*, there are documentations of the diagnosis and treatment of disorders as well as other health maintenance behaviors for longevity—many of which we still practice even to this day. From generation to generation, years of accumulated medical wisdom formed the basis of traditional Chinese medicine (TCM) as we know it today. Rather than rigorous randomized controlled trials, the discipline developed on centuries of practice experience. When China finally opened its doors to the world approximately half a century ago, there has since been a gradual influx of Western medical practices into the nation (Schnell, 1987). Since then, the interchange between TCM and Western medical practice flourished, and research evidence for TCM was gradually built up by attesting TCM treatment for Western medical diagnoses. To illustrate, the efficacy of subcutaneous acupuncture for angina pectoris and bronchial asthma was already being systematically researched in the 1950s (Hesketh & Zhu, 1997) and the pharmaceutical properties of Chinese medicinal herbs were also being validated through the establishment of a number of TCM research institutes since that period.

Consequently, the prominence that TCM has been receiving in Chinese healthcare does, by no means, fall short of Western medicine. Rather than being inclined toward one over the other, health care practitioners and patients seek to integrate the East and West so as to provide the maximal health benefit. At every tier of the healthcare system, one can find partnerships between Eastern and Western medical practices, from hospitals to medical schools or research institutes, to a specialized TCM department in the Ministry of Public Health and Bureau of Public Health at both the provincial and the national levels (Hesketh & Zhu, 1997).

The enactment of the one-child policy in addition to longer life expectancies is raising concerns regarding elderly health in many parts of China. It is projected that in 2050, the largest age cohort for men would be aged from 60 to 64 and for women, 80 and above (World Health Organization, 2004). Disability and chronic diseases requiring long-term medical attention will likely consume a significant proportion of national medical expenditures. As preventive and curative health models for the aging population are being national level and among the people, there is much wisdom that can be learnt from their integrated East–West model of health.

Eastern and Western Models of Health and Aging

Definition of Health and Aging

The World Health Organization defines health as "a state of complete physical, mental and social well-being and not merely the absence of disease or infirmity" (World Health Organization, 2011). The definition highlights an important paradigm shift in the Western conceptualization of health. For centuries, the Western biomedical model was established based on scientific evidence and observable clinical trials. Treatment for physical illnesses adopted an allopathic approach, with the goal of finding the cause of the pathology and providing treatment to contravene the illness-causing agent. Psychiatry is seen as a distinct branch in medicine, and such disorders were virtually untreatable until the discovery of their underlying physical neurological basis.

The Social Philosophies of the Conventional Western Medical Model

The Western biomedical model is heavily influenced by a linear understanding of mind–body, where the mind is a manifestation of physical neurological activity. When Descartes (1641/1990) proposed that the mind is distinct from the body, it sparked an era of positivism and expansion of scientific medicine. Personality was found changed after brain damage as in the infamous case of Phineas Gage, while fear and other emotions can now be mapped out in detailed neurological pathways. Aspects of psychosocial health, spirituality, or religion remain scientifically unexplainable and their values are undermined in medicine. This has been exacerbated by the societal inclination toward individualism which promotes self-reliance and personal responsibility for one's own health. As individualism is subtly creeping over to Eastern cultures, little attention is being given to the psychosocial aspects of health even to this day, simply because of the difficulty in objectively measuring psychosocial health and a lack of convincing evidence of its relevance to physical well-being. Stifled by the scientific rigor of the current biomedical model, there is little room for promoting holistic view of health (physical and psychosocial aspects), or in the promotion of wellness rather than the absence of illness.

Besides this, Western allopathic medicine developed based on reductionism, which essentially is an approach to understanding health by reducing them to constituent parts. This reductionist empiricism has given rise to the many specialties of medicine, had has helped uncover the root causes of many diseases. The body is broken down into discrete parts for precise examination and targeted treatment that tends to be mechanistic, overspecialized, and oblivious to the macro-perspective that underlies some health issues. More often than not, the root causes of illnesses are not chemical imbalance or cell mutation, but the underlying stressors and interpersonal problems the individual is facing that resulted in the bodily dysfunction. Conversion disorder, which categorizes illnesses with no known neurological cause, is a classic example of how a macro-perspective to health may be beneficial at times. Reductionist medicine also fails to be conductive to health promotion and illness prevention especially in terms promoting healthy lifestyle or social environment.

Holistic health concepts are especially relevant to the elderly population. Not only are they one of the major users of health care services, physical deterioration is a normal part of the aging process which means that the biomedical ideology of illness eradication may not be as relevant to the elderly population as it may be for younger patients. The notion of *active aging* was introduced by the WHO (World Health Organization, 2002) calls for active participation of the elderly individual to improve their health and their qualities of life. Perhaps what is more important to an elderly person is not perfect physical health, but the maintenance of good psychosocial well-being, functioning, and being able to take responsibility in managing their health statuses.

Eastern Health Models

One of the most influential health concepts of the *Huangdi Nei Jing* is the idea of "nurturing one's life," or *Yang Sheng*. The name alone highlights the very feature that differentiates between Eastern health concepts and the Western biomedical approach. It is not that TCM is not concerned about the prevention and treatment of illnesses, but the eradication of diseases is only one of the benefits of a life (*Sheng*) well taken care of (*Yang*). Yang Sheng is a life-long commitment, which does not begin only when illness strikes nor is it considered to have failed when the body succumbs to disease. It is the ability of the body to adapt and continue to function well despite of illnesses.

Traditional Wisdom of Yin–Yang Balance and Health

A unique concept toward health in TCM is the adaptability of our bodies and our psyches toward adverse health. The bedrock of one's abilities to embrace impingements to our health, or our lives, can also be found deep within cultural beliefs and philosophical teachings. Daoism, in the classical teachings by Lao Tzu and Chuang Tzu, fathers of the doctrine, understands life as an intricate balance between the forces of Yin and Yang. Yin, like water, connotes shadow, dimness, downward flowing and feminism, whereas Yang is fire, representing brightness, energy, upward heading, and masculinity. The two polarized forces are by no means antagonistic, but rather, complementary to one another. When imbalance between the Yin and Yang occurs, each acts to promote or repress one another so that equilibrium is attained.

Table 10.1 Table of five elements

	Metal	Water	Wood	Fire	Earth
Functional system	Lung, large intestine	Kidney, bladder	Liver, gall bladder	Heart, small intestine	Spleen, stomach
Associated emotions	Worry and grief	Fear	Anger	Joy	Contemplation

Under the Yin–Yang theory, health is simply a balance of the Yin and the Yang energies in the body (Pachuta, 1989). Imbalance, however, is not seen as pathological, but simply the result of the natural dynamics of Yin and Yang. The concept bears similarities to Western medical understanding of homeostasis or our bodily circadian rhythms of day and night. The cycle of equilibrium restoration means that neither too much nor too little of one thing is good for the body. Ultimately, the body has a natural power to restore this balance. Nurturing the body supports this balance, while medical treatment, if needed, serves to support the body's innate ability to return to equilibrium.

Law of Five Elements: The Body–Mind Connection

Another major underpinning of TCM is the Law of Five Elements, which states that our bodies constitute of five natural elements affecting the five bodily systems respectively (Table 10.1). Each with their own functions supporting life, the systems bears much resemblance to the Western medical understanding of the various functional systems in the body. Each of these systems is affected by one or two of the seven primary emotions in Chinese medicine, namely, worry/grief, fear, anger, joy, and contemplation (Chan, Ho, & Chow, 2001). There are reciprocal relationships between the system and its respective emotion. Excessive worry or grief affects the lung. Excess anger harms the liver. Overanxiety irritates the spleen and the stomach. Fear affects the kidney. According to the Yin–Yang theory, the excess of even positive emotions can be problematic which is why over-happiness has adverse effect on our hearts or blood pressure.

The law of the five elements bears important implications on the way Chinese people conceptualize both physical and emotional health. In fact, the connection between the functional systems (body) and the seven emotions (mind), describes the appreciation for a holistic view of health. From holistic lens, TCM diagnoses and treats illnesses as a system consisting of both bodily and emotional imbalance, rather than targeting treatment to the specific pathological area as Western medicine seeks to do. Similarly, the nurturing of life requires the nurturing of the body as well as the mind. It is no use to only care for our physical health while being constantly worried about when illness would strike, as the worry will gradually damage your internal systems.

Integrating the East and West: Evidence for the How's and Why's of Body–Mind Connection

There is surprising much universality in the way our emotions are intertwined with our bodily systems. Much evidence can be found even in the English language where the way we describe our emotions is often associated with our organs, including words like *gut-feelings, heavy hearted, heartbroken, liverish,* or *butterflies in the stomach.* In fact, it is just as easy to observe how our bodies react to our feelings, for instance, feeling thirsty or having to urinate frequently when in fear, or palpitations and increased blood pressure when feeling extreme joy or fear. The body–mind connection is by no means restricted to only the Chinese population.

The Link Between Stress and Health

There is accumulating scientific evidence demonstrating how psychological well-being affects health and longevity. Illnesses including cardiovascular disease, gastrointestinal problems, diabetes, obesity, osteoporosis, and cancer have been linked to various psychosocial stressors (Vitetta, Anton, Cortizo, & Sali, 2005). The development of psychoneuroimmunology and psychoendocrinology provides convincing evidence for this body–mind connection that has long laid the foundations of the five elements theory. By explicating the pathways through which our mental states impacts bodily functions, it offers some preliminary insight into the missing *how's* and *why's* of TCM (Schnell, 1987). Psychological stress elicits stress responses in the body which involves a cycle of changes to the neuroendocrine system through the sympathetic-adreno-medullary system and the hypothalamic-pituitary-adrenal axis. Through a series of endocrinal response, cortisol (stress hormone) and cytokines (proteins) are released, which serves to protect the body in times of stress by regulating immunity. Yet, prolonged stress strains and impairs the protection from this stress response, thus increasing the susceptibility to infections and deter healing (Vileilyte, 2006).

Stress and Aging

Developments in chromosomal research are also offering new insight to body–mind medicine. At the ends of chromosomes are specialized DNA sequences known as telomeres. Telomeres are responsible for health cell replication while protecting against cell abnormalities, which can lead to cancer. As cells divide, telomeres gradually wear out with time and such degeneration results in aging.

The levels of psychological stress and its chronicity can influence cell aging and their lifespan by heightening oxidative stress, lowering telomere enzyme activity, and ultimately leading to shorter telomere length (Epel et al., 2004). Healthy premenopausal women in higher stress have telomeres shorter than their counterparts with lower stress. The difference in length represents 10 years of aging. That is, the telomere length of a 40-year-old woman under high stress may be 10 years older than somebody of the same age who experience less stress. Whether such shorter telomeres can eventually lead to adverse health outcomes or premature mortality is still to be proven but its proven protection against cancer and cancer progression will no doubt affect longevity (Hornsby, 2007). Telomere length is also associated with a number of illnesses affecting the elderly population, including coronary heart disease, dementia, amongst others (Woo, Tang, Suen, Leung, & Leung, 2008). The role of stress to aging and associated chronic illnesses demonstrates the intricate relationship between the psychological status and physical health.

Body–Mind Medicine

While the body–mind connection has been practiced by the Chinese population for centuries, the growing scientific appreciation received in Western countries has led to the development of various mind–body therapies (MBTs). One of the earliest applications in the 1960s was the use of biofeedback which has since been found effective in treating problems including tension headaches in the Caucasian elderly population (Arena, Hannah, Bruno, & Meador, 1991) and anxiety disorders among the Chinese elderly (Ma, 2010). In more recent decades, techniques from traditional Asian medicine were systematically studied as a treatment for specific illnesses through randomized controlled trials. More commonly used techniques include relaxation techniques and meditation (Astin, Shapiro, Eisenberg, & Forys, 2003). Other techniques like guided imagery, hypnosis, tai-chi, yoga, qigong, acupuncture and acupressure, amongst others have also been shown to have salutary effects to various ailments or diseases. What is common among the many techniques is that their goals are to achieve mental relaxation or changes to cognitive states as well as improvements to vitality or illness prevention. The dual goals of mind–body practices can be exemplified in Tai-chi, a popular form of sequential movement exercise for the elderly, aims to improve balance control and flexibility while cultivating tranquility and mental relaxation. Even guided imagery, which involves mostly mental relaxation and visualizations, is expected to be able to exert an effect on physical health through cognitive expectations of positive health outcomes (such as imagining that toxins or diseases are being released to the body through the hands and the feet). Similarly, meditation capitalizes on the psychophysiological state of mental control and relaxation through which personal cultivation and spiritual transformation can be attained.

In Caucasian populations, MBTs are mostly adopted to maintain health and vitality for illness prevention or as an adjunct to conventional allopathic medicine (Eisenberg et al., 1998). Less commonly are they being used as a complete alternative to conventional medicine. The situation in China, on the other hand, varies considerably across cities. While in rural villages, TCM mind–body practices has even higher predominance in health protection (Hesketh & Zhu, 1997), the attitudes and adoption of mind–body health and treatment among Chinese elderly is embedded both within their traditional culture, while balancing the benefits of conventional Western medicine. In the Chinese city of Hong Kong (HKSAR), the older population, particularly those who were female and with lower socioeconomic statuses with chronic conditions causing discomfort tend to be more skeptical of Western medicine (Chan et al., 2003). There is a tendency to seek to first conceive ailments and illnesses based on traditional medical concepts (Lam, 2001). Sore throats, for instance, may be attributed to *heatiness* (preponderance of *yang* resulting in yin-yang imbalance) resulting from poor dietary habits or poor sleep rather than viral infections. Consequently, rather than seeking Western medication, some Chinese may look for remedies by modifying their dietary or other lifestyle habits. Traditional remedies are believed to be able to slowly cure the root of the illnesses without the side effects from conventional Western medication (Lam).

Besides greater faith in traditional health practices, Chinese elderly are also concerned about their health being a burden to their children, which ultimately contributes as a driving force toward self-care and mind–body practices (Pang, Jordan-Marsh, Silverstein, & Cody, 2003). In making healthcare decisions, Chinese elderly tend to hold themselves accountable for their own health; hence leading to a greater use of traditional home remedies, particularly food and exercise, to enhance immunity. Conventional Western medicine serves as an alternative, only when all other possibilities have been exhausted. Therefore, while there is an increasing use of mixed health remedies from both the East and the West, Chinese elderly, particularly elderly women, both in China and even those residing overseas (Lam, 2001; Pang et al., 2003), understands health and illnesses in the context of their own culture and traditional health beliefs, particularly the belief that the body has an innate healing capacity through the connected body and mind.

Mind–body Approach to Physical Illnesses for Chinese Elderly

The prevalence of physical illnesses among the Chinese elderly has experienced a rapid shift from communicable diseases to chronic diseases (Wang, Kong, Wu, Bai, & Burton, 2005). In 2005, 80% of deaths in China were due to chronic diseases including cancer, cardiovascular diseases, cancer, chronic respiratory illnesses, and diabetes (Chinese Centre for Disease Control and Prevention, 2005). Living with the chronicity of physical illnesses and related symptoms signify that beyond illness prevention and health interventions, is a need to also maintain a reasonable level of physical functioning and quality of life by minimizing illness burden and treatment side effects. This very need to arrive at a balance between the quick and aggressive interventions that characterizes conventional Western treatment and reduced quality of life as a result of treatment side effects is one of the major reasons why patients would choose to opt out from medical treatment to seek alternative interventions (Shumay, Maskarinec, Kakai, & Gotay, 2001). Some cancer patients believe that allopathic treatments of chemotherapy and radiotherapy hinder recovery or even cause death by weakening the body and immunity. Mind–body medicine, which is believed to be less assaulting to the body, is thus turned to as an alternative measure. Recall that the fundamental of nurturing life (Yang sheng) in TCM is the ability to strike equilibrium in our daily lives. Accordingly, the modification of lifestyle habits is one of the ways which the Chinese population readily turns to for health promotion and illness treatment.

Lifestyle Choices

Lifestyle choices can be a major illness trigger (Vitetta et al., 2005), and so it is crucial for the rectification of disease and enhancement of health. This is not unique of mind–body health approaches. Evidence for health-promoting practices pertaining to physical activity, nutrition, and the control of alcohol and tobacco intake are already well-established with non-Chinese populations. In a scale validation study of the Health Promotion Lifestyle Profile in Taiwan, factors common to both Caucasian and Chinese populations in terms of health promotion lifestyles are nutrition, exercise, responsibility toward one's own health and spiritual growth, though the actual meaning behind each of these lifestyles varies considerably from culture to culture (Teng, Yen, & Fetzer, 2010). What particularly defined the Chinese population was how the importance of interpersonal support and stress management were inextricably connected to one another in the process of managing one's health. The larger meaning of this can be that the maintenance of a physically healthy lifestyle for Chinese needs to be discussed in the context of interpersonal relationships and psychological health.

Food and Nutrition

There is an old Chinese saying that "illnesses come from what we put into our mouths." While Western medicine advises that calorie and cholesterol intake has major manifestations on coronary health and longevity, the understanding of food and nutrition among the Chinese population is a complex mind map of food characteristics and dietary restrictions. The hot or cold energies of each type of food work to alter the yin–yang balance in the body. When attacked by a cold or flu, hot energy food, like ginseng, fosters healing. Another example is how an "acidic" diet consisting of meat, is believed to help cancer cells proliferate faster within the body.

Interestingly, the health-promoting alimentation habit of tea consumption among the Chinese elderly had positive impacts extending beyond physical well-being to mental health as well as social functioning (Zhou et al., 2010). There is no proven mechanism for such associations, but a possible justification is that the Chinese consumption of tea is often associated with social gatherings or meals with families.

The contribution of social support to physical health is further supported by studies showing how poor social support and feelings of loneliness is correlated to higher resting blood pressure, particularly in older adults (Uchino, 2004; Vitetta et al., 2005). Loneliness is often coupled with feeling threatened. Blood pressure, being predominately associated with dietary habits, may be related to loneliness through the stress response, which causes a rise in blood pressure (Takkouche, Reguieira, & Gestal-Otero, 2001). This discussion offers a multilevel perspective that the mind–body health practices of Chinese elderly is more than an alternative set of lifestyle promotion behaviors, but cultural practices which are deeply ingrained within relational aspects.

Integrative Body–Mind Health Promotion

In view of this, mind–body health promotion needs to be an integrative lifestyle plan with ways to promote physical health, mental well-being, and enhancing social support. Indeed, such an integrative mind–body intervention for low-risk prostate cancer patients, which addressed patients' diets, exercise, stress management practices, and group support, had the effect of raising the immunity of patients after 3 months of intervention (Sagar & Lawenda, 2009). Telomerase (the enzyme responsible for the maintenance of telomeres) in health cells were increased, which may promote longevity among patients. In addition to that were reductions in low-density cholesterol levels as well as psychological distress.

Holistic intervention programs as such distinguishes themselves from what is commonly referred to as MBTs like meditation, mindfulness, relaxation techniques, acupuncture, etc. Many MBTs tend to target treatment either via the physical body or via the mind, rather than adopting a multifaceted approach of intervention. There is insufficient evidence to ascertain whether multicomponent holistic interventions would be more beneficial than single component MBTs, but the work by our team has found that an integrated Body–mind–spirit health intervention model (I-BMS) produces greater physiological and psychological benefits compared to an intervention that address solely the emotional concerns for Chinese breast cancer survivors (Chan et al., 2006). Founded upon the TCM concept of holism, the I-BMS model seeks to simultaneously promote physical health through simple MBTs, psychological well-being through traditional Chinese philosophical teachings combined with provision of alternative media for expression, and finally, spiritual growth through a search of meaning despite the suffering of illnesses. When conducted with a group of patients, the combined effect of mutual support has led to the reduction of stress hormone levels (salivary cortisol), psychological distress, emotional control, and negative mental adjustment. Participants further reported greater positive social support after the intervention. In contrast, the group intervention focusing predominately on emotional expression and regulation reported no significant improvements.

Therefore, the rapidly expanding body of research demonstrating the effectiveness of single modality MBTs on specific disorders and symptoms has important implications on what can be included in the design of holistic interventions. A number of chronic conditions the elderly commonly suffer from can be effectively relieved with MBTs. Astin's review of best MBT practices found relaxation and hypnosis helpful in dealing with cancer or treatment-related symptoms, improving mood, quality of life, and disease coping (Astin, Shapiro, Bishop, & Cordova, 2005). Acupressure can provide short-term alleviation of cancer pain, while relaxation techniques, biofeedback, tai-chi, qigong, yoga, and expressive arts therapies help relieve pain, insomnia, hot flashes, and nausea (Bardia, Barton, Prokop, Bauer, & Moynihan, 2006; Elkins, Fisher, & Johnson, 2010).

Besides cancer, cardiovascular diseases can also benefit from relaxation and emotional and stress management skills (Astin et al., 2005; Luskin et al., 1998), and pain, including low back pain or osteoarthritis pain, can be alleviated through stress management, coping skills training, cognitive restructuring, progressive muscular relaxation, and guided imagery (Astin et al., 2005; Morone, Greco, & Weiner, 2007). Bone mineral density in postmenopausal women can be enhanced through tai-chi practice which helps build musculoskeletal strength and reduce fall frequency (Wayne et al., 2007).

Mind–Body Approach to the Mental Health of Chinese Elderly

The prevalence of mood disorders among the elderly China (55 years old and above) is 10.56% where 3.82% suffer from clinical major depressive disorder, and 7.97% has anxiety disorders (Philips et al., 2009). The manifestations of mood disturbances can often be rather pervasive. Insomnia, which can be caused by depression, is a frequent disturbance to about a third of Chinese elderly people where almost 10% of the elderly population consequently experiences daytime symptoms (e.g., sleepiness, fatigue, and decreased mental clarify) (Liu & Liu, 2005). In the study, it was also found that two of the greatest risks of insomnia were advanced age as well as poorer perceived health. Ironically, Chinese people tend to perceive health related symptoms as physical health issues rather than having a psychiatric origin (Li, Logan, Yee, & Ng, 1999). The failure to detect psychiatric problems is compounded by the widespread societal stigma makes it difficult for mental health treatment to be adequately provided to Chinese communities.

A growing body of scientific evidence veritably demonstrates that mental health is indicative of bodily alterations or even pathologies. Structural brain imaging studies revealed that increased stress is associated with a reduced anterior hippocampal volume (Narr, Woods, Thompson, Szeszko, Robinson, Dimtcheva, et al., 2006). Hippocampal volume is also found to be less in patients with unipolar depression in a meta-analysis of 12 studies (Videbech & Ravnkilde, 2004). Depression also impairs the treatment of idiopathic urinary incontinence as a result of altered serotonin levels (Zorn, 1999). Similarly, having depression increases the risk of cardiac death by 84% for patients with chronic artery disease (Steffesnet, O'Connor, Jiang, Pieper, Kuchibhatla, Arias, 1999). The studies collectively conjure profound connection between mental health and physical well-being.

In a similar vein, traditional Chinese health concepts stress the inextricable segregation of mental health problems from physical disorders. Bridging East and West medicine involves first an appreciation that prevention and treatment for mental health disturbances among the elderly requires remedies to both the mind and the body, rather than enforcing a psychosis label and treatment into what may be more than just a psychological disturbance. The second issue involves the need to scrutinize how mental health disorders, which at the present moment, is diagnosed based on the Diagnostic and statistical manual of mental disorders 4th edition (text revision) (DSMIV-TR), may be more justifiably adapted for the Chinese elderly population.

The Chinese Concept of Negative Emotions

The DSM-IV introduced a list of *culture bound syndromes* which describes mental disorders specific to Chinese and other ethnic populations. In spite of that, there remains scrutiny in the way the other DSM diagnostic criteria can adequately describe the mental health vulnerabilities of Chinese patients. This calls for a need to describe more indigenous mental health diagnostic criteria. Depression, for instance, translates to *yiyu* in Chinese which describes a phenomenon of being *stagnated, stuck, or entangled* (Ng, Chan, Ho, Wong, & Ho, 2006). The sense of being "depressed" or "yu" among Chinese differentiates from the feeling of down and low described in clinical depression. It arises from the stagnation of one of the seven emotions described in the five elements theory. Emotional stagnation stifles the normal flow of *qi* (balancing energy) in the body which ultimately leads to a number of physical and mental symptoms such as preoccupation, anger, poor sleep, fatigue, indigestion, amongst others. Treatment, therefore, is not so much to eliminate the physical symptoms or to foster positive emotions, but to first relieve the stagnated feelings locked up inside.

Consequently, understanding psychological disturbances within the cultural context is necessary in the provision of mental health interventions. The TCM five elements theory describes the seven basic emotions as worry, grief, fear, anger, joy, and contemplation. Each of the seven emotions mutually promotes or restrains one another. Excessive contemplation leads to worry, for instance, and worry creates fear. Fear, on the other hand, is the cause of anger. Overindulgence in one's own anger leads to feelings of joy. Too much joy will lead to contemplation and the cycle continues. The Eastern perspective on mental health, therefore, is not an absence of negative emotions, since too much positive emotion can eventually lead to undesirable psychological well-being. The goal is to obtain a balanced level of emotions, and not being overly fixated in one.

Traditional Chinese philosophy states that too much good brings suffering while at the end of suffering are harmony and hope. The meaningless strive for desires will only impair health. In Daoism, longevity is the relinquishing of all yearnings and allowing everything to occur at its own natural course. Unfortunately, elderly Chinese people tend to have a myriad of emotional concerns stemming from various desires or the inability to relinquish the health they once had, such as poor health conditions, debilitating symptoms such as declining vision or memory, incontinence, frequent hospital admissions, functional disabilities leading to an inability to complete housework. Other concerns include poor social support network resulting in fears that there is few people they can turn to in times of need or when illness strikes, as well as concerns about living conditions and living expenses (Tsai, Chung, Wong, & Huang, 2005; Woo et al., 1994).

The same preoccupations with their deteriorating health, or the fear of it, the inability to carry forth the same functional or social roles as before or other unmet goals or unresolved issues from their youth haunt the elderly on a day to day basis. Such failure to live in the present and let go of obdurate attachments prove to be a great challenge to sustain a healthy balance of emotions according to the traditional Eastern medicine. In the treatment of psychological disturbances for the elderly, the root imbalance-inducing emotion first needs to be identified and treated (Lee, Ng, Leung, & Chan, 2009). To illustrate, an elderly widow expressing constant anger toward her children for not visiting enough may, in a deeper subconscious level, have a great fear of abandonment or an over-contemplation over an ailing prominent family role which she had once enjoyed.

Once the root emotion has been identified, it is essential to realize that the core cause of the emotional disturbance usually arises from the failure to assimilate virtues such as:

1. The impermanence of all worldly entities, including good health, functional abilities, or social roles.

    In the treatment of depression, yiyu, for Chinese patients, a crucial component is learning to let go of over-attachments to various worldly things (Ng et al., 2006). Often, feelings of grief and anxiety arise from being overly fixated on the past or on what may happen in the future. Illness and degeneration will inevitably lead to losses of social and functional roles. Equally prominent among the elderly are needs to cope with death anxiety, the loss of loved ones or other unresolved pains in life. Meanwhile, some are also constantly worried about the future—be it their health, financial conditions, or being a burden to their children—Being immersed in the grief and anger or the past or the anxiety over their future indicates failure to fully live in the present. Analogous to our contemporary psychological understanding of late adulthood, living in the present means finding quiescence with the past, accepting the present, while readapting to the new situation or new roles in life.

    Acceptance of advancing age and related health manifestations does not need to be a fatalistic process. Indeed, contemporary medical treatment can often make patients confused and feel unable to control their health statuses. In reality, elders can actively manage their own mental and physical well-being through a number of mind–body techniques. Controlled studies have demonstrated that massage and meditation can reduce anxiety (Mamtani & Cimino, 2002). Insomnia can be relieved via muscle relaxation (Astin et al., 2003). Twelve sessions of acupuncture was effective for putting 64% of the study participants into full remission of their depression symptoms (Allen, Schuyer, & Hitt, 1998). Although acupuncture cannot be administered by the elderly themselves, simple acupressure techniques of applying pressure to acupuncture points can also vitalize the flow of *qi* in the body, thus preventing illnesses and pain (Chan et al., 2001). Through these means, the elderly can actively engage in their lives while promoting quality of life.

2. Cultivation of personal virtues as a way of social healing
   Influenced by Confucianism, the mental health of the Chinese population is largely affected by harmonious interpersonal relationships (Hsiao, Klimidis, Minas, & Tan, 2006). It was found that guilt and shame arises from perceived inability to perform familial obligations. Conversely, expectations on family members to provide support in times of need can also create emotional disturbances among Chinese elderly (Teng, Lin, Tsai, Kwan, & Chen, 1994).

   Subsequently, psychotherapeutic counseling for Chinese elders needs to emphasize the maintenance of relational harmony (Hsiao et al., 2006). Interpersonal virtues taught in Chinese philosophical ideals, such as humility, tolerance, compassion, and forgiveness, among others, has been empirically explored in recent development of positive psychology. Positive psychology advocates the emphasis of human strengths and prosocial behavior rather than pathologies. Among the various virtues, studies on forgiveness received the most scholastic attention (Luskin, 2004). Forgiving other people and oneself helps sustain harmonious relationships through understanding and acceptance of not being able to fulfill expectations. Mental well-being is consequently improved as a result. In particular, anxiety and depression can effectively be reduced through only a brief 8-week forgiveness training (Human Development Study Group, 1991).

3. Spiritual transcendence
   The sage is symbolic of a wise Chinese elder who appears to have transcended above the rules of secularity. They become attune with themselves and the world as they no longer allow themselves to be psychologically bound by desires, others' influence, and human limitations (Yip, 2004). They are in full acceptance of the unpredictability of suffering and blessings in life where they are able to find gains and a sense of blessing within their sufferings. When the Western culture strives for the self-actualization of potentials, Chinese people hope to attain a state of transcendence where achievements are far less important than the pursuit of peacefulness and finding meaning in their lives. Transcendence can be promoted through practices including meditation, prayer or mantra affirmations, or tai-chi (Luskin, 2004). Practices as such require mental control and deep reflection and their benefits exceed just the spiritual level. Certain forms of meditation was found to have positive effects on telomere length and telomere maintenance by reducing stress, stress arousal, and enhancing positive states of mind (Farzaneh-Far, Lin, Epel, Lapham, Blackburn, Whooley, 2009).

Conclusion

In this chapter, we discussed Eastern and Western perspectives in promoting health for the Chinese elderly. There are tremendous challenges faced in the integration of both approaches, including a lack of rigorous randomized controlled studies and explanatory pathways for many mind–body interventions. Secondly, the way in which Eastern and Western medical approaches can be safely integrated and administered remains unclear. For instance, the clashing of Western medication with herbal remedies can be dangerous (Tsen, Segal, Potheir, & Bader, 2000). Meanwhile, patients are often unwilling to disclose their own seeking of MBTs to their medical doctors (Eisenberg et al., 1998). Besides patient education, physicians will also have to be adequately informed about mind–body practices so as to be able to offer meaningful and appropriate advice to patients (Mamtani & Cimino, 2002).

Despite difficulties in effectively integrating the two approaches, there is a noted global trend in pushing forward the research and application of holistic mind–body health approaches. The National Center for Complementary and Alternative Medicine (NCCAM) was established in 1998 within the National Institute of Health in the United States, offering scientific evidence and biological basis for TCM. In China, residential care facilities for the elderly are beginning to recognize the drawbacks of focusing on the provision of physical care. Socioemotional and spiritual needs are being integrated as part of routine care (Zhan, Liu, Guan, & Bai, 2006).

To conclude, promoting health for Chinese elderly should be a process of Yang sheng, the nurturing of the various integrated domains of our well-being. While it is less crucial to become disease- and symptom-free, it is necessary to flexibly readapt to the changing health status, and striking a balance between physical and psychological health.

References

Allen, J. J. B., Schuyer, R. N., & Hitt, S. K. (1998). The efficacy of acupuncture in the treatment of women with major depression. *Psychological Science, 9*(5), 397–401.

Arena, J. G., Hannah, S. L., Bruno, G. M., & Meador, K. J. (1991). Electromyographic biofeedback training for tension headache in the elderly: A prospective study. *Biofeedback and Self-Regulation, 16*(4), 379–390.

Astin, J. A., Shapiro, S. L., Bishop, S. R., & Cordova, M. (2005). Mindfulness-based stress reduction for health care professionals: Results from a randomized trial. *International Journal of Stress Management, 12*(2), 164–176.

Astin, J. A., Shapiro, S. L., Eisenberg, D. M., & Forys, K. L. (2003). Mind-body medicine: State of the science, implications for practice. *The Journal of the American Board of Family Practice, 16*(2), 131–147.

Bardia, A., Barton, D. L., Prokop, L. J., Bauer, B. A., & Moynihan, T. J. (2006). Efficacy of complementary and alternative medical therapies in relieving cancer pain: A systemic review. *Journal of Clinical Oncology, 24*(34), 5458–5464.

Chan, C. L. W., Ho, P. S. Y., & Chow, E. (2001). A body-mind-spirit model in health: An Eastern approach. *Social Work in Health Care, 34*(3/4), 261–282.

Chan, C. L. W., Ho, R. T. H., Lee, P. W. H., Cheng, J. Y. Y., Leung, P. P. Y., Foo, W., & Speigel, D. (2006). A randomized controlled trial of psychosocial interventions using the psychophysiological framework for Chinese breast cancer patients. *Journal of Psychosocial Oncology, 24*(1), 3–15.

Chan, M. F., Mok, E., Wong, Y. S., Tong, T. F., Day, M. C., Tang, C. K. Y., & Wong, D. H. C. (2003). Attitudes of Hong Kong Chinese to traditional Chinese medicine and western medicine: Survey and cluster analysis. *Complementary Therapies in Medicine, 11*, 103–109.

Chinese Centre for Disease Control and Prevention. (2005). *Chinese health statistical digest 2005*, from http://www.chinacdc.cn/en/. Accessed June 15, 2011.

Descartes, R. (1990). *Meditations on first philosophy: Meditationes de prima philosophia* (G. Hefferman, Trans.). Notre Dame: University of Notre Dame Press. (Original work published 1641)

Eisenberg, D. M., Davis, R. B., Ettner, S. L., Appel, S., Wilkey, S., Van Rompay, M., & Kessler, R. C. (1998). Trends in alternative medicine use in the United States, 1990–1997. *Journal of the American Medical Association, 280*, 1569–1575.

Elkins, G., Fisher, W., & Johnson, A. (2010). Mind-body therapies in integrative oncology. *Current Treatment Options in Oncology, 11*(3–4), 128–140.

Epel, E. S., Blackburn, E. H., Lin, J., Dhabhar, F. S., Adler, N. E., Morrow, J. D., et al. (2004). Accelerated telomere shortening in response to life stress. *Proceedings of the National Academy of Sciences of the United States of America, 101*(49), 17312–17315.

Farzaneh-Far, R., Lin, J., Epel, E., Lapham, K., Blackburn, E., & Whooley, M. A. (2009). Telomere length trajectory and its determinants in persons with coronary artery disease: Longitudinal findings from the heart and soul study. *PloS One, 5*(1), e8612. doi: 10.1371/journal.pone.0008612.

Hesketh, T., & Zhu, W. X. (1997). Health in China: Traditional Chinese medicine: One country, two systems. *British Medical Journal, 315*, 115.

Hornsby, P. J. (2007). Telomerase and the aging process. *Experimental Gerontology, 42*(7), 575–581.

Hsiao, F. H., Klimidis, S., Minas, H., & Tan, E. S. (2006). Cultural attribution of mental health suffering in Chinese societies: The views of Chinese patients with mental illness and their caregivers. *Journal of Clinical Nursing, 15*(8), 998–1006.

Human Development Study Group. (1991). Five points of the construct of forgiveness within psychotherapy. *Psychotherapy, 28*(3), 493–496.

Lam, T. P. (2001). Strengths and weaknesses of traditional Chinese medicine and Western medicine in the eyes of some Hong Kong Chinese. *Journal of Epidemiology and Community Health, 55*, 762–765.

Lee, M. Y., Ng, S. M., Leung, P. P. Y., & Chan, C. L. W. (Eds.). (2009). *Integrative body-mind-spirit social work: An empirically based approach to assessment and treatment*. New York: Oxford.

Li, P. L., Logan, S., Yee, L., & Ng, S. (1999). Barriers to meeting the mental health needs of the Chinese community. *Journal of Public Health Medicine, 21*(1), 74–80.

Liu, X., & Liu, L. (2005). Sleep habits and insomnia in a sample of elderly persons in China. *Sleep, 28*(12), 1579–1587.

Luskin, F. (2004). Transformative practices for integrating mind-body-spirit. *Journal of Alternative and Complementary Medicine, 10*, S15–S23.

Luskin, F. M., Newel, K. A., Griffith, M., Holmes, M., Telles, S., Marvasti, F. F., et al. (1998). A review of mind-body therapies in the treatment of cardiovascular disease. Part 1. Implications for the elderly. *Alternative Therapies in Health and Medicine, 4*, 46–61.

Ma, X. (2010). Clinical study of biofeedback therapy in treating anxiety disorder in the elderly. *Chinese Journal of Convalescent Medicine, 19*(6), 534–536.

Mamtani, R., & Cimino, A. (2002). A primer of complementary and alternative medicine and its relevance in the treatment of mental health problems. *The Psychiatric Quarterly, 73*(4), 367–381.

Morone, N. E., Greco, C. M., & Weiner, D. K. (2007). Mindfulness meditation for the treatment of chronic low back pain in older adults: A randomized controlled pilot study. *Pain, 134*, 310–319.

Narr, K. L., Woods, R. P., Thompson, P. M., Szeszko, P., Robinson, D., Dimtcheva, T., et al. (2006). Relationships between IQ and regional cortical gray matter thickness in healthy adults. *Cerebral Cortex, 17*, 2163–2171. doi: 10.1093/cercor/bhl125.

Ng, S. M., Chan, C. L. W., Ho, D. Y. F., Wong, Y. Y., & Ho, R. T. H. (2006). Stagnation as a distinct clinical syndrome: Comparing "yu" (stagnation) in traditional Chinese medicine with depression. *British Journal of Social Work, 36*, 467–484.

Pachuta, D. M. (Ed.). (1989). *Chinese medicine: The law of five elements. Healing east and west ancient wisdom and modern psychology.* New York: Wiley.

Pang, E. C., Jordan-Marsh, M., Silverstein, M., & Cody, M. (2003). Health-seeking behaviors of elderly Chinese Americans: Shifts in expectations. *The Gerontologist, 43*(6), 864084.

Philips, M. R., Zhang, J. X., Shi, Q. C., Song, Z. Q., Ding, Z. J., Pang, S. T., & Wang, Z. Q. (2009). Prevalence, treatment and associated disability of mental disorders in four provinces in China during 2001–05: An epidemiological survey. *The Lancet, 33*, 2041–2053.

Sagar, S. M., & Lawenda, B. D. (2009). The role of integrative oncology in a tertiary prevention survivorship program. *Preventive Medicine, 49*(2–3), 93–98.

Schnell, J. A. (1987). *The merging of traditional Chinese medicine and western medicine in China: Old ideas cross culturally communicated through new perspectives,* Retrieved from http://www.eric.ed.gov/PDFS/ED306161.pdf.

Shumay, D. M., Maskarinec, G., Kakai, H., & Gotay, C. C. (2001). Why some cancer patients choose complementary and alternative medicine instead of conventional treatment. *The Journal of Family Practice, 50*(12), 1067.

Steffesnet, B. C., O'Connor, C. M., Jiang, W. J., Pieper, C. F., Kuchibhatla, M. N., Arias, R. M. (1999). The effect of major depression on functional status in patients with coronary artery disease. *Journal of the American Geriatrics Society, 4*, 319–322.

Takkouche, B., Reguieira, C., & Gestal-Otero, J. J. (2001). A cohort study of stress and the common cold. *Epidemiology, 12*(3), 345–349.

Teng, H. L., Yen, M., & Fetzer, S. (2010). Health promotion lifestyle profile-II: Chinese version short form. *Journal of Advanced Nursing, 66*(8), 1864–1873.

Teng, J. K., Lin, J. F., Tsai, L. M., Kwan, C. M., & Chen, J. H. (1994). Acute myocardial infarction in young and very old Chinese adults: Clinical characteristics and therapeutic implications. International Journal of Cardiology, *44*(1), 29–36.

Tsai, Y. F., Chung, J. W. Y., Wong, T. K. S., & Huang, C. M. (2005). Comparison of the prevalence and risk factors for depressive symptoms among elderly nursing home residents in Taiwan and Hong Kong. *International Journal of Geriatric Psychiatry, 20*(4), 315–321.

Tsen, L. C., Segal, S., Potheir, M., & Bader, A. M. (2000). Alternative medicine use in presurgical patients. *Anesthesiology, 93*, 148–151.

Uchino, B. N. (Ed.). (2004). *Social support and physical health: Understanding the health consequences of relationships.* New Haven: Yale University Press.

Videbech, P., & Ravnkilde, B. (2004). Hippocampal volume and depression: A meta-analysis of MRI studies. *The American Journal of Psychiatry, 161*, 1957–1966.

Vileilyte, L. (2006). Stress and wound healing. *Clinics in Dermatology, 25*(1), 49–55.

Vitetta, L., Anton, B., Cortizo, F., & Sali, A. (2005). Mind-body medicine: Stress and its impact on overall health and longevity. *Annals of the New York Academy of Sciences, 1057*, 492–505.

Wang, L., Kong, L., Wu, F., Bai, Y., & Burton, R. (2005). Preventing chronic diseases in China. *The Lancet, 366*, 1821–1824.

Wayne, P. M., Kiel, D. P., Krebs, D. E., Davis, R. B., Savetsky-German, J., Connelly, M., & Buring, J. E. (2007). The effects of Tai Chi on bone mineral density in postmenopausal women: A systematic review. *Archives of Physical Medicine and Rehabilitation, 88*, 673–680.

Woo, J., Ho, S. C., Lau, J., Yuen, Y. K., Chiu, H., Lee, H. C., & Chi, I. (1994). The prevalence of depressive symptoms and predisposing factors in an elderly Chinese population. *Acta Psychiatrica Scandinavica, 89*(1), 8–13.

Woo, J., Tang, N. L. S., Suen, E., Leung, J. C. S., & Leung, P. C. (2008). Telomeres and frailty. *Mechanisms of Ageing and Development, 129*, 642–648.

World Health Organization. (2002). *Active aging: A policy framework.* Geneva: World Health Organization.

186 R.T.H. HO et al.

World Health Organization. (2011). Health for all, from https://apps.who.int/aboutwho/en/health-forall.htm. Accessed June 15, 2011.
World Health Organization. (2004). *World population prospects*, from http://www.who.int/countries/chn/en/. Accessed June 15, 2011.
Yip, K. S. (2004). Taoism and its impact on mental health of the Chinese communities. *The International Journal of Social Psychiatry, 50*(1), 25–42.
Zhan, H. J., Liu, G., Guan, X., & Bai, H. G. (2006). Recent developments in institutional elder care in China: Changing concepts and attitudes. *Journal of Aging & Social Policy, 18*(2), 85–108.
Zhou, B., Chen, K., Wang, J., Wang, H., Zhang, S., & Zheng, W. (2010). Quality of life and related factors in older rural and urban Chinese populations in Zhejiang province. *Journal of Applied Gerontology, 30*(2), 199–225.
Zorn, B. H. (1999). Urinary incontinence and depression. *The Journal of Urology, 162*, 82–84.

Chapter 11
Family Caregiving and Impact on Caregiver Mental Health: A Study in Shanghai

Vivian W.Q. LOU and Shixun GUI

Abstract This chapter analyzes the aging population and related policy in present-day China by pointing out that long-term care needs are growing but China still lacks a national long-term care policy. The case of Shanghai is reviewed, with an observation that there was a missing element in the long-term care model—family caregivers. An empirical study on family caregiving and impact on caregivers' mental health is described. Findings show that about one-fifth of the caregivers were at risk of depressive symptoms, a significant number, which deserves policy responses. A national level long-term care policy is suggested so as to clarify the definition of long-term care and policy objectives. Moreover, there is a need to integrate needs of family caregivers into the long-term care policy and service model. Financial needs, health needs, and knowledge/skills needs are recommended to be taken into consideration for policy and service development purposes.

Keywords Long-term care • Shanghai • China • Informal caregiving • Depressive symptoms of caregiver

Population Aging and Long-Term Care Needs

China has been experiencing a considerable increase in population aging in the past half century. According to the United Nation's estimation, one out of five older people worldwide are Chinese (United Nations Department of Economic and Social

V.W.Q. LOU (✉)
Department of Social Work and Social Administration, Sau Po Centre on Ageing,
University of Hong Kong, Pokfulam Road, Hong Kong
e-mail: wlou@hku.hk

S. GUI
Institute on Population Studies, East China Normal University, Shanghai, China

S. Chen and J.L. Powell (eds.), *Aging in China: Implications to Social Policy of a Changing Economic State*, International Perspectives on Aging,
DOI 10.1007/978-1-4419-8351-0_11, © Springer Science+Business Media, LLC 2012

Table 11.1 Key economic and social development indexes

	Eastern provinces	Central provinces	Western provinces	Northeast provinces	National total
GDP per capita (RMB)	40,800	19,862	18,286	28,566	25,575
Government expenditure/GDP (%)	10.0	20.0	30.0	20.0	20.0
Number of hospitals and health centers	16,728	15,488	22,562	5,140	59,918
Number of hospital beds per 1,000 persons	3.1	2.9	3.1	3.8	3.1
Licensed (assistant) doctors per 1,000 persons	1.9	1.6	1.6	2.1	1.7

Source: National Bureau of Statistics of China (2011b)

Affairs Population Division, 2008). According to the recent census report (National Bureau of Statistics of China, 2011a), the total number of people aged 60 or above has reached 17.8 million and accounted for 13.26% of the total population. Among them, 11.9 million people were aged 65 or above, which comprised 8.87% of the total population. Assume that the fertility rate increases to 2.0 until 2010 and remains stable, and the life expectancy also increases to 77 and 81 for men and women, respectively, then there will be 457,590,000 older people in China, which will compose 31.2% of the total population in 2050 (Gui, 2009).

China is a big country in which both the economic and social aspects of development experienced by people living in diversified geographic areas varies (Tables 11.1 and 11.2). In a general sense, there is more advanced economic development in the eastern provinces, but similar healthcare resources, as compared to provinces located in other geographic areas. Parallel to the above economic and social development context, the percentage of the aging population varies from province to province, ranging from 6.66% in Xinjiang to 14.08% in Shanghai (Table 11.2). Moreover, old dependency ratio also ranged from 9.23 in Xinjiang to 17.23 in Shanghai. In this regard, we can tell that Shanghai is at the top end of municipal cities and provinces with respect to economic development. Shanghai experienced the highest level of aging population, old dependency ratio, and life expectancy, which calls for an aging policy and services that can fulfill the existing and future needs of an older population.

As guided by the "Six Haves" national aging policy framework, China has enacted People's Republic of China Law on the Protection of the Rights and Interests of Older People in 1996 (Steering Committee of the National People's Congress of the People's Republic of China, 1996). In the past few decades, there has also been the development of income maintenance insurance, medical insurance, and poverty alleviation policy. The aim is to: (1) ensure that older people have been taken care of, (2) have health care, (3) have an updated knowledge about social development and the aging policy, (4) have social participation, (5) have learning opportunities, and (6) have happiness in late life. This policy framework can be regarded as an umbrella policy framework that targets older people with diversified needs along their aging process, taking the holistic needs of the older population into consideration.

Table 11.2 Population aging across municipal cities and provinces in China

| | (RMB) | Aging population (65 or above) (1000 in 2009) | | | Life expectancy (2000) | |
		Frequency	Percentage	Old dependency ratio	Male	Female
National average	*25,125*	*113,199*	*9.72*	*13.24*	*69.63*	*73.33*
Beijing	70,234	1,524	10.10	12.62	74.33	78.01
Tianjin	63,395	1,156	11.04	13.99	73.31	76.63
Hebei	24,583	5,512	8.86	11.87	70.68	74.57
Shanxi	20,779	2,457	8.09	10.83	69.96	73.57
Inner Mongolia	37,287	1,818	8.46	10.94	68.29	71.79
Liaoning	34,193	4,419	11.50	14.85	71.51	75.36
Jilin	25,906	2,164	8.89	11.26	71.38	75.04
Heilongjiang	21,593	2,951	8.66	10.96	70.39	74.66
Shanghai	*77,205*	*2,368*	*14.08*	*17.97*	*76.22*	*80.04*
Jiangsu	43,907	8,225	12.03	16.20	71.69	76.23
Zhejiang	44,895	5,035	11.07	14.75	72.50	77.21
Anhui	16,656	5,545	10.15	14.38	70.18	73.59
Fujian	33,106	3,220	10.03	13.76	70.30	75.07
Jiangxi	15,921	3,165	8.08	11.55	68.37	69.32
Shandong	35,893	8,171	9.74	13.06	71.07	76.26
Henan	21,073	7,460	8.88	12.35	69.67	73.41
Hubei	22,050	5,175	10.17	13.50	69.31	73.02
Hunan	19,355	6,351	11.18	15.61	69.05	72.47
Guangdong	39,978	6,356	7.48	9.95	70.79	75.93
Guanxi	16,576	3,993	9.31	13.41	69.07	73.75
Hainan	18,760	667	8.77	12.39	70.66	75.26
Chingqing	20,219	2,927	11.58	16.53	69.84	73.89
Sichuan	17,289	8,838	12.20	17.28	69.25	73.39
Guizhou	9,214	2,794	8.27	12.36	64.54	67.57
Yuannan	13,687	3,478	8.60	12.30	64.24	66.89
Tibet	15,294	179	7.00	9.56	62.52	66.15
Shananxi	20,497	3,315	9.89	13.32	68.92	71.30
Gansu	12,882	1,952	8.34	11.46	66.77	68.26
Qinghai	18,346	345	6.99	9.71	64.55	67.70
Ningxia	19,642	374	6.80	9.47	68.71	71.84
Xinjiang	19,119	1,264	6.66	9.23	65.98	96.14

Source: National Bureau of Statistics of China (2011b)

According to the 2006 national survey of the aged population in urban and rural China, the percentage of older people who need daily assistance varies from 5.7 to 15% in sampled provinces (Guo & Chen, 2009). Overall, around 15% of adults aged 65 or above were unable to use the toilet and ambulate; followed by 12.6% who could not dress themselves and around 9% who could not eat without assistance (Fig. 11.1). When using the four indicators of Activities of Daily Living (ADL), including eating, locomotion, toileting, and dressing for detailed analysis, it is observed that ADL deficit rates increased with aging (Figs. 11.2–11.5). Older adults aged 85 or above had 1.5–2 times the deficit rate of inability to perform these ADL

Fig. 11.1 Older adults aged
65 or above unable to
perform daily activities (%)

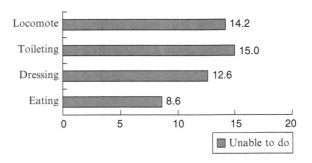

Fig. 11.2 Eating ability by
age group

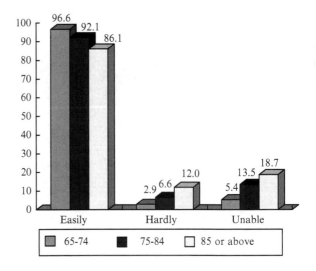

Fig. 11.3 Locomotion ability
by age group

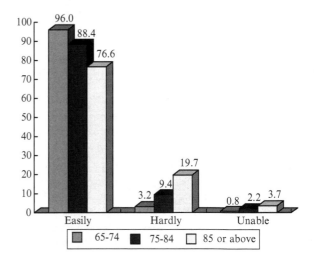

Fig. 11.4 Toileting ability by age group

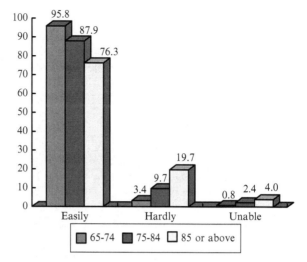

Fig. 11.5 Dressing ability by age group

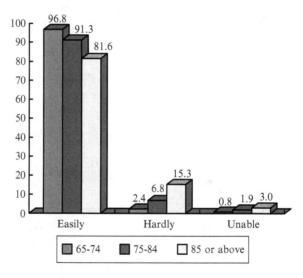

as compared to those between 75 and 84 years old, and around 4 times as compared to their counterparts between 65 and 74 years old.

According to a recent nationwide study conducted by the China Research Centre on Aging (2011), it was estimated that a total number of 33 million people aged 60 or above in China have suffered deficits in ADL, which consists of about 19% of the total aging population. In urban China, on average, about 5% of the aging population had severe deficits in ADL. In the same study, it was estimated that both total numbers of older people suffering from ADL deficits and the percentage of frail older people would continue to increase in the coming half century in China. Thus, undoubtedly, long-term care needs and policy responses are a critical concern at the national level.

In facing such a large group of aging population that has long-term care needs, an important question is who is taking care of them? Family certainly plays a significant role in taking care of frail older adults in China. According to recent statistics, there are about 4,000 elderly homes in China, which are housing around 2,109,000 older adults. Among them, only 17% have long-term needs. Hence, the majority of frail older adults are being cared for in families, within which spouses, children, and children-in-law are the three major family caregivers (China Research Centre on Aging, 2011; Ministry of Civil Affairs of the People's Republic of China, 2010).

In response to such newly emerged long-term care needs along with population aging, long-term care needs have for the first time been emphasized in the most recent guiding policy document "The twelfth five-year plan for aging policy in China" as one of the distinguished policy priorities entitled "developing an elder care system" (China National Committee on Aging, 2011). The contents covered by this policy direction is in line with the "aging in place" principle, which stated that community resources and community services be strengthened, and residential care facilities are expected to be enhanced and secured for long-term care purposes (Burgess & Burgess, 2006; China National Committee on Aging, 2011).

The Shanghai Long-Term Care Model

As discussed in the previous part of this chapter, Shanghai is one of the municipal cities in China with the highest rate of aging population. Up to the end of 2010, Shanghai has 14,120,000 people with *HuKou* registration, among which 2,260,000 were aged 65 or above, composing 16.0% of the population (Shanghai Research Center on Aging, 2011). It is estimated that the proportion of people aged 65 or above will increase to 33.8% in 2050. According to the 2006 national survey, the component of surveyed sample in Shanghai that needs daily assistance was 12.73%, which stood at the higher end among all the municipal cities and provinces in China (Guo & Chen, 2009). Estimates are that by 2050, Shanghai will have around 494,000 people aged 65 or above needing long-term care (Table 11.3) (Gui, 2008).

Since the central government of China adopted a decentralization approach in developing social policies, policy initiatives can be developed and trailed at various administrative levels (Chan, Ngok, & Phillips, 2008). Shanghai, as one of the earliest aging cities in China, experienced four stages in developing its long-term care policy (Table 11.4).

Table 11.3 Estimation of long-term care needs in Shanghai (10,000 persons)

	2000	2010	2020	2030	2040	2050
60 or Above	15.61	23.08	29.81	43.22	56.55	50.15
65 or Above	15.20	22.33	28.67	42.51	55.82	49.41

Source: Gui (2008)

Table 11.4 Shanghai long-term care model development stages

Stage	Level of administration	Policy	New initiatives
1980s, The initial stage	Street level	NIL	Community center
1990s, The residential care development stage	Municipal city level	Administrative rules governing residential facilities for the aged in Shanghai	Encourage investment in building residential facilities for the aged
Early 2000s, The community care development stage	Municipal city level	Opinions on community-based long-term care in Shanghai	Community-based long-term care system
2006, The long-term care policy model stage	Municipal city level	Shanghai 9073 model: Standard of community-based long-term care services for the elderly	Balanced development in residential and community services

In the late 1980s, the first community center providing long-term care services was established in *LuWan* district in Shanghai. At that time, meal service and assisted bathing-at-home service were initiated. Later on, day-care service was developed. However, due to lack of policy support from the municipal city level, community-based long-term care was in its initial stage. At that period of time, community-based elderly care was discussed and proposed by academics as a supplement to family care (Lin, Bao, & Sun, 1999).

Until late 1990s, Shanghai initiated a policy on regulating the development of residential facilities for the aged (Shanghai Government, 1998). At that time, the focus was on how to encourage different social agencies including enterprises and individuals to investigate and manage residential facilities for the aged. It is noteworthy that this local policy was enacted 1 year before national interim measures on regulating welfare organizations by the Ministry of Civil Affairs (1999). From the early 2000s, Shanghai moved into a stage of developing community-based services. In 2001, the Shanghai Civil Affairs Bureau published a policy document entitled "An opinion on a full-scale development of community-based services." Then, a citywide project "supporting aging in place project" was imitated, which aims to establish four core elements of community-based services, which consisted of day-care, personal care, housekeeping service, and psychological care. From 2001 to 2004, Shanghai was able to achieve 4 times the number of its community-based service recipients—up to 12,000 persons per year (Shanghai Civil Affairs Bureau, 2004).

Shanghai's 2006 11th five-year plan, set up a long-term care model of "9073," which stated that 90% of older adults would depend on family care (including paid full-time and/or part-time home helpers, *Bao Mu*), 7% of older adults would use community-based services, and 3% of older adults would benefit from institutional care (Shanghai Committee on Aging, 2006). Under this framework, Shanghai developed (1) a comprehensive assessment tool as a single entry point for eligibility assessment; (2) a sliding pay scale for government subsidy; (3) multiple channels for financing including funding from welfare, medical insurance, employment insurance, and retraining; (4) a consumer-directed voucher system for service delivery; (5) community-based long-term care service centers, which are responsible for administration and quality insurance; and (6) service standard for both residential and community-based long-term care services (Shanghai Civil Affairs Bureau, 2009). The setup of the "9073" model signified that Shanghai has entered into a balanced development stage in long-term care. Up to 2010, Shanghai has 625 residential facilities, which provided 97,841 beds (3% of the older population in Shanghai). A total number of 303 day-care centers have been established, together with 233 community elderly assistant centers, which provided home-based services for 252,000 older people annually. In addition, there are 404 community places that provide meal services, which serve around 40,000 older persons on a daily basis (Shanghai Research Center on Aging, 2011).

Family Caregiving And Its Impact on Caregiver Mental Health

After reviewing long-term care needs in China and long-term care model development in Shanghai, we argue that while the Shanghai long-term care model set a milestone and benchmark for the future development of long-term care policy in China, it still has room for improvement. The missing block in the Shanghai model is on policy and/or services targeting family caregivers. In developed countries such as the United States and Australia, family caregivers become policy targets and are integrated into aging policy in general and long-term care policy in particular (Congress of the United States of America, 2000; Department of Health and Ageing and Australian, 2010).

Family care is and will be dominating provision of long-term care in China. Recently, a draft on amending the Law on protecting the rights of older people tried to add emotional care from children toward their elders as mandatory (Jin, 2011), revealing the government's intention to further strengthen the institutional power of a family's responsibility for taking care of older adults. However, we would like to highlight that family caregivers, in particular those who provide long-term care for elders, need various types of support so that they can continue to provide care. Otherwise, premature institutionalization and/or quality of life of frail elders would be affected (Anngela-Cole & Hilton, 2009).

In response to this, the authors of the chapter conducted a study in Shanghai, which aims to examine family caregiving conditions and their impacts on mental health of caregivers.

Sampling

The sample consisted of 716 frail elder-caregiver dyads which were selected by multistage random sampling. At stage one, six urban districts were randomly selected. At stage two, random sampling was adopted to select one Street Office (*Jie Dao*) in each selected district, followed by selecting up to six neighborhood committees under each Street Office with a sequenced list. At stage three, sampling was conducted in the selected neighborhood committees starting from the first committee on the list, until at least 120 potential participants were identified from each community of which the first 120 listed participants consisted of 2/5 males and 3/5 females. Two to six neighborhood committees were finally selected from each Street Office.

Cadres from the Street Office and selected neighborhood committees, who were familiar with the residents, consulted the entire resident registry, made preliminary assessment on each resident aged 75 or above for his/her eligibility to the current study and prepared a list of potential interviewees. The inclusion criteria behind this preliminary assessment included (1) elders aged 75 or above, (2) elders who need support in ADL, and (3) elders who have at least one main family caregiver who takes care of him/her. The total number of potential frail elder-caregiver dyads finally identified from the 6 districts was 819 dyads. Surveys were conducted with the potential interviewees starting from the first dyad listed on the list to the 120th listed dyad while the remaining listed participants were used for replacement purpose until 120 valid cases in each district were interviewed.

The last stage of sampling was done by interviewers who screened the elder and his/her caregiver participating eligibility prior to the interview. To be eligible as a frail older person in this study, the elder has to be more than 75 years old and have at least two ADL deficits equivalent to a score of 90 or less (Cox, 1993). To be eligible as a main caregiver defined in this study, respondents have to be at least 18 years old, while providing support and/or supervision for a community-dwelling frail old-old person. The caregiver is receiving no money from this care provision. A total of 716 dyads were successfully interviewed and are included in the present analysis.

Measures

Measures on elder respondents included functioning health, cognitive functioning, and demographic characteristics. Measures on family caregivers included depressive symptoms, care provided by caregivers, knowledge and utilization of community-based services, living arrangement of caregivers, competence of caregivers, and demographic characteristics including relationship between caregivers and care receivers.

Functioning health of older adults was measured by the Chinese version of the Barthel Index (BI), Activities of daily living (ADL), and Lawton Instrumental Activities of Daily Living (IADL) (Chi & Boey, 1993; Lawton & Brody, 1969; Mahoney & Barthel, 1965). The BI contained ten items: dressing and undressing, walking on a level surface, ascending/descending stairs, bathing self, getting on and off toilet, doing personal toilet, feeding, moving from chair to bed and return, continence of bowels, and controlling bladder. These were measured on a 3-point scale (0 = fail to do independently, 5 = need partial assistance or need external devices, and 10 = can do independently). The scale score was the sum of the ten scores. The theoretical range of the scale score was 0–100, with a higher score indicating higher independence. The reliability of the BI in the present study was satisfactory, as indicated by Cronbach's alpha equals 0.89. The IADL was a 7-item scale used to assess the level of dependency on more complex daily living activities among elderly people in the previous week (food preparation, housekeeping, handling finance, handling medication, shopping, using telephone, and using transportation). In each item, the respondent was asked to what extent he/she needs assistance in performing the mentioned task in the past week. Ratings were based on a 4-point scale (0 = totally independent, 1 = need partial assistance and able to do most, 2 = need total assistance but elder still participates in the task, 3 = carry out by others, and 8 = no such happening). Summing up, the score of all items is the scale score with a theoretical range from 0 to 56. The higher the scale score, the lower the level of independence in the IADL. The reliability of ADL in the present study was satisfactory as indicated by Cronbach's alphas equals 0.83.

Cognitive functioning of older adults was measured by the Chinese version of the Short Portable Mental Status Questionnaire (SPMSQ), which is a 10-item scale developed by Pfeiffer (1975) and validated in the Chinese older population (Chi & Boey, 1993; Ngan & Kwan, 2002). This test assesses the respondent's cognitive functioning from aspects including orientation, personal history, remote memory, and calculations. Incorrect answers received 1 point each with the scale score calculated by summing up all ten items. The total scale score ranged from 0 to 10, with a higher score associated with a poorer cognitive function. The internal consistency of the Chinese SPMSQ scale in the present study was 0.78, which indicated a satisfactory reliability.

Demographic characteristics of older adults were measured by gender, year of birth, marital status, education attainment, living arrangement, and religious belief.

Depressive symptoms of caregivers were measured by a 20-item Chinese version of the Centre on Epidemiologic Studies Depression Scale CES-D. (Rankin, Galbraith, & Johnson, 1993; Zhang & Norvilitis, 2002). CES-D assesses depressive symptoms from aspects of depressive mood, positive emotion, somatic syndromes, and relationship difficulties. In each item, the respondent is asked to report the frequency of experiencing the described feelings or situations in the past 7 days on a 4-point scale (0 = less than 1 day, 1 = 1–2 days, 2 = 3–4 days, and 3 = 5–7 days). The sum of all items score gives the scale score with a theoretical range from 0 to 60. Higher scale score is associated with greater depressive symptoms. The Cronbach's alpha for this scale in the present study was 0.89, suggesting satisfactory internal consistency reliability.

Care provided by caregivers was measured by the caregivers' frequency of performing different caregiving tasks in the past 3 months consisting of 11 items. Caregivers were asked to report the frequency of providing each care item on a 5-point scale (0 = never, 1 = rarely, 2 = occasionally, 3 = sometimes, 4 = always). These 11 care items can further be categorized into 4 caregiving aspects, which are

ADL/IADL care (5 items), care on social aspect (3 items), emotional support (2 items), and financial support (1 item). Examples of the exact items included in these four care types are "feeding, dressing, toileting," "cooking, clothes washing," "arranging social activities," and "providing financial assistance," etc. The theoretical range of subscale score of each care aspect was 0–20, 0–12, 0–8, and 0–4 for ADL/IADL care, care on social aspect, emotional support, and financial support, respectively. The Cronbach's alphas for the first three subscales with more than one item ranged between 0.70 and 0.78, which was satisfactory. Moreover, a single item was used to measure the daily amount of time that the caregiver provided care in the previous week. The response has five categories (0 = less than an hour, 1 = 1–3 h, 2 = 4–6 h, 3 = 7–9 h, 4 = 10 h or more). Caregiver's length of providing care to the care recipient was measured by a single item on a 5-point scale (0 = less than half year, 1 = half year to 1 year, 2 = 1–5 years, 3 = 6–10 years, 4 = over 10 years).

Knowledge and utilization of community services of caregivers were measured by asking caregivers "Have you heard of this service?" "Have you ever used this service in the past 12 months?" and "In the past 12 months, how many hours have you received the service each week?" Respondents answered the first two questions on a Yes = 1 or No = 0 pattern, and they reported the number of hours in answer to the third question. A list of community-based services for family caregivers was developed based on literature review and information provided by elderly service agencies in Shanghai.

Living arrangement of caregivers was measured by three questions. The first question asked "Are you sharing the same household with the care recipient?" (0 = Not the same household, 1 = the same household). If caregivers were not sharing the same household with the elder, they proceeded to answer the question "By what means do you usually go to the care recipient's home?" on a 3-category answer (1 = on foot, 2 = public transportation, 3 = private car/motorbike or cycling). In the third question, traveling time was used as an indicator of distance (Joseph & Hallman, 1996), and caregivers answered the question "By the usual means, how much time is needed to travel to the care recipient's home?" on a 4-category answer (0 = no more than 15 min, 1 = 15–30 min, 2 = 31–60 min, 3 = more than 1 h).

Competences of caregivers were measured by three aspects: competence on knowledge and skills, health competence, and financial competence. Competence on knowledge and skills was measured by asking "How would you rate your adequacy of knowledge and skills on caregiving?" on a 5-point scale (0 = far less than adequate, 1 = less than adequate, 2 = just so-so, 3 = more than adequate, 4 = much more than adequate). Health competence was measured by asking "How would you rate your overall health before you started taking care of the care recipient?" with a 5-point Likert response (0 = very poor, 1 = poor, 2 = fair, 3 = good, 4 = very good). Financial competence was measured by asking "Do you think that you have enough money to cover your daily expenses?" on a 5-point scale (0 = very inadequate, 1 = inadequate, 2 = just enough, 3 = adequate, 4 = more than adequate).

Table 11.5 Demographic of frail elders ($N = 716$)

	Frequency	Percentage
Gender		
Male	272	38.0
Age (mean = 83.1, SD = 5.1)		
75–79	201	28.1
80–84	255	35.6
85–89	178	24.9
90 and Above	82	11.5
Education		
No education	200	27.9
Primary school graduate	258	36.0
Secondary schooling	216	30.1
Tertiary education	41	5.7
Marital status		
Currently married	360	50.3
Currently not married/widowed	353	49.3
Religious belief		
Yes	166	23.2
No	549	76.7
Living arrangement		
Living alone (without domestic helper)	68	9.5
Living with spouse only (with or without domestic helper)	237	33.1
Living with (nucleus) family members (with or without domestic helper)	383	53.5
Living with friends or relatives only (with or without domestic helper)	17	2.4
Living with domestic helper only	11	1.5

Number of missing cases ranges from 1 to 3

Demographic characteristics of caregivers included relationship between caregivers and care receivers, age, gender, education attainment, marital status, religious beliefs, and employment status.

Procedure

This study was approved by the Human Research Ethics Committee of the University of Hong Kong. The developed questionnaire was initially piloted, and then the research team recruited interviewers from Street Offices. A one-day training workshop was provided to all interviewers to ensure that each interview followed the same procedure. Interviewers conducted face-to-face interviews with participants at the participants' home. Six graduate students from a University in Shanghai were recruited to help with the interviewer training and coordinate with the interviewers during the field work. The data collection period was carried out between April and July 2010.

Table 11.6 Demographic of caregivers ($N=716$)

	Frequency	Percentage
Gender		
Male	283	39.5
Age (mean = 63.2, SD = 12.6)		
49 or Below	77	10.7
50–59	287	40.1
60–69	110	15.4
70–79	138	19.3
80 or Over	104	14.5
Education		
No education	66	9.2
Primary school graduate	83	11.6
Secondary schooling	463	64.7
Finish tertiary education or above	104	14.5
Marital status		
Currently married	622	86.9
Currently not married/widowed	94	13.1
Religious belief		
Yes	135	18.9
No	581	81.1
Employment status		
Employed	132	18.4
Not employed	584	81.6
Monthly income (RMB)		
<500	32	4.5
501–2,000	520	72.7
2,001–5,000	147 (20.5)	20.5
>5,000	16 (1.9)	2.2

Number of missing cases equals was 1 in monthly income

Findings

Profile of participants: The participating frail elders consisted of 272 (38.0%) males and the mean age was 83.1 (Table 11.5). Most of the elders attained primary school education (36.0%), followed by secondary education (30.1%) and no education (27.9%). Around half of the elders were married (with spouse alive) (50.3%), and over three-fourths of them have no religious affiliation. Concerning their living arrangement, over half of the elders were living with family members, and around 30% of them were living with spouse only. Only 9.5% were living alone.

When it comes to caregivers, over half of them are female (61.5%) (Table 11.6). The mean age of caregivers was 63.2 and around two-fifths of them belonged to the age group of 50–59. A bit less than 80.0% of the caregivers have attained secondary or higher education, and most of them were married (86.9%). Like frail elders, most caregivers were not religious (81.1%). Concerning their employment status, only 18.4% of them were employed. Regarding monthly income, 501–2,000 RMB

Table 11.7 Caregiving features ($N = 716$)

	Frequency	Percentage
Relationship with care receivers		
Spouse	249	34.8
Son	183	25.6
Daughter	201	28.1
Children-in-law	59	8.3
Grandchildren/spouse of grandchildren	7	1.0
Others (siblings, neighbors, relatives)	17	2.4
Living arrangement		
Co-residence with frail elder	568	79.3
Traveling time to frail elder's home (for those not co-residing with frail elder, $n = 147$)		
Not co-reside, less than 15 min travel	68	46.3
Not co-reside, 15–30 min travel	33	22.4
Not co-reside, 31–59 min travel	28	19.0
Not co-reside, over an hour travel	18	12.2
Daily time providing care in the past 1 week		
Less than an hour	57	8.0
1–3 h	283	39.5
4–6 h	147	20.5
7–9 h	73	10.2
10 h or more	156	21.8
Duration of taking care of the frail elder		
Less than a year	157	22.0
1–5 Years	351	49.0
6–10 Years	119	16.6
More than 10 years	89	12.4

Number of missing case was 1 in living arrangement

accounted for 72.7% of all caregivers while the second biggest group was 2,001–5,000 RMB (20.5%).

Care needs of frail older respondents: Among surveyed older respondents a mean score of 75.47 was obtained on ADL with a standardized deviation of 20.3. Their IADL score was 14.28 (mean), with a standardized deviation of 9.9. Their score on cognitive ability was 2.55 (mean), with a standardized deviation of 2.4. About 62% of the respondents were severely deficit in visual ability while only 4.5% had hearing ability that was severely affected.

Family caregiving is shown in Table 11.7. About one-third of the caregivers were spouses of the frail older person; while children or children-in-law made up about 60%. Family caregivers, other than spouses and children, cared for less than 5% of the respondents. The majority of the family caregivers live together with the frail older persons. For those who did not co-reside, more than 60% reported travel time to the care recipient as within half an hour; and 40% of them travel by foot, followed by one-third by public transport and another one-third by bicycle or car.

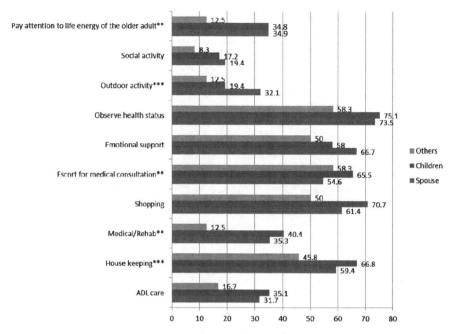

Fig. 11.6 Family caregiver by different caregivers. *Notes*: ***$p<0.001$, **$p<0.01$

Table 11.8 Mental health of caregivers ($N=716$)		Spouse	Children	Others
	CESD	12.1 (8.8)	11.4 (8.8)	8.3 (8.1)
	CESD (>=16)	23.3	23.7	16.7

Concerning care intensity, about 30% of the caregivers reported that they provided care for more than 7 h on a daily basis. Around 30% had provided care for more than 5 years.

What kind of care do family caregivers provide? Results showed (Fig. 11.6) that caregiving activities varied between different caregivers. Spouses are involved most often in emotional support, escort for medical consultation, and financial management. Children helped more in housekeeping, escort for medical consultation, and providing emotional support. Those caregivers other than spouse and children tended to be involved most often in escort for medical appointment, followed by providing emotional support and housekeeping. Gender differences were identified among four out of ten caregiving activities. Female caregivers were more likely to provide emotional support ($\chi^2=9.56$, $p<0.05$), help in housekeeping ($\chi^2=23.65$, $p<0.0001$), and ADL care ($\chi^2=13.12$, $p<0.01$) as compared to male caregivers; while male caregivers were more likely to escort older care receivers for medical consultation ($\chi^2=10.28$, $p<0.05$).

Table 11.9 Competence indicators of caregivers

	Frequency	Percentage
Competence on knowledge and skills		
Very insufficient/insufficient	220	30.7
Just meet the needs	288	40.2
Sufficient/very sufficient	208	30.0
Competence on self-rated health		
Very poor/poor	39	5.5
Fair	460	64.2
Good/very good	217	30.3
Competence on financial resources		
Very insufficient/insufficient	184	25.7
Just meet both ends	294	41.1
Sufficient/more than sufficient	238	33.2

Table 11.10 Multiple regressions on CES-D

	Beta	SE	T score
Constant		2.78	10.22***
Caregivers' religious belief (Yes = 1)	−0.06	0.76	−1.89[a]
Care needs of care recipients			
SPMSQ	0.10	0.14	2.49*
ADL	−0.10	0.02	−2.33*
IADL	−0.08	0.04	−1.89[a]
Care provided			
ADL/IADL care	0.10	0.08	2.51*
Social care	−0.10	0.10	−2.66**
Emotional support	−0.04	0.25	−1.01
Financial support	0.04	0.21	1.03
Competence			
Knowledge and skills	−0.14	0.38	−3.93***
Health	−0.20	0.52	−5.85***
Financial adequacy	−0.18	0.34	−4.73***

[a]$p < 0.10$, *$p < 0.05$, **$p < 0.01$, ***$p < 0.001$

Mental health of caregivers is shown in Table 11.8. Concerning the negative aspect of mental health, CESD score was reported with a mean of 11.53 and standard deviation 8.8. No significant differences were found among three groups of caregivers concerning mean scores of CESD. When a score of 16 or greater was used to identify at-risk depressive symptoms among caregivers, 23.3% of spouse caregivers, 23.7% of children caregivers, and 16.7% of others caregivers were identified to fall in the at-risk groups. No gender differences were found in regard with CESD at-risk groups (male 23.7% at risk vs. female 23.1% at risk).

Competences of caregivers are shown in Table 11.9. About one-third of the caregiver respondents perceived that they had sufficient knowledge and/or skills in taking up caregiver roles; while another one-third perceived they were lacking in care giving skills and knowledge. Only 5.5% of the caregiver respondents reported that they had poor or very poor health before they took up caregivers' roles. As far as perceived financial adequacy, around one-fourth of the caregiver respondents reported that their financial resources were insufficient or very insufficient. Gender differences were found in two out of three competence indicators. Female caregivers were more likely to perceive themselves as having sufficient/very sufficient knowledge and skills ($\chi^2 = 10.32$, $p < 0.05$), and financial resources ($\chi^2 = 10.13$, $p < 0.05$) as compared to male caregivers.

Associates of caregivers' mental health were computed by multiple regression analysis (Table 11.10). Who are those caregivers showing high risk of having depressive symptoms? They were those caregiver respondents without religious beliefs, who were taking care of older respondents with greater needs (e.g., poor physical functioning as indicated by ADL and IADL scores, and cognitive functioning). They provided higher frequency of ADL/IADL care, and showed a lower level of competence in taking up of the caregivers' role (e.g., perceived themselves having inadequate knowledge, poorer health, and inadequate financial status). On the contrary, caregiver respondents who provided higher frequent social care to older respondents were less likely to suffer from depressive symptoms.

Discussion

Based on our analysis on aging policy and its recent development in China, we advocate a national level long-term care policy be in place so that municipal cities and provinces can follow its recommendations. A national level policy can contribute to setting up guiding principles and long-term goals. Observations on recent development of long-term care policies in various municipal cities and provinces showed two different routes. Tianjin and Zhejiang province seemed to follow the Shanghai model by establishing long-term care eligibility criteria and community-based services (Tianjin Civil Affairs Bureau, 2008; Zhejiang Civil Affairs Bureau, 2010). Shanghai's model, mirrored by *Tianjing* and *Zhejiang* province, focuses more on long-term care of older people and mainly put this policy agency under the policy and administration of the Civil Affairs Bureau. Even though long-term care needs are mostly required by older people, it should not rule out people in other age groups including those with physical and/or mental disability occurring at birth or at a young age. According to the definition of long-term care, it basically is needed by those who "have become dependent on assistance with basic activities of daily living due to long-standing physical or mental disability" (Organization for Economic Co-operation and Development, 2005). From this perspective, the Shanghai long-term care model does not target those younger aged groups of people with long-term care needs, and might in the long run need policy reform and/or integration.

In 2009, Beijing developed its long-term care policy named "9064 model" that included people of various ages with long-term care needs. It targeted both elderly people and people with disabilities (Beijing Civil Affairs & Beijing Disabled Persons' Federation, 2009). It aims to achieve the goals that 90% of eligible adults are to be cared for by families, 6% by community-based services, and 4% by residential care. As compared to the Shanghai model, the Beijing model has the advantage of targeting various-aged people with long-term care needs, and has integrated resources from the rehabilitation sector. It is foreseen that other provinces could mirror the Beijing model in the near future. In this regard, it is crucial to develop a national government policy on long-term care.

Findings of the empirical data in Shanghai showed a significant proportion of family caregivers who were at risk of suffering psychological distress in general, and depressive symptoms in particular. Shanghai is a pioneer in developing community-based long-term care policy and services. However, such a figure was not surprising and in line with findings from other countries (Robison, Fortinsky, Kleppinger, Shugrue, & Porter, 2009). In general, it is hypothesized that high care needs generate stress and then will trigger psychological distress (Pearlin, Mullan, Semple, & Skaff, 1990). The findings of our study supported such hypothesis, which showed a positive relationship between care needs (e.g., ADL care, IADL care, and cognitive functioning level of care receivers) and depressive symptoms of the caregivers. Moreover, the findings of the present study also revealed that perceived competence, including health, finance, and knowledge/skills were negatively associated with caregivers' depressive symptoms.

Gender differences identified in caregiving activities performed by caregivers are? consistent with gender role differences—female caregivers were more likely to engage in providing in house care (e.g., daily care and emotional support); while male caregivers were more likely to provide care outside the household (e.g., escort services). Female caregivers also showed higher levels of competence in regard with knowledge/skills and financial sufficiency as compared to male caregivers. However, gender was not identified as a significant associate with depressive symptoms of caregivers. Further studies are suggested to examine gender and caregivers' mental health.

Such findings could generate three implications regarding policy and service. First, long-term care needs' fulfillment would generate extra expenses for the family, which may include medical consultation, transportation to medical consultation, buying rehabilitation aids, etc. Perceived financial constraints of the caregivers would definitely put extra stress on caregivers (Wilkins, Bruce, & Sirey, 2009). In this regard, financial needs of the family with members having long-term care needs should be attended and special allowance considered when targeting those families with financial difficulties.

Second, perceived health status of caregivers also played a significant role in predicting their depressive symptoms. It is understood that providing daily care is not only financially demanding but also physically demanding. Helping with bathing, shopping, housecleaning, and escorting to medical consultation demand a lot of physical strength of the caregivers (Robison et al., 2009). Hence, not only health status of the frail older adults deserves attention, but the health status of the caregivers also deserves policy attention. Respite services that aim to relieve caregivers within a particular time frame could be effective and were strongly recommended (Mason et al., 2007).

Third, perceived sufficient knowledge and skills also negatively related to caregivers' depressive symptoms. Taking care of frail older family members demand specific knowledge on diseases, rehabilitations, medications, and associated complications. It is expected that in most cases caregivers may not have been equipped with the necessary knowledge before they took up the caregivers' roles (Haley et al., 2009; Martinez Martin et al., 2008). Support programs on enhancing knowledge and skills of family caregivers are suggested.

Hence, in order to avoid future policy diversity in developing and planning long-term policy, we advocate a national long-term care policy be developed. At the national level, what qualifies as long-term care, policy goals of long-term care, who should be targeted, key criteria for providing government subsidy, quality insurance, and long-term care financing should all be stated considering the demography, migration and urbanization, and social and cultural values in China (Organization for Economic Co-operation and Development, 2005; World Health Organization, 2003).

Conclusion

After analyzing the aging population and associated policy in China, it is determined that China still lacks a national level long-term care policy. Current long-term care policy and models developed at municipal city level showed diversified directions and lack of consensus on the conceptualization of long-term care. A national level long-term care policy is suggested so as to clarify the definition of long-term care and policy objectives.

After analyzing the development of long-term care policy in Shanghai, the findings based on an empirical study revealed that there is a need to integrate the needs of family caregivers into a long-term care policy and service model. Financial needs, health needs, and knowledge/skills needs are recommended to be taken into consideration for policy and service development purposes.

Acknowledgment This study was supported by General Research Fund of the Research Grants Council, Hong Kong (Ref. 448009).

References

Anngela-Cole, L., & Hilton, J. M. (2009). The role of attitudes and culture in family caregiving for older adults. *Home Health Care Services Quarterly, 28*(2), 59–83.

Beijing Civil Affairs Bureau, Beijing Disabled Persons' Federation. (2009). *A solution to community-based (disability) services for Beijing residents.* Retrieved June 4, 2011 from http://zhengwu.beijing.gov.cn/gzdt/gggs/t1097076.htm (in Chinese).

Burgess, A. M., & Burgess, C. G. (2006). Ageing-in-place: Present realities and future directions. *Forum on Public Policy: A Journal of the Oxford Round Table.* Retrieved on November 2, 2011 from http://www.forumonpublicpolicy.com/archive07/burgess.pdf.

Chan, C., Ngok, K., & Phillips, D. (2008). *Social policy in China: Development and well-being*. Bristal: Policy Press.

Chi, I., & Boey, K. W. (1993). Hong Kong validation of measuring instruments of mental health status of the elderly. *Clinical Gerontologist, 13*, 35–45.

China National Committee on Aging. (2011). *The twelfth five-year plan for aging policy*. Retrieved June 3, 2011 from http://www.yinlingguanai.org/2011/0509/240.html (in Chinese).

China Research Centre on Aging. (2011). *Press release on "A national study of frail older adults in urban and rural China" by Zhang K. D.* Retrieved June 3, 2011 from http://www.cncaprc.gov.cn/info/13085.html (in Chinese).

Congress of the United States of America. (2000). *National Family Caregiver Act*. Retrieved June 4, 2011 from http://www.cga.ct.gov/2006/rpt/2006-R-0203.htm.

Cox, C. (1993). *The frail elderly: Problems, needs, and community responses*. Westport: Auburn House Publishing.

Department of Health and Ageing, Australian Government. (2010). *National Respite for Careers Program*. Retrieved June 4, 2011 from http://www.health.gov.au/internet/main/publishing.nsf/content/ageing-carers-nrcp.htm.

Gui, S. X. (2008). Constructing general care system for the elders: A case study in Shanghai. *Population and Development, 14*(3), 78–83 (in Chinese).

Gui, S. X. (2009). Demographic of China in the 21st century and sustainable development of social welfare. *Population and Family Planning, 3*, 15–17 (in Chinese).

Guo, P., & Chen, G. (Eds.). (2009). *Data analysis of the sampling survey of the aged population in urban/rural China 2006*. Beijing: Zhong Guo She Hui Chu Ban She (in Chinese).

Haley, W. E., Allen, J. Y., Grant, J. S., Clay, O. J., Perkins, M., & Roth, D. L. (2009). Problems and benefits reported by stroke family caregivers: Results from a prospective epidemiological study. *Stroke, 40*(6), 2129.

Jin, J. (2011, January 5). Law on more emotional care for the elderly. *Global Times*. Retrieved June 4, 2011 from http://china.globaltimes.cn/society/2011-01/609284.html.

Joseph, A. E., & Hallman, B. C. (1996). Caught in the triangle: The influence of home, work and elder location on work-family balance. *Canadian Journal on Aging, 15*, 393–412.

Lawton, M. P., & Brody, E. M. (1969). Assessment of older people: Self-maintaining and instrumental activities of daily living. *The Gerontologist, 9*, 179–186.

Lin, G., Bao, S. M., & Sun, X. M. (1999). Establishing an elderly protection system by intergrating family and community services. *Population Research, 23*(2), 55–60 (in Chinese).

Mahoney, F. I., & Barthel, D. W. (1965). Functional evaluation: The Barthel Index. *Maryland State Medical Journal, 14*, 61–65.

Martinez Martin, P., Arroyo, S., Rojo Abuin, J. M., Rodriguez Blazquez, C., Frades, B., & de Pedro Cuesta, J. (2008). Burden, perceived health status, and mood among caregivers of Parkinson's disease patients. *Movement Disorders, 23*(12), 1673–1680.

Mason, A., Weatherly, H., Spilsbury, K., Golder, S., Arksey, H., Adamson, J., et al. (2007). The effectiveness and cost-effectiveness of respite for caregivers of frail older people. *Journal of the American Geriatrics Society, 55*(2), 290–299.

Ministry of Civil Affairs. (1999). *Interim measures for the administration of welfare organizations*. Retrieved June 3, 2011 from http://www.mca.gov.cn/article/zwgk/fvfg/shflhshsw/200809/20080900019764.shtml (in Chinese).

Ministry of Civil Affairs of the People's Republic of China. (2010). *2009 Statistics yearbook on civil affairs*. Retrieved June 3, 2011 from http://cws.mca.gov.cn/article/tjbg/201006/20100600081422.shtml (in Chinese).

National Bureau of Statistics of China. (2011a). *2010 Census key statistics released on April 28, 2011 (no. 1)*. Retrieved June 2, 2011 from http://www.stats.gov.cn:82/tjgb/rkpcgb/qgrkpcgb/t20110428_402722232.htm (in Chinese).

National Bureau of Statistics of China. (2011b). *China statistics yearbook 2010*. Retrieved June 3, 2011 from http://www.stats.gov.cn/tjsj/ndsj/2010/indexch.htm.

Ngan, R., & Kwan, A. (2002). The mental health status and long-term care needs of the Chinese elderly in Hong Kong. *Social Work in Health Care, 35*(1), 461–476.

Organization for Economic Co-operation and Development. (2005). *Long-term care for older people*. Paris: OECD.

Pearlin, L. I., Mullan, J. T., Semple, S. J., & Skaff, M. M. (1990). Caregiving and the stress process: An overview of concepts and their measures. *The Gerontologist, 30*(5), 583.

Pfeiffer, E. (1975). A short portable mental status questionnaire for the assessment of organic brain deficit in elderly patients. *Journal of the American Geriatric Society, 23*, 433–441.

Rankin, S. H., Galbraith, M. E., & Johnson, S. (1993). Reliability and validity data for a Chinese translation of the Center for Epidemiological Studies-Depression. *Psychological Reports, 73*(2), 1291–1298.

Robison, J., Fortinsky, R., Kleppinger, A., Shugrue, N., & Porter, M. (2009). A broader view of family caregiving: Effects of caregiving and caregiver conditions on depressive symptoms, health, work, and social isolation. *The Journals of Gerontology Series B: Psychological Sciences and Social Sciences, 64B*(6), 788–798.

Shanghai Civil Affairs Bureau. (2001). *An opinion on a full-scale development of community-based services*. Retrieved June 3, 2011 from http://www.shmzj.gov.cn/gb/shmzj/node8/node15/node55/node230/node278/userobject1ai7918.html (in Chinese).

Shanghai Civil Affairs Bureau. (2004). *An introduction of community-based services for the elderly*. Retrieved June 3, 2011 from http://www.shanghaigss.org.cn/news_view.asp?newsid=547 (in Chinese).

Shanghai Civil Affairs Bureau. (2009). *A notice about strengthening community-based services in Shanghai*. Retrieved June 3, 2011 from http://www.shrca.org.cn/big/3513.html (in Chinese).

Shanghai Committee on Aging. (2006). *The eleventh five-year plan for aging policy in Shanghai*. Retrieved June 3, 2011 from http://www.shmzj.gov.cn/gb/shmzj/node8/node15/node55/node231/node263/userobject1ai22366.html (in Chinese).

Shanghai Government. (1998). *Administrative rules governing residential facilities for the aged in Shanghai*. Retrieved June 3, 2011 from http://www.ocan.com.cn/shfl-shgl.htm (in Chinese).

Shanghai Research Center on Aging. (2011). *2010 Shanghai aging population statistics*. Retrieved June 3, 2011 from http://www.shrca.org.cn/4113.html (in Chinese).

Steering Committee of the National People's Congress of the People's Republic of China. (1996). *People's Republic of China Law on the Protection of the Rights and Interests of Older People*. Retrieved June 3, 2011 from http://fss.mca.gov.cn/article/lnrfl/zcfg/200903/20090300027331.shtml (in Chinese).

Tianjin Civil Affairs Bureau. (2008). *Opinions on further development of community-based long-term care*. Retrieved June 4, 2011 from http://www.tjmz.gov.cn/zcfg/system/2008/09/02/010001206.shtml (in Chinese).

United Nations Department of Economic and Social Affairs Population Division. (2008). *World population ageing 1950–2050*. New York: United Nations Department of Economic and Social Affairs Population Division.

Wilkins, V. M., Bruce, M. L., & Sirey, J. A. (2009). Caregiving tasks and training interest of family caregivers of medically ill homebound older adults. *Journal of Aging and Health, 21*(3), 528.

World Health Organization. (2003). *Long-term care in developing countries—Geneva: World Health Organization*.

Zhang, J., & Norvilitis, J. M. (2002). Measuring Chinese psychological well-being with western developed instruments. *Journal of Personality Assessment, 79*, 492–511.

Zhejiang Civil Affairs Bureau. (2010). *A notice on trial of long-term care needs assessment in Zhejiang province*. Retrieved June 4, 2011 from http://www.zjmz.gov.cn/il.htm?a=si&key=main/ 12/shfl&id=4028e4812c33ef74012c9ba6874a056a (in Chinese).

Chapter 12
Housing Stratification and Aging in Urban China

Bin Li and Yining Yang

Abstract This chapter investigates a key aspect of aging in place with important impact on family care, that is, housing. There has been a disconnect between aging and housing in China research. The chapter approaches the topic by linking it with the study of social stratification in urban China. By using national statistical and survey data, policy recommendations are made to support aging in place by improving housing provision, increasing geriatric nursing training, etc.

Introduction

Urban housing reform was once considered to be the last battlefield of the marketization reform in China. In 1998, the Chinese central government decided that the "pay-as-you-go/own" housing distribution system should take the place of the old welfare-type housing distribution system in all major cities throughout the country, setting a milestone in urban housing reform. From 1978 to 1998, China constructed a new housing system consisting of: commercial housing, housing distributed by working units or *Danwei*, public housing provided by the government, and settlement/low-rent housing. The government implemented eight major policies, one after another, including: rent raise system, public housing sales policy, economical and affordable housing policy, housing reserve fund policy, cooperative housing construction policy, housing monetization policy, commercial housing policy, and low-rent housing policy. These led to the marketization of housing in major cities throughout China (Li, 2009).

B. Li (✉)
Central-South University, Changsha, Hunan, China
e-mail: libin@mail.csu.edu.cn

Y. Yang
Pace University, New York, NY, USA

S. Chen and J.L. Powell (eds.), *Aging in China: Implications to Social Policy of a Changing Economic State*, International Perspectives on Aging,
DOI 10.1007/978-1-4419-8351-0_12, © Springer Science+Business Media, LLC 2012

China has made great strides in its housing provision and reform, which can be summarized as follows:

1. Per capita living space in urban areas and the rate of homeownership have increased rapidly. In 1978 when China started reform and opening up, the average living space of urban and rural dwellers used to be 6.7 m²; it rose to 18.7 m² 2 decades later. In the year of 1998, the central government implemented monetized housing allocation to take the place of housing allocation in kind; meanwhile, it cultivated and developed a real estate market targeting residential housing. In 2006, the average living space of urban and rural dwellers was 27 m² and reached 28 m² in 2007; at the same time, the homeownership rate of urban and rural dwellers reached 83%.

2. Housing policies are improved gradually. Since the inception of housing reform, the central and local governments tried to build a housing system to attribute housing according to the structural characteristics of urban dwellers. Although the policy showed various flaws in the process of implementation, a housing supply system had come into shape in many urban cities by 2007 under a series of Macroeconomic Regulation and Control policies, which varied in accordance to the housing consumption ability of different social classes. Based on the correlation between household income and housing consumption ability, the urban and rural households in China can fit into three major categories: no housing consumption ability, incomplete housing consumption ability, and complete housing consumption ability. Under these three major categories, there are five subcategories, which are: extremely destitute households, destitute households, rental housing unaffordable households, households with insufficient housing purchasing ability, and households with sufficient housing purchasing ability. For instance, most of the households with none housing consumption ability are supplied with low-rent housing; households with complete housing consumption ability are mostly supplied with condominium; households with incomplete housing consumption ability are supplied with government-subsidized housing, which includes affordable housing and price-limited condominium. In 2007, the government established an urban and rural housing supply system made up of low-rent housing, affordable housing, price-limited condominium and general condominium, in response to three types of guarantee which were called relief guarantee, supportive guarantee, and cooperative guarantee. In practice, however, the outcomes of these policies were far from satisfactory since too much emphasis had been put on the general condominium policy.

3. Housing has become a "constant property" for urban dwellers. Eighty-three percent of urban housing has been and it has come to be a "constant property" for urban dwellers. On March 16 2007, the Property Law of People's Republic of China went into effect right after the fifth meeting of the tenth National People's Congress (NPC). The Property Law gives legal protection to residential housing and clarifies that residents have multiple rights toward their housing such as occupancy, inhabitation, investment, and disciplines. The "constant property" has given "constant belief" to urban and rural dwellers in the long-run development of the country. Some urban dwellers begin to invest on their housing and make profits from it.

Housing Stratification in Urban China

Urban housing stratification refers to the process in which urban dwellers of same social class tend to cluster in a certain type of living space while people of different social classes tend to live in different types of space, or in other words, the social stratification leaves an impact on urban housing stratification.

We can observe urban housing stratification in the five following areas:

1. Previous members of a community began to differentiate, disaggregate, and recombine with each other, people of the same or similar social status come to cluster in a certain living space. There are two different situations in this case. One is that people with ascending social status initiate to move out of the previous community to the community that fits their current social status, and leave the previous community to those who have fairly lower social status. One obvious characteristic of urban housing reform in China is the work units or *Danwei*. Staff of the same work unit tend to live in a relatively concentrated area, no matter what his or her occupation, position or income was. However, the situation has changed since the urban housing reform. Those who with higher income obtained high-quality condominium or houses through market mechanism, and thus left their previous community. The other is that people with descending social status cannot afford to live in their previous community and move out of it so as to find a new place to live. The urban renewal and development in China has left its low-income residents with no alternative way but to find a much cheaper place to live.

2. The newly built communities, varying greatly in price and taste, attract people of different social status and thus form housing stratification. Some cities in China are encouraging this tendency. For instance, Beijing and Shanghai offer condominium with the price ranging from ¥10,000 to 150,000/m². Undoubtedly, people of different social status tend to live in communities with different price offer.

3. Those who live in the same community gradually fall into the same social class and show the tendency to share similar cultural quality and social status.

4. Many junior labor force entered the city and clustered in the connecting areas between urban and suburbs. In those connecting areas, there always are poor community facilities, overpopulated living condition, and high crime rate. Besides, the residential structure is complex in those areas. People there have their own subcultures and often call themselves "city vagrants." The majority of off-farm or migrant workers in some cities in China serve as a good example.

5. "Yiju." It is a phenomenon that undergraduates with low-income cluster to live in the connecting areas of urban and suburbs. They cohabit like ants do: six or seven people share a single room and spend 2 h/day on bus to go to work. They have a monthly income of about ¥1,000 while the bunk will cost them about ¥300. They have poor living conditions, lack of social security, suffer from mood fluctuations, have psychological issues like anxiety, are not willing to tell their family the truth, and connect to the social world mainly by the Internet. Different from the fourth situation stated above, they show strong motivation to finally fit

into the city. Nowadays, the social phenomenon of "Yiju" exists in some first-tier cities such as Beijing, Shanghai, and Shenzhen and attracts more and more public attention.

Polarization and segregation will be two extreme cases in urban housing stratification, which means urban planning and changes in housing market will make the poor and the rich start gathering into separate zones. This phenomenon takes place for the following reasons: (a) the widening income gap has become the prerequisite for urban housing stratification; (b) the housing price determined by the land location has made urban housing stratification possible; (c) the commoditization and marketization of housing has served as a motivational mechanism in the process of urban housing stratification; (d) the breakdown of the work units policy and the increasing social mobility has accelerated the process of urban housing stratification.

Generally speaking, during the 30 years' Reform and Opening-up Policy, social stratification has come with income stratification. Under the impact of housing commoditization, stratification of social status and capacity of assets influence peoples' choice on housing to a great extent, thus accelerating the process of urban housing stratification. Therefore, the government of China should take measures to keep urban housing stratification under control. These two strategies which worth further consideration, are as follows:

First, providing complex housing and encouraging communication among different social classes. With government intervention into both economy and social fields, the strong correlation between housing and income can be weakened by such means as financial aids, improved education, and healthcare system. However, the objectives of government activities always encounter potential difficulties during implementation. The outcome of welfare policy is to make housing affordable to those who with low income. However, the housing stratification has not yet been improved and poverty-stricken households are still limited to a certain areas. In 1970s, a new strategic planning was innovated in the USA by introducing people of different social status into complex housing. The government intended to gather people with different income into the same neighborhood so as to build a complementary community, where the poor households could obtain heterogeneous resources. In this kind of complex housing community, by developing public housing and commoditized housing at the same time, there could be both for-profit and not-for-profit housing. This strategy aims to reach the balance between profits of real estate market and non-for-profit social goals through market itself and government intervention.

Second, raising the quality of poor communities. To relieve urban poverty by renewing poor communities helps to realize social integration. In late 1980s, most developed countries raised the quality of their residential areas by the following means: improving the living environment and public services in poor communities; building more shopping malls, recreation zones, traffic system, and relaxation facilities. These approaches were adopted so as to improve the living conditions for residents and to advocate humanistic management. By renewing the poor communities, the gap between the poor communities and the rich communities was diminished. Through cultural development can poor communities have new opportunities.

Many major cities in China implemented clearance policy on the renewal program of the old communities in urban areas. Under the impact of marketization, this kind of renewal programs put too much emphasis on rewards. The offered compensation for demolition is far below the market price while the supply of affordable housing and low-rent housing cannot meet the need of low-income households. Therefore, these renewal programs always turn the previous poor communities located in city center into an exclusive community; meanwhile, the mass majority of low-income households cluster in the connecting areas between urban and suburbs. Thus, it forms a social structure of "inner rich and outer poor." It does not only increase the transportation cost for low-income households but also increase the poor's hostility to the rich, which harms social stability.

To make urban housing stratification under control, some scholars have raised several proposals as follows: first, to adopt the general principle of complex housing. To realize combination and differentiation by developing different communities in the same area and setting them apart by one another. Second, to keep the community scale in control, especially the scale of clustered low-income housing. As in China, the number should be about 1,500 households per community. Third, to diminish the difference of social welfare among all social classes by renewing communities. The community renewal policy is an institutional arrangement to reduce the current resource allocation pattern, which can encourage residents' initiative toward community life.

Aging in China

As the world's most populous country, China had elderly population (age 60 and above) 178 million in 2010, accounting for 13.26% (National Bureau of Statistics of China, 2011), with an annual increase rate of nearly 10 million.

In 2006, China Research Center on Aging, on behalf of China National Committee on Aging (CNCA), carried out a research project on the elderly population in China and had following findings: till the end of 2006, those who aged from 60 to 64 years old occupied 30.3% of the total elderly population, and those who aged from 65 to 69 years old held 25.5%, the 70–74 years old made up 20.6%, the 75–79 years old accounted for 13.1%, and those who aged 80 or above occupied 10.5%. Of the total urban elderly population, 85.4% could fully take care of themselves, 9.6% could partially take care of themselves, and 5.0% could hardly take care of themselves. Of the total urban elderly population, 9.9% thought they needed to be taken care of. In daily life, 6.7% of those who aged 79 or below and 33.1% of those who aged 80 or above thought that they needed to be taken care of (CNCA, 2007). The 2011 CNCA Conference was held in Yunnan on February 25. Liguo Li, the minister of Civil Affairs and the director of CNCA, announced that during the period of the 12th five-year plan, the aging problem would become more serious in China, which featured issues like aging and "empty nests" and showed new characteristics. The elderly population age 60 or above will reach 216 million by 2015, accounting for 16.7% of the total population. With an annual increase of more than 8 million, it will exceed the total newborn population. The number of elderly people age 80 or above will reach 24 million and occupy 11.1% of the total elderly population, with an annual increase of 1 million. The number of elderly people age 65 or above living in "empty nests" will be more than 51 million and make up one fourth of the total elderly population, which puts elderly care service an issue on the agenda. The nation faces three major problems with regard to aging. First, half of its older people do not have social security to meet their basic needs including medical care. Second, many cities in China have grown overcrowded, with insufficient community facilities for the elderly to use. Third, long-term care has become an outstanding issue. How to deal with the challenge of rapid aging has become a major issue for China in the twenty-first century.

According to traditional Chinese culture, people favor to live in extended family, in other words, the elderly tend to live with their adult children. However, the rapid aging has left the elderly no choice but to live on their own. The elderly those who do not live with their adult children reached 49.7% of the total elderly population in 2006, among which 8.3% of them live independently, 41.4% of them live with their couple, and 50.3% of them live with relatives other than their adult children (CNCA, 2007). The ways and patterns the elderly choose to live differ by factors such as age, health conditions, and family members. Nowadays in China, there are three major ways people choose to live when they are aging: home-housed aging, community-based aging, and institutionalized aging.

Home-Housed Aging

In 1991 the UN Assembly adopted "The United Nations Principles for Older Persons," including the following statements: "Older persons should have access to food, water, shelter, clothing, health care, work and other income-generating opportunities, education, training, and a life in safe environments." "Older persons should have access to social and legal services and to health care so that they can maintain an optimum level of physical, mental and emotional well-being." Most of the old people live in home when they are aging in both developed countries and developing countries around the world. In USA, the percentage of home-housed aging reaches 95%; in the Netherland it is 91.4%, in Japan it is 96.9%, in the Philippines it is 83%, in Singapore it is 94%, in Thailand it is 87%, in Vietnam it is 94%, in Indonesia it is 84%, and in Malaysia it is 88% (Zhao, 2005). In January 2008, CNCA, National Development and Reform Commission, and Ministry of Civil Affairs announced "Suggestions on Improving Services for Home-housed Aging." The government put emphasis on the home-housed aging and implemented specific policies on the issue. The truth is the vast majority of elderly people in China prefer to live at home when they are aging while only 4% of them want to stay at nursing homes (Li & Chen, 2010).

Home-housed aging means staying and receiving public services at home. The advantage of it is that the elderly do not need to leave the environment they are familiar with. Open public services can be provided for the elderly while taking their living space as kind of residential resource. This home-housed aging service system, with public services provided, is supposed to be the best solution according to the willingness of the elderly. The city of Qingdao pioneered to take it into practice. By the end of 2009, about 10,000 old people took home-housed aging services provided by the government and about 5,200 of them enjoyed it free. The city government of Qingdao plans to provide a complete home-housed aging service by 2012.

Besides government service, home-housed aging requires the community to have certain integrity, which means the community cannot show too much stratification in income, age, occupation, education, and so forth, especially in age and income. The 30 years' urban housing marketization has worsened the living space of the elderly. Table 12.1 shows that few of the elderly age 60 or above live in the newly built high-quality community; most of them still live in the old community built during the planned economy period. It indicates that old people tend to cluster in traditional communities in most cities in China.

Chinese culture always features family values in aging issues. On one hand, the old people in China do not have many alternatives other than home-housed housing. On the other hand, the government should have done more. Table 12.2 shows that although people still have family values in aging issues, they have increasingly made appeal for government services.

Community-Based Aging

The percentage of "empty nests" households is high in both urban and rural China. In some cities, it reaches about 70%. Under this circumstance, the government provides great financial support for aging policies such as social security, pension, supplementary policies, health insurance in rural areas, and so forth. However, the nursing problem for the elderly has just come into public eye. So far, there are more than 20 million old people age 80 or above needing nursing service. These old people do not have any children or their children cannot take care of them due to various reasons. Therefore, government should take care of them by legislation and providing special service. This kind of service and system is now so-called community-based aging.

This kind of aging service is based on community, supplied by all kinds of resources from the community, and carried out with combined effort of the government, the community, and the households. It is a type of aging service to provide the elderly with both facilities and services like nursing, medical care, counseling, recreations, and so forth. First, it helps to release the burden both for the adult children and the elderly, especially the financial burden for the adult children. Second, it

Table 12.1 Types of urban community the elderly live in

| | | Urban community types | | | | | | | | | | |
		Squatter settlement	Old/traditional community which has not been renewed	The community of factories, mines, and other enterprises	The community of government departments and institutions	Affordable housing community	General community with condos	Exclusive community	Urban community which transferred from rural community	Migration community	Others	Total
Age group 60 and Below	Frequency	57	982	938	449	668	1,227	17	259	2	1	4,600
	Percent	1.2	21.3	20.4	9.8	14.5	26.7	0.4	5.6	0.0	0.0	100.0
Above 60	Frequency	9	155	182	105	71	164	0	28	0	1	715
	Percent	1.3	21.7	25.5	14.7	9.9	22.9	0.0	3.9	0.0	0.1	100.0
Total	Frequency	66	1,137	1,120	554	739	1,391	17	287	2	2	5,315
	Percent	1.2	21.4	21.1	10.4	13.9	26.2	0.3	5.4	0.0	0.0	100.0

Notes: (1) Sources from CGSS2006 (http://www.cssod.org/cgss/login.php); (2) Pearson Chi-Square Test Sig=0.000

Table 12.2 Who should take care of medical, health care, nursing, and basic needs of the elderly

Who should take the responsibility?	Medical, health care, and nursing			Basic living needs		
	N	Percent	Cumulative percent	N	Percent	Cumulative percent
Government should take all the responsibility	283	8.8	8.8	199	6.2	6.2
Government should take most of the responsibility	611	19.0	27.9	341	10.6	16.8
The responsibility should be taken by both government and individual/households	1,294	40.3	68.2	948	29.6	46.4
Individual/households should take most of the responsibility	654	20.4	88.6	1,069	33.3	79.7
Individual/households should take all the responsibility	366	11.4	100.0	651	20.3	100.0
Total	*3,208*	*100.0*		*3,208*	*100.0*	

Note: Sources from CGSS2006 (http://www.cssod.org/cgss/login.php)

shows the traditional values of taking care of the old. Third, it reduces government's budget on building nursing homes. Fourth, it helps to create more jobs and decrease unemployment rate. The community-based aging is a "win-win" strategy. Although cities like Shanghai have already brought the service to the public, it is a new kind of service that needs further improvement in resource allocation, evaluation, supervision, and so forth.

Social and housing stratification have negative impact on the implementation of the service since it is community-based. The basic needs of old people should be met by the community and by local service providers. If the elderly cluster in a certain community, their accessibility to the service will be reduced. Besides sufficient financial budget, it requires government invention to equip the community with more service providers. Many cities are now training social workers to provide such aging service. Besides, there should be a service center for the elderly in each community to solve problems they may be faced with. Once, there was a news about an old person living alone whose body was found long time after his death. This kind of cases should be avoided and the elderly living alone should be taken care of.

Institutionalized Aging

Aging institutions are institutions that provide housing and nursing for the elderly, like nursing homes and etc. There are some major problems existing in institutionalized aging in China, which are as follows:

(a) There is a huge gap between demand and supply. The facilities and service we now provide cannot meet the need of the elderly let alone the increasing rate of aging. In 2010, the total population of the elderly age 60 or above is about 2 billion. By the end of 2009, however, there are 38,060 aging institutions across the country with 2.662 million beds accommodating 2.109 million people. Among these aging institutions, 5,291 are urban aging institutions with 0.493 million beds accommodating 1.73 million people; 31,286 of them are rural aging institutions with 2.088 million beds accommodating 1.73 million people; 1,401 of them are Honor Resthomes (Guangrongyuan), with 67,000 beds accommodating 46,000 people; 47 of them are Disabled Soldier Recovering Hospitals, with 8,000 beds accommodating 4,000 people; and 35 of them are Demobilized Soldier Nursing Homes, with 6,000 beds accommodating 4,000 people. The number of beds compared to the population of the elderly is far below the international standard of 5–7%. Some rural aging institutions located in poverty-stricken areas or some private aging institutions cannot meet the various needs of the elderly due to lack of financial support and resources.

(b) In market economy, the potential clients of the institutionalized aging service are always not those who need the service most badly since they cannot afford the long-term caring service at all. With the lack of legislation, evaluation system, and insufficient resource allocation, many aging institutions provide their service under market rules which compromises social welfare in aging service in China. Many old people with no affordability have no access to aging institutions.

(c) The conflict between the open policy and the relatively conservative manage system is hard to reconcile. With rapid aging process, it is almost impossible for the government to shoulder all the responsibility to provide aging services. There are some private organizations that enjoy policies like tax reduction by providing such services. However, it is quite common that local governments do not implement policies advocated by the central government. The public aging institutions provide service of poor quality while the private aging institutions cannot provide sufficient facilities due to lacking financial support.

(d) When it comes to professional and special nursing services, it is also hard to reconcile the conflict between the elderly and the service providers, especially for those who cannot take care of themselves and need long-term caring. They have higher demand for long-term caring and recovery services when they are getting older. Therefore, the professional aging institutions should have been their best choice by providing professional nursing skills and knowledge. However, for those who cannot take care of themselves and need long-term caring, the service provided cannot meet their need. The aging institutions are short of professional staff. "The National Standard for Nursing the Elderly" has been advocated since 2001. However, there are only 20,000 people who have received certificates for providing nursing services for the elderly, with a shortage of about ten million professionals.

Housing for the Aging in China

As stated above, more than 95% of the elderly population plans to live where they used to spend their whole lifetime when they are aging. Therefore, home-housed aging and community-based housing should become two major approaches. It is of profound importance to control housing stratification so as to make these two approaches effective. The elderly living in community with high integrity can not only receive more respect and become more self-realized but also can contribute to the development of the community. Thus, the youth population in the community can have someone to provide their service to and also look up to.

As a result, a community center should be built in any community with more than 1,000 residents. Special sectors should be set to provide medical service, recreational facilities, and nursing service to the elderly in the community. All members in the community should take care of each other. To be more specific, we should accomplish the following:

1. To staff with professions: in each community with more than 1,000 residents, there should be at least one social worker with professional training in nursing the elderly.
2. To gain financial support: 80% of the total expense should be granted by government while 20% covered by the community. The elderly could also purchase additional services with affordability.
3. To build a service system as follows: first, an information system to get timely information of the living conditions and needs of the elderly; second, a system to timely distribute our service to the elderly; third, to provide more recreational facilities and programs to make their live more fulfilled.

References

CNCA. (2007). *A tracking survey of elderly in China*. Retrieved February 16, 2011 from http://www.china.com.cn/news/txt/2007-12/17/content_9392818.htm.

Li, B. (2009). *Housing policies in differentiation: An evaluation study on urban housing reforms*. Beijing: Social Science Academic Press.

Li, B., & Chen, S. (2010). Aging, living arrangements, and housing in China. *Aging International, 35*(4).

National Bureau of Statistics of China. (2011). *The sixth population census of China*. Retrieved February 16, 2011 from http://www.stats.gov.cn/tjdt/gjtjjdt/t20110429_402722652.htm.

Zhao, L. (2005). A study to elderly living status between China and the developed countries. *Academic Exchange, 12*, 168–170.

Chapter 13
Institutional Elder Care in China

Heying Jenny Zhan, Baozhen Luo, and Zhiyu Chen

Abstract The authors begin with conceptual issues surrounding institutional care and offer a working definition for the discussion in this chapter. The development of institutional care is reviewed by placing it within the context of Chinese elder care system including family support and social provisions. Elder care institutions in Nanjing and Tianjin and their residents are profiled to provide specific examples. Empirical data from a survey conducted in Zhenjiang City, Jiangsu province, are used to examine the willingness of elderly Chinese for institutional care. Implications of the findings are discussed and policy recommendations are made.

Defining "Institutional Care"

Numerous research articles have been published about institutional care in China (Chu & Chi, 2008; Shang, 2001; Zhan, Liu, & Bai, 2005, 2006), yet no article appears to have established a clear definition of the actual meaning of this practice. Part of the reason, as our earlier research has shown, is that China's institutional care still remains captive to the processes of professionalization and specialization. Many scholars use the American concept of "nursing home" to translate the Chinese elder care homes or "yang lao yuan 养老院." The authors of this chapter strongly disagree with this translation due to the major differences in connotation between nursing homes in the U.S. and elder care homes in China. Therefore, we see a definite need to clarify the concept of institutional elder care in the Chinese context.

H.J. Zhan (✉) • Z. Chen
Department of Sociology, Georgia State University, Atlanta, GA, USA
e-mail: heyingzhan@gmail.com

B. Luo
Department of Sociology, Western Washington University, Bellingham, WA, USA

S. Chen and J.L. Powell (eds.), *Aging in China: Implications to Social Policy of a Changing Economic State*, International Perspectives on Aging,
DOI 10.1007/978-1-4419-8351-0_13, © Springer Science+Business Media, LLC 2012

By institutional elder care, we mean, residential long-term care for older adults with or without any disabilities affecting one or more counts of their activities of daily living (ADL) or instrumental activities of daily living (IADL) in institutions managing major spheres of life for its residents, whether or not these institutions are equipped with medical staff or facilities.

By institution, we borrow in part from Goffman's (1961) concept of "total institution," by which he meant "a place of residence and work where a large number of like-situated individuals, cut off from the wider society for an appreciable period of time, together lead an enclosed, formally administered round of life" (p. xiii). In such institutions, human needs, such as food, sleep, and daily activities, are handled "by the bureaucratic organization of whole blocks of people" (p. 6). Using this concept of institution, nursing homes, assisted living facilities, and some hospice may fit into the category of an institution. Elder day-care centers, short-term stay centers, and senior centers are, therefore, excluded from the discussion of institutional long-term care in this chapter. Furthermore, because institutional long-term care continues to be an evolving concept in contemporary China, processes of professionalization and specialization range unevenly throughout China. As earlier research demonstrated, the Chinese concept of elder care institutions (yang lao yuan 养老院) encompasses all types of institutions for elders with varying degrees of disability. Using notions similar to practices in the United States, a yang lao yuan in China could mean anything from a nursing home, to an assisted living facility, a continuing care community, a retirement home, or even a hospice care center. In major cities, such as Beijing, Shanghai, and Tianjin, specialization and professionalization have resulted in a separation of these services into different types of institutions. But in most medium-sized or small cities together with all rural areas of China, elder care institutions are still a lump-sum notion or concept of do-it-all service for elders living in various types of elder care residential facilities.

Demographic, Social, and Policy Change

China's population is aging at a much faster pace than all industrialized societies due to its enormous population size. It took most industrialized countries around 100 years to double their elderly population from 8 to 16%; such a process is predicted to take only 27 years in China. Compared to Western societies such as the U.S., where industrialization precedes population aging, China will become an "old" society before the economic development is ready for such a rapid demographic transition. From 2000 to 2008, the 65+ population in China increased from 7 to 8.9% of the total population (The World Fact Book, 2011). It is projected that such a number will increase to 22.6% in 2040—an estimate of 371 million people at 65 and above (Kinsella & He, 2009). This number is going to be comparable or potentially exceeding the estimate of the entire population of the United States in 2040. In particular, the oldest old (80+), the fastest growing aging population, will increase from 1.4% in 2010 to 5.0% in 2040 (Kinsella & He).

The average life expectancy in 2010 for the total population is 75 years old, with women (77) living 4 years longer than men (73) (The World Fact Book, 2011). However, the morbidity rate among elderly women is higher than elderly men, which means women live longer but are more likely to be immobile or disabled than men. The magnitude of population aging presents a critical challenge for Chinese society—who and how will the nation take care of such a large number of its elderly people?

Declining Family Care

Traditionally, elder care has been provided by adult children at home under the cultural norm of *xiao* (or filial piety), the central value of Chinese family relationships. The practice of *xiao* is instilled among children to foster obedience and respect toward their parents; adult children are expected to provide financial, physical, and emotional support to elderly parents (Johnson, 1983). The Chinese Communist Party (CCP) further institutionalized *xiao* by codifying it into the law. The constitution of 1954 stated that "parents have the duty to rear and educate their minor children, and the adult children have the duty to support and assist their parents." In 1980, the penal code of 1980 decreed that children could be imprisoned to a maximum of 5 years for neglecting their parents. In 1996, CCP passed the *Law of Protecting the Rights of the Elderly of the People's Republic of China*, which officially and legally spelled out adult children's obligations to respect and take care of their aging parents physically, financially, and emotionally. The law formally regulated adult children's provision for aging parents in terms of housing, medical care, property protection, and so on. It is important to note that the legal regulation of filial piety by the CCP is closely related to the lack of comprehensive social welfare programs for the majority of elderly population in China, especially those in rural areas.

The implementation of one-child policy and rapid socioeconomic transformation, however, have led to much doubt on whether family, especially adult children, will be able to care for such a large number of elderly population. In 1979, the CCP enacted the one-child policy to control the rapidly rising fertility rate since the Mao era. Although it controlled the population growth in a short period of time, the one-child policy created unintended consequences as the one-child cohort become adults and their baby boomer parents enter old age and are in need of elder care. Chinese family in urban areas is gradually emerging into a pattern of "4-2-1"—four grandparents, two adult children, and one grandchild—inverted family structure (Zhan, 2002). According to previous studies, although adult children are not necessarily less indoctrinated to the cultural tradition of *xiao*, it is financially and physically too difficult and constrained for them to care for their elderly parents with dual demands of work and caring for their own offspring (Jiang, 1995; Zhan, 2004). Compared to their urban counterparts, Chinese elders in rural areas are in a much more vulnerable position due to the massive scale of rural–urban migration. Since the enactment of economic reform in 1979, tens of millions of rural young laborers have been migrating to urban areas to meet the demand for cheap labor. Consequently, large numbers of elderly parents are left behind at home in impoverished villages. At the societal level, adult children are becoming more and more unavailable to fulfill their elder care responsibilities. Although studies have shown that migrated children were able to provide more financial support to their elderly parents back in the village; physical care, which was mostly needed for a frail elderly parent, has simply been diminishing (Du, 1997; Li, 2001; Luo, 2009).

Lack of Social Welfare

Will the government step in to help the family share some of the elder care responsibility? Currently, China does not have a comprehensive health insurance program such as Medicare in the U.S. to provide financial support for medical or long-term care for the elderly population. However, in the last decade or so, in order to sustain its economic development, Chinese government initiated a series of policies to address the emerging concern for the needs of the old (Liang, 1995; National News Office of the PRC, 2006). The basic principle is to set up a collaborative support system involving four parties—the state, the community, the family, and the individual. The government also established five principles to ensure the wellbeing of the elders—"older people should be supported, have medical care, be contributory to the society, be engaged in lifelong learning and live a happy life" (National News Office of the PRC, 2006). These initiatives have resulted in the growth of home and community-based services for elders—such as senior housing, recreational facilities, adult day care programs, etc. (Wu, Carter, Goins, & Cheng, 2005). However, these programs were neither comprehensive nor did it address the most emerging need of the elders—medical and nursing care—that would facilitate the elders to function independently.

With the unavailability of care from adult children and lack of social welfare support, how is China as a nation going to provide care for the aging baby boomers? Institutional care is anticipated to be one of the answers.

Historically, institutional care for elders has been rare due to the long-lasting influence of the tradition of *xiao* and its expectations of family care. In 1992, the Chinese government published the White Paper on the Development of China's Social Services which briefly described the history of institutional care development in China.

Before 1950, some institutional care facilities existed under the Kuomingtang government. These elder care facilities were mostly foreign-operated charity organizations (Barlett & Phillips, 1997). After the establishment of People's Republic of China in 1949, the CCP took over these facilities and used them for the care of the most needy—the Three No's, those with no children, no income, and no relatives (Chen, 1996; Ikels, 1993). Between 1950s and 1980s, although more institutional care facilities were established by the governments to care for the elders with no family, the number continued to be minimal. Most of these facilities were social welfare institutions where childless elders were cared along with mentally retarded and orphans. In the mid-1960s, only 819 of them were in urban areas (Barlett & Phillips, 1997). In 1988, such a number increased to 870 and 46,837 older adults in the entire nation were served (Chen, 1996, p. 115).

In the 1990s, China's welfare system underwent a dramatic structural shift under the principles of decentralization and a market economy. Funding for welfare services were cut substantially (from 0.58% of GDP in 1979 to 0.19% of GDP in 1997). As a result, welfare institutions were pushed to the market where they had to search for new revenue by gathering funding from both public and private sectors. Depending on the major source of revenue, generally speaking, three types of institutional care facilities emerged in China—government, nongovernment, and privately owned. In major cities, there has been rapid growth of elder care homes.

Recent Developments in Institutional Care

For the discussion of recent developments of institutional long-term care, this chapter will draw upon three data sets to illustrate key points of change in the status of institutional care. The first data was collected in 2009 in Nanjing China; the second, in 2010, Tianjin, China; the third, in 2007, Zhenjiang city, Jiangsu Province. The first two research projects are funded by the U.S. Fogarty International Research Center and National Institute of Aging (NIA); the goal of the research was to understand the general profile of recent developments in institutional care in these two major Chinese cities. The third dataset is drawn from a survey about intergenerational relationships and age models collected in Zhenjiang by the Zhenjiang Population and Family Planning Committee in 2007. The first two datasets are identical in methodological approach—a mixed method of collecting aggregated quantitative data from all elder care institutions in both cities about major institutions' and residents' characteristics, and conducting qualitative focus group discussions with three different populations (elderly residents, family members, and facility managers). The third data explored intergenerational relationships among family members in a medium-sized Chinese city of Zhenjiang. This chapter only uses the part of the data that focuses on family resources and issues of elders' willingness to enter institutional care, using Zhenjiang as a case study.

Nanjing is an ancient capital city in south central China, roughly 150 miles west of Shanghai. Nanjing is an ancient capital and is a "traditional" Chinese city. It is located slightly inland where elder care practices have not experienced dramatic changes due to out-migration or industrialization. Nanjing has experienced a rapid pace of population aging. while the total population of Nanjing rose from 5.09 million in 1990 to 6 million in 2000, the population of 60 plus surged from 0.5 million in 1990 (10.2%) to over 0.7 million (14.4%) in 2000. Instead of taking 27 years to double the elderly population at the national level, as projected, it may take Nanjing only 22–23 years to do so. In 2025, one in every four people in Nanjing is projected to be older than 60 (Nanjing Gerontology Office, 2006).

Tianjin is one of the four autonomous[1] cities in China. It has a registered population of 9.1 million in 2000; in 2010, the population has reached 12.9 million (People's Network, 2011). As early as in the 1980s, Tianjin became the second city, next to Shanghai, in China to experience population aging. In 1985, those 60 and over had reached 1.3 million or 13.9% of the total population. By the end of 2003, the 60 plus reached 14.5% of the total population (Northern China, 2003). In 2010, the elderly population has reached 17.9% (China Daily, 2011).

Profiles of Elder Care Institutions in Nanjing and Tianjin

Based on our research, both Nanjing and Tianjin have experienced rapid growth in elder care institutions (see Table 13.1). Prior to the 1990s, there were only 27 (19.3%) elder care homes in Nanjing and 12 (7.64%) in Tianjin, all government-owned. During the 1990s, both cities had seen growth in elder care institutions, 41 (26.11%) in Nanjing, and 25 (17.9%) in Tianjin. This growth in 1 decade exceeds the total number in the 4 decades (1950–1990) of facilities in both cities. Yet, real growth did not take off until after 2000. In both cities, over 60% of all elder care facilities, 88 (62.9%) in Nanjing and 104 (66.25%) in Tianjin, were established after 2000.

Another observation from the studies of these two cities is that the distribution of ownership types of these elder care institutions have changed dramatically over time. Prior to the 1990s, ownership of elder care institutions was exclusively government in both cities. The recently established institutions were almost all "private" or corporative in both cities. As shown in Table 13.1, private ownership is the mode in both cities, 61 (43.57%) in Nanjing, and 104 (73.24%) facilities in Tianjin, were "privately owned." In addition, ownership of facilities has seen diversification in both cities. Among government-owned elder care facilities, multi-tiered government-owned facilities are shown to be in nearly a pyramid shape, few (1 in Nanjing, 3 in Tianjin) at the top of the municipal level, several, 15 (10.7%) in Nanjing, and 7 (4.47%) in Tianjin, at the district level (similar to the County level in the U.S.), and many at the street or community or neighborhood level, 41 (29.28%) in Nanjing, and 10 (6.36%) in Tianjin. Another variance of government ownership is its contractual relationship with individuals. In Nanjing, for instance, eight facilities are reported to be government-owned, but contracted to individuals to manage. No similar case is found in Tianjin.

The strong hold of the former "work-unit" seems to be losing its grip in the development of elder care facilities. Only 10 (7.14) in Nanjing, and 6 (3.82%) in Tianjin were collectively owned, mostly by former "work-unit." In the meantime, the cooperative or enterprise or chained elder care facilities are emerging.

---

[1] The major findings of this study used in this chapter are a synthesis of Chen, Zhiyu's M.A. thesis (2011) directed by the senior author of this chapter. It is titled, *"Stigma and Knowledge—Chinese Older Adults' Willingness to Accept Institutional Elder Care."* For further details and tables, please see his thesis.

Table 13.1 Characteristics of elder care facilities in Nanjing and Tianjin

	Nanjing ($n = 140$)	Tianjin ($n = 157$)
Year established		
Prior to 1980	3 (2.14%)	3 (1.91%)
1980–1990	24 (17.14%)	9 (5.73%)
1990s	25 (17.9%)	41 (26.11%)
2000s	88 (62.9%)	104 (66.25%)
Ownership		
Government	61 (43.6%)	20 (12.74%)
Nongovernment	79 (56.4%)	137 (87.26%)
Types of ownership		
Government		
Municipal	1 (0.007%)	3 (1.91%)
District/county	15 (10.7%)	7 (4.46%)
Street/community	41 (29.28%)	10 (6.36%)
Collective (work-unit)	10 (7.14%)	6 (3.82%)
Private (individual)	61 (43.57%)	115 (73.24%)
Private (partnership)	12 (8.57%)	8 (5.09%)
Corporation/enterprise	9 (6.43%)	
Government-owned, privately managed	8 (5.71%)	
Charity		3 (1.91%)
Number of beds in facilities	($n = 140$)	($N = 157$)
1–30	24 (17.14%)	5 (3.2%)
31–50	44 (31.42%)	21 (15.4%)
51–100	39 (27.85%)	68 (50%)
101–150	23 (16.43%)	36 (26.47%)
151–200	4 (2.86%)	12 (8.82%)
200+	6 (4.28%)	15 (11.03%)
Number of residents in facilities		
1–30	53 (37.86%)	24 (15.28%)
31–50	46 (42.86%)	34 (21.65%)
50–100	27 (19.29%)	66 (42.04%)
101–150	9 (6.4%)	20 (12.73%)
151+	2 (1.43%)	15 (9.55%)
Elderly residents characteristics (in %)		
Independent	47.1	34.48
Need assistance with eating	18.4	35.99
Need assistance with dressing	40.1	25.17
Need assistance with walking	40.8	24.64
Have dementia	23.2	26.46
Payment source (number of residents in %)		
Private pay (out of pocket)	35.2	16.88
Private pay with pension	61	74.62
Welfare (government pay)	16	6.20

In terms of the size of the care facilities, Nanjing seems to have mostly small- or medium-sized facilities with over 70% of facilities having less than 50 residents; nearly 20% have 50–100 residents. In Tianjin, on the other hand, facilities tend to be larger, 66 (42.04%) of facilities are medium-sized, having between 50 and 100 clients. A little over a third, 58 (36.94%) have less than 50 elders. Maybe due to the larger size of the elderly population, Tianjin also has larger-sized facilities, 20 of them have between 100 and 150 clients; and 15 facilities have over 150 clients. These larger facilities are either government-owned or corporate enterprises.

Profile of Elderly Residents in Institutions

What kind of elders tends to move into an elder care facility? By looking at the elders' health conditions, we found that actually a large proportion of elders do not have problems in ADL—47% in Nanjing, 34% in Tianjin. These elders are mostly able to take care of themselves on a daily basis. Many elders entered elder care homes as a result of a stroke or simply aging and disability. Elders who had trouble walking or dressing were most likely stroke patients. These disabilities, of course, could overlap. According to survey findings in Nanjing, roughly 40% of elderly residents had problems with walking or dressing themselves; in Tianjin, this proportion was roughly one in four. Elderly residents who needed feeding were evidently bed-ridden and needed care 24/7. These residents were less than 20% (18.4%) in Nanjing and over a third (35.99%) in Tianjin. Roughly a quarter of the residents in both cities had dementia (23% in Nanjing; 26% in Tianjin).

As regards to financial resources, study results revealed that government funded welfare patients were a small minority. Only 16% of elders in Nanjing, 6.2% in Tianjin are paid by government welfare. The majority of older adults in institutions were paying the fees themselves using their own pensions (61% in Nanjing; 74.62% in Tianjin). Percentages in payment sources exceeding or being below 100% are due to overlapping in categories of out of pockets or paying with children's help. They are not mutually exclusive categories.

Why Do Elders Enter Elder Care Institutions?: An Analysis of Willingness

Earlier qualitative studies have revealed various reasons for older adults' entry into an institutional care setting. Among these reasons, three came consistently to the top: elders' disability or health conditions, adult children's unavailability, and housing problems or issues, such as crowded housing, having no elevator, or reconstruction and relocation. These qualitative findings were based on focus group studies or interviews with older adults already in elder care institutions and/or their family members. Few studies have examined community elders, or young elders who are still in the so called "third age," 55–70, who are not yet physically dependent to need long-term care. The following section will use data collected from a survey over intergenerational relationships and age models in Zhenjiang City, Jiangsu province, China, to examine the issue of willingness for institutional care.[1]

Zhenjiang city had a population of three million in 2007 and about 10.93% of them were aged 65 and over. The survey randomly selected 2,000 urban residents from its seven administrative districts in Zhenjiang city. All participants were asked to fill out a questionnaire concerning intergenerational relationships. A total of 1,612 respondents completed the questionnaire. This chapter will only use data from participants aged 55 and over (N = 628, including 310 males and 318 females). Fifty-five was selected as an age-criterion because it is the legal age for female workers and the median age for male workers to retire in China.

Characteristics of the Respondents

According to descriptive data analysis, among the 628 respondents, 310 (49.4%) were male, 318 female. Their age ranged from 55 to 91. Most of them (88.2%) were under age 75; their mean age was 65.6. About four fifths (78.7%) of them were married. The vast majority of them (81.8%) had less than middle school education. Only 2.5% respondents were childless and most of them (81.5%) had two or more children. As for the financial status, over a quarter of the respondents (26.9%) had no pension and 87.3% of them had medical insurance. Regarding their self-rated health, the majority of respondents (93.1%) did not report having any health problems. Accordingly, only 11.1% of the respondents had ADL problems, 17.8% reported some IADL problems. With regard to living arrangements, 85.9% of the respondents lived with their children or had children living nearby. Most respondents (87.4%) had regular or more frequent contact with their children and 74% of the respondents were confident that their children would provide them with familial care in their late lives. More than half (67.8%) of the respondents knew little or hardly anything about elder care homes. Nearly half of the respondents (49.1%) had good or very good impressions over elder care homes. Out of all the respondents, only 8.7% of them answered they were willing to use institutional care, 13.7% somewhat willing, 28.9% "does not matter or hard to say," 19.3% not very willing, 29.4% not willing.

Willingness for Institutional Care

Using a multivariate regression analysis on older adults' willingness toward institutional elder care, we find that Chinese elders' gender, elders' confidence in availability of familial care, knowledge about and impression of elder care homes' services were related to their willingness to use institutional care.

Gender was related to Chinese elders' willingness to use institutional care. Male elders were less likely to report "willing to accept institutional care" compared to their female counterparts. Chinese elders' confidence in their adult children's provision of familial care had a negative correlation ($B = 1.88$) with their willingness to move into elder care homes. Chinese elders who were more confident in their children's provision of familial care expressed lower levels of willingness to move into elder care homes.

Chinese elders' knowledge about elder care homes was positively related to their willingness to accept institutional care. Older adults who had more knowledge, including personal visits, having relatives who have visited or stayed in facilities, increased older adults' willingness to accept institutional care. Chinese elders' impression on elder care homes was also positively associated with their willingness to accept institutional care. Older adults who had a good impression about elder care institutions expressed higher levels of willingness to accept institutional care in their old age.

Discussion

Institutional care has not been an alternative for families with physically dependent elderly parents until very recently. Earlier literature has shown that China has traditionally emphasized familial elder care following the Confucian tradition of filial piety (Chen, 1996; Ikels, 1993; Zhan & Montgomery, 2003). Under the communist system, filial piety continued to be a central principle organizing family and intergenerational relationships. Only childless elders were cared for by the government in welfare institutions. Elders who entered those welfare institutions were often viewed with sympathy and attached with stigma. Older adults who had adult children were expected to be cared for by these adult children and their family members. This expectation was explicitly spelled out in the Chinese family law. Recent demographic changes of population aging and the shrinking of family size as well as changes toward a market economy have increased the pressure on the middle generation adult children in Chinese families. Though these adults, or the Chinese baby boomers, are still likely to have more siblings to share care responsibilities, they often have found themselves busy with work and raising their own families. Our earlier research in Tianjin in 2001 showed that over half of the elders (55.8%) entered care institutions because of their adult children's unavailability (Zhan et al., 2006). In our research in 2009 and 2010 in Nanjing and Tianjin, the percentages of all elders who were childless or were on welfare were small, 16% in Nanjing and 6% in Tianjin. Vast majority of the elders in contemporary China entered an institutional care facility due to health crisis and/or the lack of adult children's direct care as a result of their busy work schedule. As more and more facilities are made available, more and more elders are entering care facilities. This research also finds major changes in organizational patterns among elder care institutions.

Organizational Changes Among Institutions

As findings in Table 13.1 have shown, elder care institutions have changed in their ownership type, service population, and organizational structure. Prior to the 1980s, there were only three elder care homes in both Nanjing and Tianjin, caring for exclusively childless elders in poverty. These homes used to be referred to as "Home of Respect for the Aged" or jinglaoyuan 敬老院. In the 1980s–1990s, there was a slow growth in both cities in elder care homes. Some elder care homes started to admit older adults whose adult children were either abroad or out of town. In these cases, families or adult children almost always paid for the cost. The rapid growth of elder care institutions caring for elders with adult children did not occur until late 1990s or after 2000. As shown in Table 13.1, over 60% of all elder care homes in 2009 or 2010 in Nanjing and Tianjin were established after 2000. The same is true in most other cities of China as shown in other research (Feng et al., 2011). Is it the increasing consumer demand or the change of government policy that propelled such rapid growth in elder care facilities? To address this issue, we have to look at the major organizational changes in elder care homes.

As shown in Table 13.1, both cities of Nanjing and Tianjin had only three elder care homes prior to 1980s. By cross-tabulation of ownership, we found that all of these three elder care homes were government-owned welfare institutions. The growth of elder care institutions in the 1980s and 1990s were mostly lower levels of government-owned elder care homes, such as street or district level of care homes (see Table 2 of Feng et al., 2011). A few work-unit or collective elder care homes also emerged. After 2000, the rapid growth with elder care institutions has been accompanied by the diversification of ownership of these care institutions. Our research in Nanjing and Tianjin clearly indicates that government policy changes have played a significant role in encouraging the growth of nonprofit and nongovernment institutions in elder care services. In Nanjing, for instance, provincial as well as municipal governments offered subsidies for nonprofit and nongovernment elder care institutions for each occupied bed, in an amount ranging from 200 to 400 yuan per month differing by location of the elder care homes. There was also a lump-sum compensation in reimbursement for newly constructed or remodeled care homes. These favorable policies and financial incentives have played a significant role in propelling the growth of private nonprofit nongovernment elder care institutions.

Changing Profile of Consumers

Who are the people who enter these elder care homes? By looking at Table 13.1, we can clearly tell that welfare patients or elders are the minority (16% in Nanjing, 6% in Tianjin). Vast majority of consumers of these elder care homes in both Nanjing and Tianjin were elders with children and families. Over half of elders who entered care institutions suffered from health crisis, thus had disabilities in ADL or IADL. Roughly a third of the consumers were dementia patients. There were a significant percentage of elders who need feeding (18% in Nanjing and 36% in Tianjin). Curiously, a large proportion of elders were able to take care of themselves on ADL items, roughly 47% in Nanjing and 35% in Tianjin. This finding is consistent with earlier research findings that elder care institutions in China are more than just nursing homes; they are also assisted living facilities, or retirement homes, taking care of older adults in a range of health conditions (Zhan et al., 2005, 2006). Some elders entered a facility mainly due to inconveniences of climbing stairs to reach an apartment home or simply because living alone in an apartment caused worry for adult children.

Another dynamic that showed major social change is the source of payment. From studies in Nanjing and Tianjin, a large percentage of elders are paying for institutional elder care using their own pension. Some elders may need some assistance from their adult children. But increasingly, more and more elders are able to use their own pension to pay for institutional care; 61% of Nanjing and 75% of Tianjin residents used their pension to cover some or all of their cost. Earlier qualitative studies have also showed that elders with pensions are more likely to emphasize independence in decision-making, less likely to expect or rely on children's direct care (Zhan et al., 2011). While some elders may continue to need financial assistance from their adult children, as time progresses, more and more retiring Chinese elders, especially in urban China, are likely to have worked under the Communist system, and are more likely to draw a pension. When they reach the age of physical dependency, their financial ability to pay for services is likely to play a role in deciding whether or not they would choose institutional care as their elder care option.

Cultural Baggage and Willingness for Accepting Institutional Elder Care

What kind of elders are more likely to be willing to accept institutional elder care? The third dataset helps shed some light to this question. Findings for studies of elders living at home or in the community revealed that elders who have more confidence in adult children's provision of care were less likely to express willingness to accept institutional care. This finding suggests that the psychological and cultural reliance on familial care continues to be strong and real. Less than 10% (8.7%) of elders expressed willingness to accept institutional care; nearly half (48.7%) expressed, "not very willing" or simply "not willing" to consider institutional care. In the meantime, intergenerational bond seems to be still strong. Elders and adult children (87%) reported regular and frequent contact between the generations, and the majority of elders (74%) expressed confidence about their children's filial piety in provision of elder care. These findings suggest that familial care is likely to be the main source of elder care in some time to come. Even though families will increasingly feel the strain of direct care for older adults, the strong cultural expectations of filial piety are likely to continue, institutional care is likely to be only an option, or alternative, for familial care.

These study findings suggest that increasing older adults' knowledge about elder care institutions will be important in the future to increase the level of comfort for using institutional care. Knowing more about elder care institutions and having better impressions about care institutions increased elders' levels of acceptance for institutional care. These findings shed light on the importance of marketing and improvement of the quality of care in institutional settings. To allow institutional care to become a viable option, government as well as elder care institutions may have to increase their outreach programs and educate the elders in the community that institutional care can be and is an option for long-term care.

Institutional Care as an Evolving Concept and Practice

Using Goffman's (1961) definition of "total institution," most of the facilities we have studies in China seem to fit into the category well. Chinese elders, like many in the U.S., due to health, disability, or children's unavailability, come to "a place of residence…where a large number of like-situated individuals, cut off from the wider society for an appreciable period of time, together lead an enclosed, formally administered round of life." The quality of life in some institutions can be truly bleak; in others, however, can be more satisfying and enjoyable than staying at home alone. In some small-scale, neighborhood or community-based elder care homes, however, the level of bureaucratic handling of elders seems to be less systematic, more interactional.

To what extent institutional care in China is going to adapt to the native culture of filial piety by de-bureaucratizing or humanizing the handling of human needs will be important for scholars as well as intellectuals to explore. When institutional care become less "institutional," more flexible or humane, maybe more Chinese older adults are likely to choose such an option in the future.

Policy Implications

Our studies of institutional care in both cities of Nanjing and Tianjin revealed a rapid growth of nonprofit and nongovernment organizations in provision of elder care services. We also have discovered that there is a lack of governmental regulation and quality control during this rapid pace of development and growth. The lack of regulation leads to blind competition in the construction of elder care homes as well as the lack of guidelines for professionalization and specification. In addition, the lack of quality control is likely to make institutional care an option for families less viable and realistic. When families place their loved ones in an institution, quality of care is what keeps the elders there and adult children at peace. So far, we have not found programs, similar to ombudsmen program in the U.S., in existence in China. Developing innovative approaches to quality control of care homes is critical for the future growth of elder care institutions. Finally, to remark on the future of institutional elder care, we would predict that the numbers of elders who enter institutional care settings are likely to continue to grow till 2040. Various factors are going to contribute to this growth, these include the shrinking family size, the increasing number of elders who need long-term care, the growing army of Chinese baby boomers who have pension and are able to afford institutional care, and the growing specification and professionalization of institutional care. Whether or not elders are willing, whether or not children intend to fulfill their filial responsibility, a proportion of elders will have to use institutional care for its professional service and convenience. Although we do not necessarily argue that institutional care is the best option for long-term care in China, we do foresee the unavoidability of this growth in several decades to come. The matter is, then, how to "deinstitutionalize" or humanize institutions or facilities caring for older adults will remain to be both a conceptual and practical puzzle for China as well as all aging societies for decades to come.

References

Barlett, H., & Phillips, D. R. (1997). Ageing and aged car in the People's Republic of China: National and local issues and perspectives. *Health & Place, 3*(3), 149–159.

Chen, S. Y. (1996). *Social policy of the economic state and community care in Chinese culture: Aging, family, urban change, and the socialist welfare pluralism.* Avebury: Brookfield.

China Daily. (2011). Retrieved June 7, 2011, from http://www.chinadaily.com.cn/dfpd/tianjin/2011-05-05/content_2519626.html.

Chu, L. C., & Chi, I. (2008). Nursing homes in China. *Journal of the American Medical Directors Association, 9,* 237–243.

Du, Y. (1997). *Leaving the villages: An analysis of rural labor migration in China.* Beijing: Economic and Science Press (in Chinese).

Feng, Z., Zhan, H. J., Feng, X. T., Liu, C., Sun, M., & Mor, V. (2011). An industry in the making: The emergence of institutional elder care in urban China. *Journal of America Geriatric Society, 59,* 733–744.

Goffman, E. (1961). *Asylums.* Garden City, NY: Anchor.

Ikels, C. (1993). Chinese kinship and the state: Shaping of policy for the elderly. In G. L. Madox & M. P. Lawton (Eds.), *Annual review of gerontology and geriatrics* (Vol. 13, pp. 123–146). New York: Springer.

Jiang, L. (1995). Changing kinship structure and its implication for old-age support in urban and rural China. *Population Studies, 49,* 127–145.

Johnson, K. A. (1983). *Women, the family and peasant revolution in China.* Chicago: The University of Chicago Press.

Kinsella, K., & He, W. (2009). *An aging world: 2008.* International population reports. Washington: U.S. Census Bureau.

Li, Q. (2001). Rural–urban migrants and their remittance in China (in Chinese). *Sociology Studies, 4,* 64–76.

Liang, H. C. (1995). The health management of the aged in China. In *Paper presented at 5th Asia Oceania regional congress of gerontology,* Hong Kong.

Luo, B. (2009). *The impact of rural–urban migration on familial elder care in rural China.* Unpublished dissertation, Georgia State University, Georgia.

Nanjing Gerontology Office. (2006). *Population statistics in Nanjing*. Accessed on March 1, 2011, from http://www.xhby.net/xhby/content/2006-02/25/content_1160416.htm.

National News Office of the People's Republic of China. (2006). Press release: The development of China's undertakings for the aged.

Northern China. (2003). Retrieved March 1, 2011, from http://news.enorth.com.cn/.

People's Network (Tianjin Division). (2011). Retrieved June 7, 2011, from http://www.022net. com/2011/4-29/50685939259215.html.

Shang, X. (2001). Moving toward a multi-level and multi-pilar system: Changes in institutional care in two Chinese cities. *Journal of Social Policy, 30*(2), 259–281.

The World Fact Book. (2011). Retrieved June 10, 2011, from https://www.cia.gov/library/publications/the-world-factbook/geos/ch.html.

Wu, B., Carter, M. W., Goins, R. T., & Cheng, C. (2005). Emerging services for community-based long-term care in urban China: A systematic analysis of Shanghai's community-based agencies. *Journal of Aging & Social Policy, 17*, 37–60.

Zhan, H. J. (2002). Chinese caregiving burden and the future burden of elder care in life course perspective. *International Journal of Aging & Human Development, 54*, 267–290.

Zhan, H. J. (2004). Willingness and expectations: Intergenerational differences in attitudes toward filial responsibility in China. *Journal of Marriage and Family Review, 36*(1/2), 175–200.

Zhan, H. J., Liu, G., & Bai, H. G. (2005). Recent development of institutional elder care in China: A reconciliation of traditional culture. *Ageing International, 30*(2), 167–187.

Zhan, H. J., Liu, G. Y., & Bai, H. G. (2006). Recent development in Chinese institutional elder care: Changing concepts and attitudes. *Journal of Aging & Social Policy, 18*(2), 85–108.

Zhan, H. J., & Montgomery, R. J. V. (2003). Gender and elder care in China: The influence of filial piety and structural constraints. *Gender and Society, 17*(2), 209–229.

Chapter 14
Changing Welfare Institution and Evolution of Chinese Nonprofit Organizations: The Story of Elder Care Homes in Urban Shanghai

Linda Wong and Na Li

Abstract Shanghai has the oldest population profile in the country. The growing pool of seniors creates a large market for residential care. The municipality has actively engaged in institution building to promote welfare nonprofit organizations (NPOs) so that they can take on the brunt of delivering care. Using selected elder care homes in the city as case examples, we attempt to trace the institutional change in the development of residential care as well as its impact on the evolution of non-state care agencies since the new millennium. Our empirical findings suggest that in Shanghai, the state indeed has strengthened its role in welfare planning, financing, and provision through formal and informal institutional arrangement. In this process, welfare NPOs serve as agents of the state and are regulated and incorporated under a system of state dominance. Therefore, it is arguable whether the expansion of non-state welfare can be taken as a success indicator for the welfare socialization policy. Rather, at the present stage, state hegemony in welfare development remains unchallenged. State dominance looks set to become a long-term feature of the new welfare economy in China.

Keywords Welfare NPOs • Social welfare socialization • Urban elder care homes

Under the state policy of "socializing social welfare (*shehui fuli shehuihua*)" promulgated since the late 1990s, Chinese welfare nonprofit organizations (NPOs) have been promoted to relieve the state burden of providing welfare to meet the unmet social needs of the population. Agencies using non-state funds and civil society resources become state partners and agents to deliver welfare services under a new cooperative framework with the state. The case of non-state elder care homes

L. Wong (✉) • N. Li
Department of Public and Social Administration, City University of Hong Kong,
Kowloon Tong, Kowloon, Hong Kong
e-mail: Linda.Wong@cityu.edu.hk

S. Chen and J.L. Powell (eds.), *Aging in China: Implications to Social Policy of a Changing Economic State*, International Perspectives on Aging,
DOI 10.1007/978-1-4419-8351-0_14, © Springer Science+Business Media, LLC 2012

is a prime example of this new provision pathway. As pioneers, non-state care homes for urban elders have grown in leaps and bounds to meet the burgeoning needs of China's senior citizens. These agencies are financially and administratively independent of the government but are regulated by an austere state regime that requires agencies to be subjected to the control and supervision by a professional sponsoring unit and registration with the civil affairs authority as non-state non-profit enterprises (*minban feiqiye danwei*). In return, such agencies are designated as nonprofit units or NPOs and are given tax exempt status and preferential policies. Nowadays, they have become the major service providers, offering over 411,000 residential beds for urban elders. Therefore, a study into the changing institutional framework and evolution of non-state residential care for urban elders is both timely and important. The development of this sector, touted as the third sector besides the state and market welfare providers, also offers insight into the success or otherwise of the state policy toward welfare socialization. In particular, the relationship between the state and NPOs sheds light on the special feature of China's transitional welfare economy.

Explosion in Non-State Welfare Organizations

The liberalization of China's command system in the last two decades spawned an explosion in NPOs. Normally, NPOs are not only formally constructed, nongovernmental and nonprofit distributing, but also voluntary and self-governing (Salamon & Anheier, 1992). Yet the framework may not be applicable while linked to the party-state institutional structure like China. As Saich (1994) said, "…this approach [emphasizing the need for organizations of civil society to have autonomy vs. the party state] underestimates the role of a competent state structure in the birth and growth of any civil society." Particularly in China, the state plays a predominant role in the birth and growth of NPOs, because it "remains keenly aware of political risks of associational actions (Ma, 2006)" while integrating NPOs in the network to deal with economic, social, and political crises. That is, official policies of the state impact NPOs in China more significantly than any in other countries. Like Yu (2006) indicates, China's civil society is a typical case of a civil society led by the government, and it obviously has both official and unofficial aspects. This kind of government dominance is manifested in the dual-management system toward Chinese NPOs. Its principles are summarized as "registration by professions, dual responsibility and multitier management." According to the *Regulations for Registration and Management of Social Organizations* and the *Temporary Regulations for Registration and Management of Non-state Non-profit Enterprise* promulgated in 1998, an NPO should secure an official/professional sponsor willing to oversee its daily management before it registers with the Ministry of Civil Affairs (MOCA) or local bureaus of civil affairs. The dual-management system brings NPOs under the supervision and monitoring of the state. Also, this restrictive registration system forces some organizations to operate without the official approval and register as business entities. Based on official statistics up to the end of 2009, 431,069 NPOs have registered under the three legal registration categories[1] recognized by MOCA (2010). Among the three, organizations that deliver social welfare services grow especially rapidly. Welfare NPOs normally register as non-state non-profit enterprises (*minban feiqiye danwei*) with MOCA or local civil affairs bureaus. At the end of 2009, the country had 190,479 non-state nonprofit enterprises, making up 44.2% of total registered NPOs (MOCA, 2010).

In China, welfare NPOs can be divided into five categories: social welfare homes (*shehui fuliyuan*), children's homes (*ertong fuliyuan*), social welfare hospitals (*shehui fuli yiyuan*), urban welfare homes for the elderly (*chengzhen laonian fuli jigou*), and rural welfare homes for destitute households (*nongcun wubao gongyang fuli jigou*) (National Bureau of Statistic of China, 2008). As said before, these agencies operate under the dual-management system just as other Chinese NPOs. While services are often run on commercial principles, profit is not the primary objective. Rather, income generated is ploughed back into expanding or improving the service and not distributed as dividends to investors. In return, Chinese welfare NPOs enjoy tax exemption and preferential policies like reduced public utility charges. Operating as self-sustaining entities also allows these agencies to enjoy management autonomy. Among these types of welfare NPOs, the development of urban old age homes is particularly noteworthy. As pioneers in non-state care, these agencies are products of the welfare socialization reform since 1998.

The rise of non-state welfare cannot be understood without revisiting the institutional and socioeconomic contexts. Along with the marketization and privatization in the economic sphere since the 1980s, the work unit system collapsed and its accompanying enterprise-based welfare arrangements gradually eroded. In the political sphere, the acceleration of economic reform also highlighted repeatedly the importance of social welfare for maintaining social stability and political authority. Meanwhile, the demographic change resulted from the population growth and the one-child policy led to a rapid process of aging and a declining dependency ratio between the work-age population and the retired, causing the increased needs for welfare services. Under these contexts, how to accommodate both the growing needs of the population and the financial constraints deriving from its reduced institutional capacities became the most significant challenge of the government (Croll, 1999).

To deal with the situation, social welfare reform has been listed in the national reform agenda since the late 1990s. The overriding aim of this reform was "socializing social welfare (*shehui fuli shehuihua*)," which could be regarded as a mixed arrangement for welfare or welfare pluralism. The reform was officially put forward in the Tenth National Civil Affairs Work Conference in 1994. In 1998, 11 ministries and commissions including MOCA co-promulgated the *Suggestions on Accelerating the Implementation of Socialization of Social Welfare*. In this document, the general

---

[1] These three legal registration categories include social organizations (*shehui tuanti*), non-state nonprofit enterprises (*minban feiqiye danwei*) and foundations (*jijinhui*).

goals of socialization were to promote "various investments, public recipients and diverse services." To sum up, the main trends of the welfare reform lied in two aspects: the first was the decentralization of social welfare from the central government to local governments; the second was the responsibility shift from the state to the market and the third sector. Due to the decentralization, local governments have become key actors in accumulating funds, redistributing resources, and establishing new welfare systems, with the central government's input only comprising a small part (Croll, 1999). Hence, local experiments and innovation presented a decentralized and diverse collage of frameworks in Chinese welfare socialization campaign. More importantly, there has been a diversification in the welfare delivery system, with the emergence of non-state welfare providers, especially the nonprofits becoming the most visible manifestation of the development trajectory. This view was supported by Shang (2001), who pinpointed the mobilization of new economic resources in the provision of social welfare products and the growth of non-state nonprofit enterprises as the main driving force of welfare delivery under the socialization policy.

China officially became an aging society in 1999. Based on government statistics, Chinese elders over 60 amounted to 167.14 million (12.5%) while people above 65 reached 113.09 million or 8.5% of the total population in 2009 (MOCA, 2010). It is estimated that in 2020, the elder population in China will swell to 248 million (China National Committee on Ageing or CNCA, 2010). Faced with the massive and rapid demands of the elderly, the weakness of the government is only too obvious. In 2000, urban bed supply was a mere 977,000, enough for only 0.8% of the target population (Jia, 2000). At the end of 2008, the country has 2.38 million residential beds for elderly, only accounting for 1.5% of the total aging population (CNCA, 2010) while the international recognized standard for developing countries is 3%. To deal with this situation, the government began to promote the practice of "making greater use of the third sector, NPOs and commercial operators to produce welfare services" (Yan, 2003). To encourage and regulate welfare NPOs, it announced the *Temporary Regulations for Registration and Management of Non-state Non-profit Enterprise* in 1998. The decree specifies strict rules on the formation, operation, administration, and dissolution of non-state nonprofit enterprises. Before they can register with MOCA, agencies must secure an official or professional sponsor willing to oversee their daily management. Besides, they must pass an annual inspection by the civil affairs authorities (Wong, 2007) before they can renew their license.

Since the late 1980s, commentators have noted that "a significant number and wide range of associations are in operation in the country [China]" (Wang & He, 2004). In 2005, MOCA published the *Opinions on Supporting Social Entities to Operate Social Welfare Institutions* to lay out the principles and preferential policies (MOCA, 2005). To the state, just as it is eager to shift the burden of delivering welfare to civil society, the growth of social organizations creates considerable anxiety. In particular, agencies that represent troublesome groups and pose a threat to state authority arouse suspicion. On the other hand, organizations engaged in economic and social development are seen as nonthreatening and useful. To the extent that government perceptions of social organizations differ, conflicting policy stance is

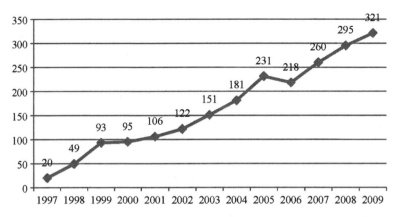

Fig. 14.1 The number of non-state elder care homes in Shanghai from 1997 to 2009. *Source*: Shanghai Civil Affairs Bureau (2010)

apparent. In practical terms, social service and commonweal organizations won the government's favor, whereas political, religious, and advocacy organizations face suppression by the state. Some scholars attribute the policy bias to state concern for social stability after the destabilizing effects of market reforms (Ma, 2006). As vehicles to deliver social care and handle the enormous social problems brought by the lack of a comprehensive social security system, welfare NPOs can repair the havoc in the social and personal life of citizens as the state withdraws from providing the old existential guarantees. This line of thinking is best summarized by Ma (2006) who argues that "[the government's social organization policy is] a response to economic, social, and political crises. The government has turned to nongovernmental institutions to shoulder responsibilities on numerous fronts, including social welfare, economic development, and disaster relief." As a result, government policy is geared toward strengthening the social service functions of NPOs since these agencies serve as agents and allies of government. Consequently, their numbers shot up rapidly since the 1990s. In elder care alone, in 2008, registered non-state care homes have exceeded 4,141 nationwide, operating a total capacity of some 411,723 beds (CNCA, 2010).

Shanghai has the oldest population profile in the country. According to official statistics, people over 60 exceeded 3.15 million in 2009 or 22.5% of the total population (Shanghai Committee on Aging, 2010). The growing pool of seniors creates a large market for residential care. As a pioneer city to implement the welfare socialization program, the municipality was keen to promote NPOs so that they could take on the brunt of delivering care. Indeed the increase in non-state services has been dramatic. In 1997, the city had 20 non-state elder care homes with 1,851 beds; in 2002, the numbers shot up to 122 homes and 10,243 beds. At the end of 2009, 321 homes registered with the civil affairs bureau, providing 47,947 beds and accounting for 53% of total supply, much more than the state sector (Shanghai Civil Affairs Bureau, 2010). In short, in about one decade, unit numbers jumped 15 times while bed numbers jumped nearly 26 times, as shown in Figs. 14.1 and 14.2.

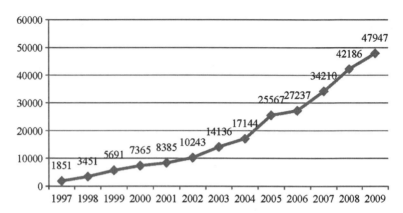

Fig. 14.2 The number of beds of non-state elder care homes in Shanghai from 1997 to 2009. *Source*: Shanghai Civil Affairs Bureau (2010)

What are the reasons behind this stunning growth? What does the numbers represent—success in the socialization policy or otherwise? What are the institutional measures that promoted NPO development? What is the situation of non-state elder care homes in Shanghai now? To answer these questions, we compare the findings of an earlier survey completed by Wong and Tang in 2002 with new investigations we conducted in April 2009.

In the 2002 survey, 137 non-state old age homes in three cities—Tianjin, Shanghai, and Guangzhou—were investigated, including 50 care homes in Shanghai. The early study explored in depth the agency characteristics, development history, operational profile, and difficulties encountered by non-state care homes from the perspective of key stakeholders (investors, administrators, staff, client, and family).

The findings of the 2002 survey have been reported in various venues. Of relevance to this chapter, five conclusions can be summarized. First, the homes had mixed auspices, comprising individual-run, community-run, enterprise-run, shareholder-run, and joint venture units. Over 81.5% of homes registered as NPOs. Second, most of the agency founders were women of middle to late middle age who had a vocational background as a managerial and technical cadre in a socialist work unit and had been made redundant or faced imminent job loss. Third, their financial status was rather weak: only 17% were making a profit (revenue exceeding expenditure), 47% could balance income and expenditure, but 36% were in deficit. Moreover, 51% of agencies had little prospect of ever recovering their capital investment. Fourth, the agencies encountered many daunting problems: high cost, unfair competition from state-run welfare institutions, state failure to enforce preferential policies, and lack of government support and trust (Wong & Tang, 2006). Hence, all agencies were unanimous in demanding government support. Fifth, four factors crucially affect the survival and development of non-state care homes: their ability to meet the needs of users, manage human resources, ensure financial sustainability, and satisfy the expectations of their investors (Wong, 2007, 2008).

Fig. 14.3 Change in institutional status of 50 non-state elder care homes in Shanghai, 2002–2009. *Sources*: Wong and Tang (2006) and Shanghai Civil Affairs Bureau (2010)

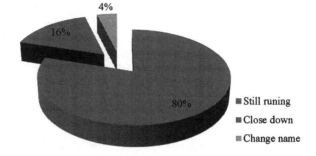

Fig. 14.4 Change in size in 50 non-state elder care homes in Shanghai, 2002–2009. *Sources*: Wong and Tang (2006) and Shanghai Civil Affairs Bureau (2010)

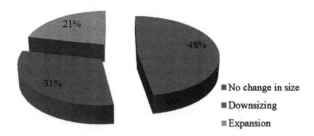

Of the 50 homes in Shanghai which were included in the first survey, eight have closed down while 42 have remained, including two that have changed their names. This means that the vast majority (84%) of the sample homes are still in operation. Among the eight defunct units, two were community-run, four were enterprise-run, and two were joint ventures of enterprises and local communities. The size of the closed units was usually small (less than 10–50 beds). Of the existing 42 homes, 20 homes have retained their original size, 13 homes have expanded (the biggest expansion being 96 beds), and 9 have down-sized (a reduction of 60 beds in the most dramatic case). Figure 14.3 shows the change in institutional status of the 50 agencies from 2002 to 2009; Fig. 14.4 shows the change in bed numbers.

In order to find out what happened to these agencies since the last survey, an initial study of five projects was carried out in April 2009 and June 2010. Without the interlocution of the municipal civil affairs bureau, we used the current list of non-state elder homes on the civil affairs Web page and contacted a number of homes in different locations. To our pleasant surprise, we readily secured the agreement of five agencies for a re-visit. We then called on the Shanghai Civil Affairs Bureau where we interviewed officials from various departments including social organization registration, social welfare services, services for the elderly, policy and research office, and Shanghai Committee on Aging affiliated to the civil affairs bureau. A review of policy documents from government and Web sources and an interview with an expert on old age care from East China Normal University complement the current study. It is envisaged that the current investigations will form a part of a bigger empirical study of Chinese NPOs to be carried out in the future.

Changing Institution of Welfare NPOs in Shanghai

An institution is a series of regulations or norms that influence human activity. North argues that institutions are a set of rules, compliance procedures, and moral and ethical behavioral norms designed to constrain the behavior of individuals (North, 1981). As far as Chinese NPOs are concerned, the impact of institution is particularly visible and important, because the genesis was traceable to the push of the government rather than the autonomous development in civil society. In China, the state still holds a dominant position vis-à-vis civil society. Social science scholars have used the terms "nascent civil society," "semicivil society," or "state-led civil society" to depict the state–society relationship (Frolic, 1997; He, 2003; Whiteet al., 1996). State control of NPOs is strong—it controls them through a tight dual-management system and regulations promulgated in national and local administrations. This comprehensive regulatory framework pervades the process of registration, agency management, fiscal aid, preferential policies, and administrative sanctions.

Shanghai has often stood at the forefront in pioneering social services. For example, the municipality is the first in China to introduce social assistance or minimum living security allowance (*dibao*) in 1993 (Leung & Nann, 1995; Wong, 1998). It is also the first to set up re-employment service centers for redundant or *xiagang* workers in 1994 (Wong & Ngok, 2006). In elder care and NPO development, the Shanghai authorities are keen to take the lead. A review of policy evolution shows a strong record of initiation and consolidation of the regulatory framework. The changing institution can be summarized in two aspects—formal and informal. The former includes law, administration regulations, and other administrative arrangement. The latter comprise officials' attitude toward non-state care homes.

Formal Institutions: Work in Progress

As a means to deliver real benefit (*zuo shishi*) to the citizens of Shanghai, the municipal government announced a plan to establish 10,000 new residential beds each year and to provide a subsidy to each newly added bed (*chuangwei butie*) in its 11th Five Year Plan (FYP) released in 2005. In its wake, the city's civil affairs authorities announced an implementation plan. As a measure of support, the municipal civil affairs bureau will grant a subsidy of 5,000 yuan to each new bed. Meanwhile, the district civil affairs department will also give the same level of support. Hopefully, the scheme will give sufficient boost to the creation of new facilities, which was absent in the early years. The implementation of the two-tier scheme is supervised at the city level. According to official report, the system came into effect in 2006. However, 1 year away from the end of the plan period, the government still has not announced further plans to increase bed supply or how to manage the existing ones. What's more, qualitative issues appear to be neglected—the quality of newly added beds and the services these provide are not included in the monitoring and supervision process.

The second measure in institution reform is the formation of a trade association (*hangye xiehui*) for the residential care industry. In 2003, Shanghai Social Welfare Association (SSWA) was set up under the instigation of the government. At the end of 2009, over 80% of non-state care homes have joined up as members. The major responsibilities of the association include the training of care staff and assisting the municipal civil affairs bureau to conduct the annual inspection. In short, it is the intention of the government to use this intermediary agency to manage the elder care homes. In one of our interviews, a government cadre used "mentoring or governing" to describe its relationship with the trade association. Actually, SSWA has been controlled by the government through fiscal and personnel allocation. Where funding is concerned, income for the association comes from state allocation and membership fees. The vice president of SSWA confessed that the association could only rely on the government input because the membership fees were "a drop in the ocean." Where human resources are concerned, the association has 13 cadres on its staff. Eight of them come from the municipal civil affairs bureau, district civil affairs departments, or subordinated public institutions (*shiye danwei*). All cadres are paid by their original or parent units. In terms of personal status, SSWA cadres are not classified as civil servants. Instead they are graded as quasi-civil servants in public institutions. Therefore, to the government, the association functions as a government agent even though it does not have formal line authority over the elder care homes. At the same time, however, the association is supposed to represent the interests of member agencies. This double identity may be a bit awkward. However, the directors we interviewed made no bones that the SSWA was a government front. As all office bearers were appointed by the government, the government naturally dominated the decision making.

The third measure is directed at perfecting the legal framework, namely the introduction of accident insurance in elder care homes. The need for this has been identified in the 2002 survey. In the past, accidents involving residents in care often ended in expensive litigation and compensation settlements. The lack of collective insurance cover drove affected homes to bankruptcy and created long standing hassle. In order to build a safety net for both residents and agencies, the government launched an accident insurance scheme at the end of 2008. In the *Notice to Implement Accident Insurance in Elder Care Homes*, the basic principles, coverage, enforcement measures, and funding arrangements are laid down. The premium comes from three sources: the municipal government, the district government, and the agency, each paying one third of the cost. The scheme was welcomed by agencies. However, as it has only 2 years' experience, the effects remain to be seen.

Informal Institutions: Questionable State Attitude

However vital, formal codes and systems are only one part of reality. Informal institutions are equally salient. Informal rules that are not found in official documents but nevertheless continue to influence the maturation and development of civil society are influential (Yu, 2006). In China, where legal norms are not well enforced and officials' attitudes and behavior can arbitrarily affect policy outcomes, the force of informal institutions cannot be ignored. Our literature review and interviews with home operators and experts suggest considerable conflict and vacillation of official attitudes toward non-state agencies. On the one hand, the state is anxious to unload the heavy burden of social welfare to the third sector. This is mainly driven by its weak capacity in delivering services, especially financial constraints. On the other hand, the government still keeps a watchful eye on civil society organizations for fear of insubordination, intransigence, or loss of state authority. Pundits say NPOs are never treated as equal partners but subordinates. Their activities and operations require supervision and monitoring.

Even though non-state homes are supposedly instruments of the welfare socialization policy, the civil affairs cadres we interviewed seldom referred to the policy. Instead, officials often stressed that the government would never give up "managing" such homes. They believed that in the transitional stage, NPOs needed strong support and guidance from the state. However, the state could not meet all their needs. In terms of financial support, for example, civil affairs funding is always short and must be used judiciously. They are even more worried about potential loss of state assets if more money passes to non-state operators. One cadre opined bluntly that Shanghai socialization was a failed project. He pointed out that the government has been spending more money on social welfare and on NPOs in the past several years. Instead of bringing in new investment and service operators to relieve the pressure on state coffers, the government was actually enlarging its role in welfare planning, financing, and management. This suggests that state–NPO relationship is one of principal and agent. The latter must operate under a strict regulatory regime.

In sum, the government endorses the function of NPOs in elder care. It encourages social organizations to shoulder the major role in service delivery and is prepared to give more support to these agencies. Nevertheless, it still believes in keeping its guidance and supervision role as its underlying attitudes are colored by both pragmatism and distrust. These attitudes are likely to produce policies that are wavering, inconsistent, and conflicting.

Evolution of Welfare NPOs: Development and Challenges

As mentioned before, the 2002 study identified four factors that affected agencies' survival and development prospects: their ability to ensure financial sustainability, manage human resources, meet the needs of users, and satisfy the expectations of their investors (Wong, 2007). The follow-up interviews also explored these issues by asking investors or managers to reflect on their evolution during these years. The interviewees consisted of home investors, superintendents, and financial managers.

Table 14.1 Features of the revisited non-state care homes in Shanghai

	Home A	Home B	Home C	Home D	Home E
Inception	1999	2002	1998	1998	1999
Auspice	State-owned enterprise	Private enterprise	Individual household	Joint venture	Joint venture
Location	*Yangpu* district	*Yangpu* district	*Hongkou* district	*Pudong* district	*Hongkou* district
Neighborhood characteristic	Old industrial area with a mass of laid-off workers	The biggest civilian residential community with many new residents	Old downtown neighborhood surrounded by warehouses and civilian housing	Emerging community with a mix of local elders and young immigrants	Traditional residential community with many indigenous Shanghai residents

Characteristics of the Revisited Homes

The interviewees spoke quite freely. To protect the anonymity of the agencies, they shall be designated as Homes A, B, C, D, and E (Table 14.1).

In terms of auspices, they fell into different types. Home A was started by a state-owned enterprise in 1999 with the initial aim to cater for the unit's retired employees and to place their redundant staff. Home B was initiated by a private enterprise owned by an individual entrepreneur in 2002. Over the years, this entrepreneur formed a string of five elder care homes and two nursing hospitals in a consortium under the district labor union as its official sponsor. Home C was founded by a female entrepreneur in 2001, who used to be a laid-off worker of a state-owned enterprise. Home D was established by an oversea religious group as one program of an entire community project in 1998. This community project was co-founded by the government, the religious group, and the local community, providing community services, community education, and cultural activities to residents. Home E was set up as a joint venture of a collective-owned enterprise and a local community organization (*jiedao banshichu*) in 1999.

Service demand and client profile is crucially affected by location. All five homes are situated in Shanghai's inner city areas: Homes A and B are located in *Yangpu* District, Homes C and E in *Hongkou* District while Home D is in *Pudong* District. However, their neighborhood characteristics are different. Home A sits in an old industrial area with a mass of laid-off workers; hence most of its residents are retired personnel or redundant staff from its parent company. Home B is domiciled in the biggest civilian residential community in the city—*Zhongyuan* Community.[2] Not a traditional neighborhood, the community houses many residents resettled from other areas after the large-scale land clearance in the 1990s. According to the data of Shanghai Civil Affairs Bureau, there are ten residential homes in the neighborhood (Shanghai Civil Affairs Bureau, 2010), so competition for customers is intense.

---

[2] According to government statistics, the population in *Zhongyuan* Community is more than 250,000.

Home C is found in the old downtown of *Puxi*[3] surrounded by civilian residences, warehouses, hospitals, and schools. Although the earliest urban renewal project was started in this area in the 1990s, most of the original residents had moved back since, thereby maintaining the traditional custom and community identity. Home D is located in the newly developed residential community of *Pudong* District. Most of the residential units were workers' housing provided by state-owned enterprises. After welfare housing distribution came to an end in 1998, the original residents moved out gradually and rented their apartments to outsiders. Nowadays, this community is a mixed neighborhood dominated by local elders and young immigrants. Home E lies in a traditional community with local Shanghainese making up the majority of the population, especially retired workers from the education bureau. The local environment is comparatively close and stable.

Financing Issues: Funding Chain, Preferential Policies, and Housing Problem

1. Ruptured funding chain

The initial capital of non-state residential homes comes from four sources: individual investment, enterprise funding, government subsidy, and social donations. Wong and Tang found that 66% of homes had individual investment, 39.6% of homes obtained startup capital from enterprises, 24.7% enjoyed a subsidy from the government, and 13.6% secured donations from society (Wong, 2007). In this chapter, the revisited agencies obtained initial funding either from their parent units or individual investors. Home A got the capital funding from its parent company, a state-owned enterprise. Home B was started by an individual investor although it was supposedly supervised by the district labor union. Home C was initiated by a female laid-off worker, who received fiscal assistance from the district civil affairs bureau and the Women's Federation, and became the manager of the home later. Home D was funded by a social group but obtained extra support from the government because it was supposed to be a "model project of community service provision." Home E was a joint venture set up by a collective enterprise and one local community. For all, the biggest financial problem was not the startup investment but the ruptured funding chain after the projects came into operation. The government was aware of this difficulty. In our interviews with civil affairs officials, one cadre emphasized that most non-state elder care homes had sufficient funding at the founding stage. Agencies often ran out of money to maintain and expand their operations.

---

[3] *Puxi* is a geographical concept that refers to the west bank of *Huangpu* River in Shanghai city except for suburbs like *Jiading* District and *Baoshan* District.

Continuing injection of capital was needed to expand the scale of operation and to enhance the amenities and services. In 2002, Wong and Tang discovered that it was rare for the government and enterprises to make further injection after the initial investment, hence most homes relied on fee charges to keep afloat (Wong, 2007). The situation has not changed since then. As mentioned before, the municipal government now provides a subsidy of 5,000 yuan to each newly added bed. Some districts are able to offer more subsidies. For example, in *Pudong* District, the district government provided an extra subsidy of 45,000 yuan for each new bed. However, after receiving the startup grant, agencies do not get any new funding to run the existing beds. The financial manager of Home A expressed her anxiety about the funding crunch during our visit. The subsidy was not enough to set up a standard new bed according to the government's guideline.

To augment the shortfall, agencies turned to three sources. The first and most important funding source was the residential charges at an average of 1,000–1,600 yuan per resident per month. Indeed, from a report of CNCA, 90% of non-state old age homes in the whole country depend heavily on the charges as the primary funding to maintain their daily operation while only 7% of them have other financial sources like social donation (CNCA, 2010). The second source came from the investors, such as the continuing operation subsidy, and the exemption or deduction of rent or management fee.[4] Funding from the local community made up the third source, with the neighborhood agency contributing funds to support destitute seniors who did not have any children. Notwithstanding this, the latter two sources are uncommon for non-state agencies. For example, in Home A, the parent company provided a subsidy for each resident who had no family members. However, for other homes, the only income resource is the fee charges. Consequently, the shortage of money meant that although agencies perceived new demands, expansion is not possible. Unable to raise new funds, Home E had to give up its expansion plan and focus on cost reduction, for example, controlling the food bill. Another illustration comes from Home C. Because of its unstable finance, it has to downsize from 130 to 70 beds over a 7-year period. Meanwhile, although the superintendent of Home B refused to admit that he obtained funding from other service units in the consortium, primarily the two nursing hospitals, to balance its finances, its residential fee was higher than other homes, ranging from 1,300 to 1,900 yuan per month.

When the issue of funding shortage was raised with the civil affairs officials, they were well aware of the issue. They felt unable and reluctant to give ongoing support. As mentioned before, the government was deeply concerned about the potential loss of state assets. To the state, spending money (albeit on a more generous level) on government-run facilities was regarded as justified. Giving a subvention to non-state agencies was not. In their view, granting public money to non-state programs was inherently dangerous. One official even suggested that it was tantamount to divesting state resources to benefit individuals and enterprises.

---

[4] Some non-state old age homes rent space from their parent companies. Meanwhile, some agencies initiated by state-owned enterprises have to submit management fees to their investors.

Another cadre admitted that, apart from the government's own financial incapacity, they did not have enough confidence to support non-state projects year by year. Besides, agencies should "run their own business." Such an attitude reflects a lack of trust in the third sector.

In view of these difficulties, NPOs began to look for alternative income sources besides fee charges. In Home A, the manager tried very hard to obtain aid from the Shanghai Charity Foundation. For years, the superintendent of Home B intimated that the income deficit was met by transfers from its sister organizations. However, he refused to confirm this point and insisted that all the units in the group had independent accounting. Raising donations from society might be a solution in the future. In 2007, the *Enterprise Income Tax Law of the People's Republic of China* considerably increased the scope for pretax deductions when companies donate money to public welfare undertakings (National People's Congress, 2007). One home manager we interviewed expressly looked to donations as the next step to augment their funding.

2. Difficulty of implementing preferential policies

On paper, the government has promulgated preferential policies to encourage non-state welfare provision. Supportive policies, from the central to the local government, include concessions on land and land costs, construction expenditure, water and electricity charges, tax exemption, and so on (Yi, 2006). In reality, implementation was haphazard. The failure to honor preferential policies has been identified by Tang and Wong. The situation has persisted to this day. For example, Home A did not enjoy half price for electricity during nighttime according to policy; the electricity bureau refused to offer discount even when the relevant notice from the civil affairs bureau was presented. For gas charges, they did not enjoy the domestic rate as promised; the actual charge was set between the levy for domestic households and that for commercial users. Home E reported another anomaly. In 2008, the home earned a profit of 10,000 yuan. As an NPO, if the profit was put back into existing services, it should be tax exempt. However, the tax bureau insisted on taxing them. Notwithstanding the generally uncooperative attitude of fellow state agencies and utility companies, exceptions are possible. For instance, homes that enjoy close connections with the authorities will stand a better chance of enjoying the preferential policies. Home D, designated as part of a "star project" is one such case. This home has multiple auspices. Its founders include the district civil affairs bureau, a street-level agency, and a Hong Kong-based charitable organization with a religious background. Although the government does not involve in its daily operation, its close link to the state and its special place in the welfare socialization campaign in Shanghai give it favored standing. The manager of Home D acknowledged this privilege: "We are luckier than most of our fellows. We enjoy everything we are entitled to from the state, sometimes even more."

Actually, the issue of enforcing preferential policies exposed the weak position of the welfare authorities. Within the government bureaucracy, the civil affairs

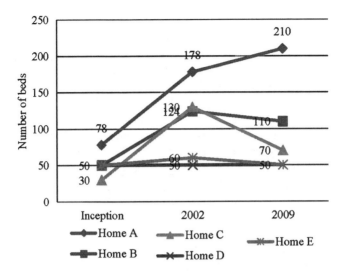

Fig. 14.5 Changing size in the revisited non-state elder care homes in Shanghai, the inception to 2009. *Sources*: Wong and Tang (2006) and Shanghai Civil Affairs Bureau (2010)

bureau is known as a marginalized agency. It has no power over finance and personnel in other departments and their subordinate agencies. When fellow bureaucrats refuse to observe the preferential codes, such as tax exemption or reduction in fee charges, the civil affairs bureau is powerless. The ambiguity of the policies themselves creates difficulties too. For example, in the *Suggestions on Supporting Social Sector to Set Up Social Welfare Institutions*, the authorities only promise to "grant proper preferential treatment to social welfare organizations on the land price," but there is no elaboration on what "proper" treatment means. Further, the government professes to give priority to welfare agencies in the use of telecommunications service, but it does not state what the preference is (MOCA, 2005). Such vagueness and ambiguity makes enforcement almost impossible. In the opinion of Yan Qingchun, vice president of CNCA, policies that are promoted by the central government are not being enforced by the local authorities. "Normally, the land allocation, government subsidies and utilities concessions exist in name only" (Yan, 2007).

3. Problem in securing affordable housing

All home operators wanted to tap into the big market for residential care in the rapidly aging city. Indeed all homes had plans to expand facilities. However, such ambition had to be put on hold due to difficulties in security land and buildings at affordable cost. As Fig. 14.5 shows, all five agencies had below 100 beds at the beginning. In the 2002 survey, four out of the five homes had grown significantly. By 2009, however, all of those expanded homes had to scale back their capacity.

Three factors hindered service expansion. First, from 2005, the government set higher building standards and strengthened the inspection of newly added beds. This increased the cost of installing new facilities. Second, the shortage of new funding curtailed their ability to expand. Finally, difficulty in acquiring affordable space through purchase or leasing had been the most critical factor. The finance manager of Home A complained that because of space limitation, they could not add extra beds even though each new bed could bring in more revenue from the government. As few agencies can afford to buy land or property, success in negotiating a long lease and keeping the rent from rising too sharply is crucial to survival. Compared with Home A's failure in expanding its facilities, the story of Home C is more pathetic. Since its establishment in 1998, it has rented space from the district education bureau. For quite a long time, Home C's relationship with the landlord was said to be "stable and close." The grace ended at year-end 2008, when the landlord demanded to double the rent from 2009. This is a big blow to the home. Its manager immediately pleaded for a new and more affordable rent. Unfortunately, the negotiation quickly broke down, with the owner refusing to sign a new lease. This caused Home C to be slapped with a serious warning in the annual inspection: "Literally, we pass this year's annual inspection, but the government refuses to stamp on our inspection form because we cannot produce an effective leasing contract to prove our housing status." The home was ordered to fix this problem before the next inspection or face failure in the subsequent round. Given the regulation that an elder care home cannot fail the inspection in two consecutive years, its future is in jeopardy with the threat of closure over its head.

The constant struggle over the space issue forces home managers to look for survival strategies. For the manager of Home E, cultivating a good relationship with the landlord is imperative. He cited another case as a lesson: one care home in the vicinity was close to collapse after the landlord raised the rent from 60,000 to 90,000 yuan in a period of 2 years. Each re-location means a loss of investment that had gone into decoration and provisioning. Home A experienced this trauma first hand. In 2005, new decoration and fitting out work was carried out. After completing the improvements, the landlord (the parent company) suddenly demanded repossession of several floors in the building. The blow to the agency's finances was severe.

To non-state agencies with modest means, buying their own premises is unthinkable. All five homes rely on renting and are unanimous in welcoming government support in securing affordable and stable housing space. One suggestion, from the superintendent of Home B, was to have the government sign a long-term contract with land/property owners and sublease the space to agencies. Another idea, from the superintendent of Home E, was for the government to provide land to agencies as incentive to start new facilities. The most favored suggestion was offering rent allowance to agencies. These suggestions were not well received. The idea of rental allowance, especially the feasibility of a unified rent allowance scheme, was dismissed as unfeasible. First, the government suffered from resource constraints. The municipal civil affairs bureau already spent 50 million yuan a year to create new beds; to give recurrent rental subsidy would not be possible. Second, due to the unstable and variable real estate prices in different districts, a single rent aid standard is impractical. Notwithstanding the caution of the municipal authorities, local districts and communities were encouraged to experiment with rent support schemes when resources permit. For example, *Changning* District started to offer non-state care homes a rental subsidy at 0.5 yuan per day per square meter from 2008. The investor of Home C, who has another non-state old age home in this district, regards the rental allowance as the greatest and most useful help from the government, "If I set up another elder care home, I will definitely put it in *Changning* District, because the rent allowance is attractive for us [investors]."

Staffing Concerns

A care home is a human service; hence the staffing issues are of crucial importance (Wong, 2007). In non-state care homes, the service team comprises administrators, doctors, nurses, care workers, and support staff. The 2002 survey identified a number of salient issues related to staffing: staff lacked training and experience in elder care, wages for most staff was low, working conditions were tough, and turnover of nurses and care workers was high.

The situation has not changed much since then. Because of the limited budget and unfavorable working environment, the homes faced problems in hiring professionals like doctors and nurses as well as care workers. For example, Home D only hired one healthcare doctor who was unqualified to prescribe. Home B did not hire any doctors or nurses itself. All the four medical professionals working in the home came from the nursing hospitals in the same consortium. They were paid by the nursing hospitals, not the home. Home A had more than 200 beds but had only four doctors and one nurse. All medical workers were former employees of the parent state-owned enterprise who had been reassigned there since its founding in 1999. However, attracting new medical and nursing staff was difficult because of the low pay. It was the same with care workers. The majority were also redundant personnel transferred from the parent company. All of them spoke the Shanghai dialect and were familiar with the community. In time, most of the first cohort has retired. Nowadays, the 50 strong care staff is dominated by migrant workers from Gansu Province.

While migrant workers provide an abundant reservoir of care staff, their hiring has brought about new issues. One problem is the high mobility and turnover. Currently, care workers are paid 1,500–1,900 yuan per month. Working conditions are disagreeable: a two-shift system, no vacations and heavy work load. Home E saw a lot of turnover among its care staff each month; the shortest stint was 2–3 months. Signing a contract was no guarantee to staff retention. To the management, the never-ending task of making new hires has become a constant headache that consumed a lot of their time. Furthermore, frequent staff changes disrupt the service and harmed work relationships with clients. Besides, migrant workers coming from villages in western China did not speak the local dialect and were unfamiliar with local customs. In our interviews, communication and cultural differences resulted in residents' complaints that care workers did not understand them and could not communicate with them. Agency administrators could do very little to solve these problems. Even worse, the new generation of migrant workers was reluctant to work in elder care homes, since it was viewed as a high-risk job with poor salary, poor working environment, and low social status. The manager of Home B complained that it was "extremely difficult" to hire new nursing workers from the labor market: "They prefer to work in restaurants or hotels, because these jobs are much easier and less pressurized." In addition, all five homes were eager to hire professional social workers. However, trained social workers are in short supply in China and where available, prefer to work in community settings rather than in elder care. As a partial remedy, some agencies took in social work students or volunteers but their short-term affiliation limits their usefulness. In some homes, existing staff undertook social work training to meet the shortfall of social workers and to lower the staff costs. The female manager of Home D is one such example.

Meeting Users' Needs and Enhancing the Quality of Care

Chinese tradition of filial piety stresses care of elderly parents in the family, and seeking residential care for a parent is frowned upon. Values have changed a lot in modern times (Peng, Liang, & Cheng, 2006; Wong, 2008; Wong & Tang, 2006). Now, placing a parent in a care facility no longer invites social disapproval. Many elders also accept residential care as a normal and professional form of old age support. Such thinking has impacts on service development. First, as residential care becomes more accepted and as income rises, better amenities and higher professional standards are in demand. Care homes have to respond by improving their amenities and care skills. Second, higher expectation on the services provided and nature of care has been evident. Previously residential care was seen as catering to the physical needs of inmates; now new professional thinking emphasizes the provision of a comprehensive service package which includes emotional tending, mental stimulation, and even hospice care. From our recent visits, however, not all the homes were well prepared to meet the new challenge. The views of the agencies on the nature of care and their vision for the future differed. The manager of Home A appeared unconvinced about the need to provide mental and psychological care. Her main concern was seeing to the satisfactory discharge of the traditional functions—the diet, facilities, cleanliness, medical treatment, and personal care. However, for the superintendent of Home B, a former army doctor and ambitious social entrepreneur who led a "group" of five care homes and two nursing hospitals, the updating of the service was his goal. Noting the increasing longevity of the residents, he gave priority to the purchase of new medical equipment to improve the inmates' health and wellbeing. The introduction of hospice care was also in the pipeline. By contrast, the superintendent of Home C felt powerless to follow the fashion of care service updating. She defined her elder care home as an institution that offered elementary accommodation, diet, and care service to low-income elders because of its limited funding, facilities, and space. Actually, among our sample homes, Home C set the lowest charges and 30% of its residents were supported by the municipal Minimum Living Security Allowance Scheme. For Home D, thanks to the social service experience and good standing of its Hong Kong sponsor, the agency is able to offer psychological counseling and various cultural activities to its inmates. The enhanced standard of care gives the home its major selling point. The manager of Home D encouraged her employees to acquire knowledge and skills in nutriology, psychology, hospice care, and related fields to meet the rising expectation of their residents. Not all homes have such aspirations. In the case of Home E, a different perspective is evident. The manager knew about the modern ideas in elder care, but disdained to follow them, regarding them as meaningless for the present. In his opinion, nutritious meals and comfortable living environment were the essence of good residential care; hence much time went into ways of improving the residents' diet.

All five homes seemed to run well since they were last surveyed. Occupancy rate was over 90% in all five. Residents and their family members were reportedly satisfied with the services provided, in particular, the food and personal care. Moreover, all agencies tried hard to control their fee charges at an acceptable level. The monthly fees[5] in Home A was 1,200–1,600 yuan; in Home C, charges ranged from 900 to 1,350 yuan per month; and in Home D, fees covered from 1,150 to 1,550 yuan. Home C had the lowest monthly fees from 600 to 1,000 yuan while Home B sets the highest monthly charges from 1,300 to 1,900 yuan.

Investors' Concerns

Wong and Tang identified four important motivations for investors/founders to start their social projects: their personal history, their need for employment and economic survival, the skills and experiences they developed in their past careers, and their entrepreneurial vision and ability (Wong, 2007). After founding the projects, almost all the founders were involved in running the business. To staff and clients, they were known as the "bosses." Running a non-state home turned out to be more testing than they imagined; "the feeling of being overwhelmed by problems and not getting help from the government was widespread among them" (Wong, 2008). In the latest visit, our conversations with the administrators revealed interesting perceptions about the challenges they faced and their vision for the future.

For managers of Homes A and E, running a care home was a task assigned by their former employers. Without relevant skills and experience, they found the work difficult and demanding. Their common complaint was a lack of support given by the government. Besides, providing care to frail elders on a 24 h basis made them tired and frustrated. Both seemed to be resigned and cautious. When asked what

---

[5] Normally, the monthly fees or charges of an old age home consist of three parts: accommodation, diet, and nursing service fee which vary according to the level of care.

their plans for the future were, their answer was somewhat surprising. Their concerns were mundane, such as decorating the premises, replacing windows and doors, and replenishing the facilities. They confessed that they were not looking for big profit, but for survival. Nevertheless, they were proud of their social contribution. They thought their work had great social value.

For the manager of Home C, in contrast, the development of the non-state agency reflected her life story and personal struggle as a former laid-off worker and a current social entrepreneur. After retiring from the state-owned enterprise at her 40s in 1998, she got fiscal support from the district civil affairs bureau and the Women's Federation to start a civilian-run elder care home. Since that time, she had received a number of awards from the authorities. The most treasured honor was being conferred the title of "National Female Pacesetters (*quanguo jinguo jiangong biaobing*)" by the All-China Women's Federation for her remarkable contribution to social welfare reform and re-employment of laid-off workers in 2003. Currently, she supervised three small to middle size elder care homes in two different districts. Despite difficulties in securing affordable housing and facilities, she was thankful that she had gone into the residential care field. As a female entrepreneur, she is still optimistic about the future of the business. She understands the constraints of the civil affairs authorities and is not overly critical of the government policy toward NPOs. At this point, her business ambition is modest: no expansion in the coming years, just focus on maintaining and improving the services incrementally.

All managers found their work draining and stressful. To Home D's manager, the pressure is keenly felt. Having served as home superintendent for 8 years, she was newly promoted to a more senior post in her agency. As a modern manager who acquired knowledge in nutriology, social work, and public administration, she attempted to bring new ideas into the operation of care homes. She strongly believed in encouraging inmates to participate in various community activities and strengthening communication with neighborhood residents. Despite her tireless efforts to reach out to the community and strengthen programming in her home, a change in government personnel and attitude deprived her of the support she used to enjoy from the state authorities. Two factors led to the falling status of her home. One was the religious background of her Hong Kong sponsoring agency, which made government leaders uncomfortable in visiting her project. A more immediate cause was the departure of the home's patron, a senior government cadre who was promoted and transferred to another government department. Hence instead of being held up as a model for government–NPO collaboration, the home lost its favored status. Much as she still felt hopeful about the future of non-state residential care, her aspirations have become more modest. More attention is now being given to internal management issues, particularly the cost control. No expansion plan is envisaged.

The superintendent of Home B was more visionary. He had been a doctor in the army before entering the elder care industry. Using his personal resources, medical knowledge, and social connections, he started his first care home in Shanghai in the early 1990s. After more than a decade of hard work, he built up a consortium of seven care establishments (five elder care homes and two care hospitals) through collaborating with other individuals and organizations. In his view, the old age care industry was full of opportunities and challenges. He professed his eagerness to develop his projects to meet the higher service requirements and advanced ideas in the near future. He also saw the need to strengthen the corporate management skills in his team. Indeed, he had the bearing and vision of a true businessman or entrepreneur. However, even for an optimist like this boss/manager, the market environment at the present time is fraught with risk and uncertainty. Government policy still forbids NPOs to set up branch organizations and operate as a consortium. In the interview, he was wary in drawing attention to his "group" of five projects although he was proud of being head of a big organization. He made no bones in saying he had little confidence in the legal environment of Chinese society. In his words, government policy might change at any time. Besides, "a tree that grows tall captures the wind"—the government might give him trouble if his business becomes too successful.

Conclusion

The development of non-state elder care homes filled a gap in social welfare provision in a rapidly aging society. In Shanghai, which has the oldest population in China, non-state elder care homes have become the major service providers. In recent years, the government has publicly endorsed the contribution of NPOs. To encourage their growth, it has actively engaged in institution building. However, policy development is very much a work in progress. In particular, translating policy into practice turns out to be inconsistent, arbitrary, and uncertain. First, there is the absence of long-term direction and detailed plan to guide the development of the industry. Beyond the plan to increase new beds, issues of quality and monitoring have not received much attention. Second, the status of the trade association is ambiguous. As an organization that purportedly represents members' interests, issues like agency involvement, autonomy, and accountability should be stressed. However, the way the association operates suggests it is another form of government organization to manage and regulate the non-state sector. Third, the attitude of the government toward NPOs is a source of anxiety to agencies. In contrast with before when NPOs were solely responsible for raising the initial capital, the government now gives a startup grant. However, there is still no policy on providing ongoing subvention to help alleviate the funding shortage. Weak support to NPOs was attributed to resource constraints and anxiety about the loss of state assets. This is not the entire story. At a deeper level, there is strong distrust in NPOs and their use of public resources. More importantly, the state is in two minds about the value of the non-state sector. On the one hand, NPOs are useful as service providers, helping to relieve pressure on the state. On the other hand, non-state players have to be kept subservient, to be used as agents and controlled as part of the new regulatory state. Government–NPO relationship is one of dominance, not equality. The idea of NPOs as partners is still a long way off.

The return visits to five non-state elder care homes reveals that problems relating to funding, staffing, and space pose big challenges to service operators, similar to the time of the first survey. Uncertainties about state policy and the nature of residential care also create obstacles to further development. Of the five homes, three appeared to lack vitality and were just coping on a day-to-day basis. Two homes are rather more forward looking and appear adroit in harnessing networks and capital to expand their operations. It now seeks to upgrade its facilities, services, and standards to capture the emerging market for quality care. All agencies are frustrated by the low level of state support and trust. The welfare market is still risky and uncertain.

Notwithstanding the explosion of non-state elder care, it is questionable whether such growth is a step forward toward welfare socialization. In the case of Shanghai, the state has strengthened its role in welfare through provision of startup grants, higher spending on welfare, standard setting, and initiation of a trade association. If welfare socialization means cultivating greater participation of societal actors in welfare funding, provision, and management, it is difficult to come to any conclusion on what expansion in non-state care means and NPOs' position vis-à-vis the state. So far the state–agency relationship is marked by tension. At the present stage in the new welfare economy, the market in social care is far from mature. The authority of the state is well accepted and agencies demand more state support to increase their chance of survival. State dominance looks set to become a long-term feature of welfare development in China.

References

China National Committee on Ageing. (2010). *Research report on the basic information of non-state old age homes in China*. Beijing: China National Committee on Ageing.

Croll, E. (1999). Social welfare reform: Trends and tensions. *The China Quarterly, 159*, 684–699.

Frolic, B. M. (1997). State-led civil society. In T. Brook & B. M. Frolic (Eds.), *Civil society in China* (pp. 46–67). Armonk: M.E. Sharpe.

He, B. (2003). The making of a nascent civil society in China. In D. C. Schak & W. Hudson (Eds.), *Civil society in Asia* (pp. 114–139). Hampshire GU: Ashgate.

Jia, X. (2000). Conflicts and choices: Some problems in socialization of welfare for elders. *Zhongguo Minzheng (China Civil Affairs), 3*, 26–27.

Leung, J. C. B., & Nann, R. C. (1995). *Authority and benevolence: Social welfare in China*. Hong Kong: Chinese University Press.

Ma, Q. (2006). *Non-governmental organizations in contemporary China: Paving the way to civil society?* London: Routledge.

Ministry of Civil Affairs. (2005). *Suggestions on supporting social sector to set up social welfare institutions*. Retrieved April 7, 2008, from http://fss.mca.gov.cn/article/cjrfl/zcfg/200711/20071100003615.shtml.

Ministry of Civil Affairs of China. (2010). *China civil affairs' statistical yearbook 2010*. Beijing: China Statistics Press.

National Bureau of Statistic of China. (2008). *China statistical yearbook 2008*. Beijing: China Statistics Press.

National People's Congress. (2007). *Enterprise income tax law of the people's republic of China*. Retrieved June 22, 2009, from http://www.gov.cn/flfg/2007-03/19/content_554243.htm.

North, D. C. (1981). *Structure and change in economic history*. New York: Norton & Company Inc.

Peng, X., Liang, H., & Cheng, Y. (2006). *The research on service system for elderly in cities*. Shanghai: Shanghai Renmin Press.

Saich, T. (1994). The search for civil society and democracy in China. *Current History, 93*, 260–264.

Salamon, L. M., & Anheier, H. K. (1992). In search of the non-profit sector I: The question of definitions. *Voluntas, 3*(2), 125–151.

Shang, X. (2001). From welfare statism to welfare pluralism: Cases of Nanjing and Lanzhou. *Journal of Tsinghua University (Philosophy and Social Sciences), 16*(4), 16–23.

Shanghai Civil Affairs Bureau. (2010). *Shanghai social welfare annual report 2009*. Shanghai: Shanghai Civil Affairs Bureau.

Shanghai Committee on Aging. (2010). *Report on the development of shanghai old age programs 2009*. Shanghai: Office of Shanghai Committee on Aging.

Wang, S., & He, J. (2004). Associational revolution in China: Mapping the landscapes. *Korea Observer, 35*(3), 1–66.

White, G., Howell, J., & Shang, X. (1996). *In search of civil society in China*. Oxford: Clarendon.

Wong, L. (1998). *The marginalization and social welfare in China*. London: Routledge.

Wong, L. (2007). The emergence of non-state welfare in China: Performance and prospects for civilian-run care homes for elders. In J. Y. S. Cheng (Ed.), *Challenges and policy programmes of China's new leadership* (pp. 367–388). Hong Kong: City University of Hong Kong Press.

Wong, L. (2008). The third sector and residential care for the elderly in China's transitional welfare economy. *The Australian Journal of Public Administration, 67*(1), 89–96.

Wong, L., & Ngok, K. L. (2006). Social policy between plan and market: Xiagang (off-duty employment) and the policy of the re-employment service centres in China. *Social Policy and Administration, 40*(2), 158–173.

Wong, L., & Tang, J. (2006). Non-state care homes for older people as third sector organizations in China's transitional welfare economy. *Journal of Social Policy, 35*, 229–246.

Yan, Q. (2003). Observations on public sector reform and social welfare socialization. *China Civil Affairs, 7*, 25–26.

Yan, Q. (2007). The major conflicts and solutions on the development of China's social organizations for aged. *Zhongguo Minzheng (China Civil Affairs), 7*, 41–42.

Yi, S. (2006). *Theories and practices of social welfare socialization*. Beijing: China Social Sciences Press.

Yu, K. (2006). *Institutional environment of China's civil society*. Beijing: Peking University Press.

Chapter 15
Ageing Policy Integrative Appraisal System in the Asia-Pacific Region: A Case Study on Macao Special Administrative Region, China

Cheung Ming Alfred CHAN, Pui Yee Phoebe TANG, and Hok Ka Carol MA

Abstract The population in Asia-Pacific Regions has moved from a state of high birth and death rates to one characterized by low birth and death rates. This transition has resulted in the growth in the number and proportion of older persons. With this rapid and ubiquitous growth, the United Nations has long promoted a re-conceptualization of how we think of older persons, what it means to age and what impacts ageing have on society. In response to UNESCAP's call to enact ageing policies in reference to the Shanghai Implementation Strategy, which was developed in the footsteps of the Madrid International Plan of Action on Ageing, the Asia-Pacific Institute of Ageing Studies at Lingnan University in Hong Kong has developed the "Ageing Policy Integrative Appraisal System" (APIAS). The APIAS acts as a comprehensive indicator of policy implementation and a validation instrument for end users' appraisal of life and service quality. This paper uses Macao Special Administrative Region (Hereafter "Macao SAR"), China as a case study for explaining the development of APIAS and its results and implications for the Macao government.

Keywords Ageing • Policy • Appraisal • Bottom-up approach • Participatory approach • Macao • Asia-pacific

Background

The Asia-Pacific Region is the most rapidly ageing region in the world. In 2002, 52% of the world's population of people aged 60 and above lived in the Asia-Pacific Region and this is predicted to increase to 59% by 2025 (United Nations Population Division, 2005a). One of the key factors driving this growth is the rising life expectancy in the

C.M.A. CHAN (✉) • P.Y.P. TANG • H.K.C. MA
Lingnan University, Tuen Mun, Hong Kong
e-mail: sscmchan@LN.edu.hk

S. Chen and J.L. Powell (eds.), *Aging in China: Implications to Social Policy of a Changing Economic State*, International Perspectives on Aging,
DOI 10.1007/978-1-4419-8351-0_15, © Springer Science+Business Media, LLC 2012

region. The data in 2004 shows that life expectancy in the region is up to 66 for males and 70 for females, and both figures are projected to continue increasing. For example, the life expectancy at birth of males and females in Hong Kong is 77 and 82 years respectively, whilst comparable figures in Singapore are 76 and 80, and 70 and 75 in Malaysia. By 2050, one out of every five people will be 60 years old or older and by 2150, it is expected that one out of every three people will be 60 years old or older. The oldest segment of the older population (80 years or older) is currently the fastest growing segment; they numbered 70 million in 2000 but are projected to increase fivefold, reaching 350 million by 2050. Since most of the developing countries are located in the Asia-Pacific Region and their paces of demographic ageing are more rapid than developed countries, they will have less time than the developed countries to adapt to the consequences of population ageing. Taking into consideration that demographic ageing is increasingly serious in the Asia-Pacific Regions, the Ageing Policy Appraisal Index (APAI) is an assessment tool for the member countries in the Asia-Pacific Regions to collect information about the ageing population, to develop a protocol for assessing their ageing policies and to offer a set of indicators (instrumental and outcome) for measuring the current ageing policies.

Ageing in the Asia-Pacific Region

The Asia-Pacific Region has the largest aged population. About 60% of the world's total aged population lives in the Asia-Pacific Region and many will live to older ages beyond 70. The longest average life expectancies from birth beyond age 80 are found in Japan, Hong Kong and some parts of China. The consequences of this means more demands on health care services, specifically health concerns associated with age and older lifestyles including high fat, salt and sugar diets; lack of physical activity; tobacco and alcohol usage; and diseases such as dementia, osteoporosis, lung cancers, arthritis, etc.

The region ages faster than any place ever recorded in human history. It took Europe about 120 years to increase its aged population from 7 to 14% (for the 60+), while the Asia-Pacific Region will only take 60 years for their aged population to grow beyond the 25% mark. There is less time for these governments to prepare and for societies to adapt to these changes. A social protection system (e.g., provident fund), professional training mechanisms and healthcare systems also need time to build and adapt. In addition to that, aged women will outnumber aged men in the region. It is a global trend that women tend to outlive men by a few years. This is greatly affecting women in the region as culturally their livelihoods are more dependent on men. The region's male-oriented culture also poses a threat to women's health, as the less-educated have less access to health services and so are more vulnerable to sexually transmitted diseases, as has been seen and documented for young women in Vietnam and Cambodia.

The Asia-Pacific population gets old before the region's wealth grows. Countries like Britain and the US, whose industrialization and urbanization came earlier than rapid population ageing, were able to accumulate adequate wealth to build the infrastructure needed for an ageing population, including things like a universal pension, training institutes and hospitals. Most Asian regions (particularly in rural areas) are still poor when faced with an ageing population, especially seen in Sri Lanka, India and China. Thus, accessibility, availability and affordability for health care services are crucial areas for development.

The huge differences between the urban and rural areas will add up challenges to the region. Many of the countries in the Asia-Pacific Region have rapidly developed cities with a lingering rural sector. Normally, able-bodied men would move to cities to earn a better living, thus leaving their older parents, wives and children at home. While cities have benefited from such a vast influx of cheap labour that has helped them grow to be more vibrant and wealthier, farmlands in the villages are left unattended or half-ploughed by the less physically able family members. This results in poorer villages with less access to healthcare and other services. Adding to the problem is the low literacy level of the villagers, as they are less able to educate themselves about maintaining a healthy lifestyle and are thus more susceptible to health risks.

Many policy-makers have paid attention to these characteristics and have thought of what kind of actions should be taken in view the unprecedented ageing phenomenon happening around the world, especially, in the Asia-Pacific Region.

The United Nations' Work for the Ageing Population

The United Nations has long been promoting a re-conceptualization of how we think of older persons and what it means to age. Its agenda is to call for "a society for all ages". This model recognizes the demographic changes that are taking place and promotes a society that is more accommodating towards an ageing world. By integrating "age" into a "society for all", holistic and multigenerational approaches are to be adapted whereby "generations invest in one another and share in the fruits of that investment, guided by the twin principles of reciprocity and equity" (United Nations, 1995b). A society for all ages has been promoted under the leadership of the UN, and the summary below highlights the UN's agenda and key actions since 1982.

In 1982, the First World Assembly on Ageing was held in Vienna, Austria, where they adopted the International Plan of Action on Ageing (United Nations, 1982). The plan, endorsed by the United Nations General Assembly, set forth 62 recommendations for action in the areas of (a) Health and Nutrition; (b) Protection of Elderly Consumers; (c) Housing and Environment; (d) Family and Social Welfare; (e) Income Security and Employment; and (f) Education (United Nations, 1982).

In 1991, the United Nations devised the *Principles for Older Persons* (United Nations, 1991). The 18 principles are divided into five categories, and all were adopted by the United Nations: (a) Independence; (b) Participation; (c) Care; (d) Self-Fulfilment; and (e) Dignity (United Nations, 1991).

In 1992, The Proclamation on Ageing was adopted by the United Nations General Assembly, paved the way for the 1999 International Year and gave further impetus to the work of the UN programme on Ageing (United Nations, 1995b).

In 1995, the World Summit for Social Development held in Copenhagen, Demark marked the beginning of the conceptual framework of "a society for all ages". With the fundamental aim of social integration, it refers to a society where "every individual, each with rights and responsibilities, has an active role to play" (United Nations, 1995b).

In 1999, the United Nations General Assembly designated the year as the International Year of Older Persons. This was celebrated by national, regional and international initiatives for its awareness raising, partnership formation and networking (United Nations, 2003).

In 1999, the Macao Plan of Action on Ageing for Asia and the Pacific (Macao POA, ESCAP, United Nations Economic and Social Commission for Asia and the Pacific (UNESCAP) 1999) was concluded at the Regional Meeting on a Plan of Action on Ageing for Asia and the Pacific at Macao Special Administrative Region (Macao SAR), China. The first regional plan devoted to issues related to older persons provided a set of concise recommendations and specific guidelines that created a framework for individual countries to set their own goals and targets for establishing an ageing policy. The seven areas of concern related to ageing and older persons are: (a) the social position of older persons; (b) older persons and the family; (c) health and nutrition; (d) housing, transportation and the built environment; (e) older persons and the market; (f) income security, maintenance and employment; and (g) social services and the community.

In 2002, the Second World Assembly on Ageing was held in Madrid and brought together 159 country representatives to develop a revised international ageing policy document, namely the Madrid International Plan of Action on Ageing (MIPAA). The MIPAA was developed to guide international policy on ageing for the twenty-first century, which has become the most important United Nations document on population ageing for the following 20 years. For the first time, all of the region's governments agreed on the need to link up ageing to other frameworks for social and economic development and human rights, thus recognizing that ageing will be the dominant and most visible aspect of world population change in the twenty-first century. Governments also decided that the promotion and protection of all human rights and fundamental freedoms, including the right to development, is essential for the creation of an inclusive society for all ages. Three important agenda items were agreed upon at the meeting: (a) ageing and development; (b) health and well-being into old age; (c) enabling a supportive environment (United Nations, 2002).

With these directions given in World Assembly, United Nations Economic and Social Commission for Asia and the Pacific (UNESCAP) also held a regional meeting in Shanghai to talk about what the region can take forward based on the MIPAA. Thus, the Shanghai Implementation Strategy (SIS) on ageing was developed in Shanghai in 2002 and has specifically covered four key Priority Areas (PA) that included (a) older persons and development, (b) advancing health and well-being into old age, (c) ensuring an enabling and supportive environment,

Table 15.1 The 16 action areas (AA) under the four priority areas (PA) of the Shanghai Implementation Strategy (SIS) (2002)

PA-I	PA-II	PA-III	PA-IV
Ageing and development	Health and well-being	Enabling supportive environments	Implementation and monitoring (national capacity)
AA-1: The challenges of mainstreaming ageing	AA-8: Quality of life for all ages	AA-10: Older persons and their families	AA-15: National mechanisms
AA-2: Protection and security	AA-9: Quality health and long-term care	AA-11: Social services and community support	AA-16: Regional and Inter-government cooperation
AA-3: Alleviation of poverty		AA-12: Housing and living environment	
AA-4: Older persons and emergencies		AA-13: Care and support to caregivers	
AA-5: Positive attitudes toward ageing		AA-14: Protection of the rights of older persons	
AA-6: Employability and workability			
AA-7: Gender specific issues: the concerns of older women			

and (d) implementation and follow-up action. These four key PAs have been extensively discussed and acted upon in many countries in the Asia-Pacific Region. In general, a total of 16 Action Areas (AA) have been included under the four key PAs after the discussion (for details, please refer to Table 15.1).

Macao SAR Government's Response to the Call of United Nations: The Development of Ageing Policy Appraisal Index

The challenges from this regional trend of population ageing are enormous, not only for individuals but also for every aspect of community, national and international life. It is generally acknowledged that ageing is a positive outcome of combined social, economic and health advances from the SIS. The challenge now faced by many countries in the region is to develop appropriate policies and take practical measures to transform this positive concept of ageing into reality. Countries that have already developed national policies on ageing will certainly have a more prepared and structured strategy in dealing with population ageing over the next 30–50 years. A national strategy for preparing society for the challenges of ageing is essential for ensuring that the goals of active ageing are achieved. While the SIS provided a direction for national governments to improve the standard of living across specific focal points of ageing, country government should think of using it as a policy evaluation. The Macao government is one of the most proactive members who used the SIS as a platform to evaluate their current ageing policy. Macao then developed an Ageing Policy Integrative Appraisal System (APIAS).

The Macao SAR government held a seminar as a follow-up to the SIS for the Madrid and Macao Plans of Action on ageing in 2004. Since then, the Macao government has been very committed to developing its ageing policy. With the consent and cooperation of different stakeholders including different government departments, community partners and the elderly, the government decided to review and evaluate its ageing policy in 2009 in reference to SIS. The review has provided an overall picture for the government of its ageing policy, specifically looking at both its policy implementation and service performance.

The 2010 population estimate for Macao is 552,300, of which the male and female population accounted for 287,000 (48%) and 265,300 (52%) respectively. As at June 2011, the overall population increased by 1.1% since the end of 2011. According to the Statistics and Census Service of Macao SAR Government, the youth population (aged 0–14) accounted for 12.5% of the total, the adult population (aged 15–64) was 79.6% and the elderly population (aged 65+) took up 7.9% in 2010. With reference to the *Macao Resident Population Projections 2007–2031*, the Macao population is projected to increase at an average annual rate of 1.9% and reach 829,000 by 2031. In regards to sex structure, the female population is projected to account for 52% of the total population in 2031. Meanwhile, fertility rates will remain low, thus leading to a continuously ageing population. The proportion of the youth population (aged 0–14) is projected to decrease constantly and the elderly population (aged 65 and over) is projected to increase steadily to 19% in 2031, making the ageing ratio to increase from 48 in 2006 to 156 projected for 2031.

In order to address the ageing population in Macao, the government decided to focus on ageing as a policy priority. So, the APIAS was developed to review its current ageing policy.

Methodology: Bottom-Up Participatory Approach

The implementation of the action plan on ageing varies from country to country given different cultural, social, political and economic development. For this reason, the measurement tool developed in the West might not jigsaw-fit the situation in the Asia and the Pacific especially in the area of ageing, so we used the bottom-up participatory approach to do the assessment. Unlike conventional approaches, the bottom-up participatory approach to policy review puts people in the centre of the monitoring and evaluation process. Traditional top-down approaches are mostly government led and lacking creditability. Lip-service has been paid to demands, while the voices of people have been relatively muted during the process, albeit older persons' action groups are increasingly vocal (Phillips and Chan, 2002). The bottom-up approach towards policy review is therefore a participatory process and serves to include various voices, especially the end users of services, which are the results of policy being implemented, in the process of review and appraisal on ageing policies. The process for review and appraisal is the monitoring process that adopts a continuous and systematic way to examine all aspects of a plan, policy, programme, project or data collection procedure. Throughout the process, different key informants are involved, such as government officials, policy-makers, stakeholders, older people in rural and urban areas, community partners from owned or privately owned enterprises, or NGOs. They are invited to participate in monitoring, and give their views on the current actions on ageing.

The process for review and appraisal is broadly defined into five stages guided by the major procedures of participatory action research. Firstly, the research team/participatory team identifies a specific programme, plan, or policy to review and to appraise, and sets the degree of priority for the action on ageing of SIS. Secondly, they develop research methods and measuring tools for monitoring and evaluating the programme with different key informants. Thirdly, they collect relevant data through different sources. This allows the voices of different parties to be heard by policy-makers. A wider picture of how programmes or policy impacts each individual and the community as a whole can be mapped. After collecting data, the fourth step involves analyzing and interpreting the data and the impacts of the programmes, and then making suggestions on actions to improve the aims or implementation methods. The final step is to continue the improvement of effectiveness and positive outcomes of the programmes or policy. Having through these stages, not solely the policy-makers, everyone who involved in the process will have a better awareness of older people's issues, supported with the distillation of findings on relevant policy and the possible adjustments of policies and programmes on ageing in accordance with the conclusions and recommendations of the review and appraisal. The APIAS is developed based on this bottom-up participatory approach.

In response to UNESCAP's call for enacting ageing policies with reference to the SIS, which was developed in the footsteps of the Madrid International Plan of Action on Ageing (MIPAA), the developed "Ageing Policy Integrative Appraisal System (APIAS)" served the purpose of having a comprehensive indicator of policy implementations and a validated instrument for end users' appraisal of life and service quality. The APAI combined two sets of indicators that measure both policy implementation and service performance, which are developed from both the SIS and MIPAA. These two indicators are the Policy Implementation Index (PII) and the Ageing Services Index (ASI), both developed from the ESCAP Background Paper (UNESCAP, 2004) and follow the 4 Priority Areas and 16 Key Areas of Action as detailed in the SIS (Table 15.1).

The PII is an instrumental indicator and is suggested as mainly a quantitative evaluation of the availability, scope and coverage of programmes and policies that have been adopted to address issues of rapid population ageing. The types of indicators were compiled on the basis of existing statistical data as appropriate and available taken from government, NGO and private sector reports. The PII is organized in the form of a checklist of 88 items that can be grouped under 16 areas of action (see Table 15.2). The content on the checklist was based on a 2005 survey of the UNESCAP, which serves to update the survey conducted in 2002 with countries in the Asia-Pacific region after the regional follow-up of MIPAA, SIS, is developed in the same year. The areas metered formed a sound standard in evaluating policy implementation in response to UNESCAP's directives.

Table 15.2 Policy
Implementation Index (PII)

	Area of action	Number of indicator
1.	Active participation of older persons	4
2.	Productive ageing	5
3.	Older persons and the family	7
4.	Older persons and the market security	3
5.	Social protection/social security	3
6.	Poverty and old age	7
7.	Social services and the community support	6
8.	Health and nutrition	9
9.	Access to health care services	4
10.	Older persons and HIV/AIDS	5
11.	Disability and mental health needs	6
12.	Housing and living environment	6
13.	Care and support for caregivers	7
14.	Neglect, abuse and violence	4
15.	Regional mechanism on ageing	6
16.	Regional and international cooperation	5
		88

The procedures for the development of the PII are as follows:

1. Using the UNESCAP survey (2005) as a base, the items serve to inquire new information on the situation of older persons, areas of concern and priority directions in national policies and programmes on ageing of the country. The PII focuses on the four priority areas of the SIS for the implementation of MIPAA and Macao Plan of Action on Ageing in Asia, and have extracted items under the four priority areas from the questionnaire used for UNESCAP 2005 survey. A total of 88 key actions grouped under 16 areas of actions from four priority areas were distilled in the PII. The adapted Chinese version was first passed on to senior government officials working in the area of ageing of Macao SAR for first review, also cross-checked by independent consultant, Dr. Edward Leung who is the adviser for many ageing related researches in Hong Kong, and finally approved by the chief investigator (Professor Alfred CM Chan), who is also a UNESCAP expert involved in the development of the SIS.
2. Demographic items and relevant information about different departments (e.g. amount of old age grants, types of health services) were added so as to produce a fuller profile of all services provided for the older persons by the Macao SAR Government.
3. The final version has 88 questions that are grouped under 16 areas of action. The PII questionnaire is appended for reference.

A checklist of 88 items with Yes/No answers to "Has your department/unit such a provision?" was formed for the department/unit's self-evaluation. Department chiefs would rate their work against each indictor (i.e., instrumental indicator) and award 1 point for "Yes" and 0 for "No" if they considered such a provision to be provided. Evidence or examples for such a provision were required to attest any positive responses. The accumulated score as a percentage of the total score of 88 items forms the PII and thus the first indicator of the APIAS in Macao, revealing the proportion of achieved policy domains. The PII, apart from being indicative of what a government has done for its ageing policies, serves to echo directly the UNESCAP member governments' implementation protocol for actions on ageing, as it has included all of the Priority Areas (4) and Areas of Action (16) prescribed by the SIS. Member governments like Macao, if they so wish, could in fact fill in the UNESCAP survey by simply extracting relevant results from the PII.

The other set of indicators used for this assessment is the ASI. It is the outcome indicator used to identify positive or negative changes in quality of life and the development of enabling and supportive environments for older people, conceptualized in terms of family relationships, community care, living environments, social protection, security, and socio-economic conditions of older people. According to the spirit of the UNESCAP's "bottom-up appraisal" in involving the service users, the ASI is an outcome indicator reflecting the quality of services (hence the efficiency and effectiveness of policies) rated by service users. The ASI is organized in the form of a questionnaire with 156 items that are grouped under 6 core domains (for details, please refer to Table 15.3). Adhering to the ultimate objective for public policies in enhancing older people's life quality, the research team has selected the World Health Organization's measurement of Quality of Life Scale for Older Adult (WHOQoL-OLD) as the blueprint for the development of ASI.

The procedures for the development of the ASI are as follows:

1. A questionnaire was constructed to cover the six domains of ageing services (i.e., general living, health, ageing services, social security, employment and training services, and personal finance) in an attempt to harmonize WHOQoL-OLD domains (i.e., physical health, psychological health, social interaction, environmental, spiritual and financial aspects) into the PII, especially as the ASI is meant to be a "bottom-up appraisal".
2. The questionnaire was then distributed to all grade officers in the Social Welfare Bureau for editing and later cross-checked by the independent consultant, Dr. Edward Leung. The first draft of the questionnaire with 68 objective items and 88 subjective items, adding up a total of 156, was accepted without much alteration.
3. A sample survey was carried out resulting in 519 successful interviews, of which 64.5% were female and 35.1% were male with an average age of 76.6 years. Eighty-three percent of the respondents attained only a primary education or lower. As per employment status, 85.2% were retirees, 6.5% were homemakers, 2.9% were working full time, 1.7% had part time jobs, while 1% claimed to be unemployed. 50.7% were widowed, 34.1% were married, 11.8% were single and the remaining 3.4% were either separated or divorced. Their living arrangements varied, as 40.8% of respondents were living alone, 17.9% were living with a

Table 15.3 Ageing Service Index (ASI)

Domain	Facets and its corresponding items
General living[a], $\alpha=0.4$	Living condition Corresponding objective item(s): 12, 13, 14, 15, 16 Corresponding subjective item(s): 17, 18, 19, 20 Elder learning Corresponding objective item(s): 21, 25 Corresponding subjective item(s): 22, 23, 24
Health, $\alpha=0.6$	Physical health Corresponding objective item(s): 30 Corresponding subjective item(s): 27, 28R, 29R Mental health Corresponding subjective item(s): 32, 35, 36, 37, 38, 39 Habits and activities for daily living Corresponding objective item(s): 40, 41, 42, 43, 44, 45 Corresponding subjective item(s): 46, 47R, 48R, 49R Participation of social and interpersonal activities Corresponding objective item(s): 65, 66 Corresponding subjective item(s): 50, 51, 52, 53, 55, 56, 57, 58, 59, 60, 61R, 62R, 63R, 64R
Ageing services, $\alpha=0.5$	Access to ageing services information Corresponding objective item(s): 67, 68 Elderly homes services Corresponding subjective item(s): 71, 72, 73, 74, 75, 76, 77, 78, 79, 80, 81 Community support services Corresponding objective item(s): 82, 83, 84, 85 Home care services Corresponding objective item(s): 86, 87, 88, 89, 90, 94, 95, 100, 101, 102, 103, 104, 105 Corresponding subjective item(s): 91, 92, 93, 96R, 97R, 98R, 99R Services and protections for abused and neglected elders Corresponding objective item(s): 106, 107, 108 Corresponding subjective item(s): 109, 110, 111
Social security, $\alpha=0.7$	Social welfare Corresponding objective item(s): 113, 114, 115, 116, 117, 118, 119, 122, 123 Corresponding subjective item(s): 120, 121, 124 Health services Corresponding subjective item(s): 125, 126, 127, 128, 129, 130, 131, 132 Corresponding subjective item(s): 133, 134, 135, 136, 137 Community and family support Corresponding objective item(s): 139, 140 Corresponding subjective item(s): 141, 142, 143
Employment and training services[b]	Employment and training Corresponding objective item(s): 144, 146, 147, 148, 149 Corresponding subjective item(s): 145
Personal finance[b]	Income and expenditure Corresponding objective item(s): 150, 151, 153, 154, 155, 156 Corresponding subjective item(s): 152

[a]Socio-demographic status contains objective figures (e.g. population demographics, employment statistics and finance) from current official records, thus reliability is only relevant to those tapping on subjective appraisals. For items that have been included in reliability tests, please refer to the corresponding subjective item(s) detailed in Table 15.3

[b]*Employment and Training Services* and *Personal Finance* has one single item in the ASI, thus reliabilities are not computed separately. In any case the objective data (e.g., Census median income) may be a more objective indicator to show adequacy. The two items are included to complete the subjective appraisal of a fuller aspect of life quality

Table 15.4 Full scale (index) and domain reliability

Domain	Reliability	Number of case	Number of item
General living	0.444	519	7
Health	0.614	519	27
Ageing services	0.536	519	21
Social security	0.652	519	11
Employment and training	0.705[a]	519	1
Income	0.695[a]	519	1
Full scale	0.7046	519	68

[a]The domain has only one item, with its reliability ascertained to the α if item deleted

Table 15.5 Correlations between composite ASI and domains (Pearson r)

	Composite ASI	General living	Health	Ageing services	Social security	Employment	Finance
Composite ASI	N.A.						
General living	0.676**	N.A.					
Health	0.400**	0.150**	N.A.				
Ageing services	0.549**	0.312**	0.128**	N.A.			
Social security	0.607**	0.361**	0.028	0.230**	N.A.		
Employment and training services	0.228	0.054	−0.158	0.577*	−0.400	N.A.	
Personal finance	0.693**	0.236**	0.172**	0.056	0.229**	0.190	N.A.

** Correlation is significant at the 0.01 level (2-tailed).
* Correlation is significant at the 0.05 level (2-tailed).

spouse, 16.4% were living with their children, 6.2% were living with their spouse and children, 2.3% were living with relatives or friends, while 14.6% were residing in elderly homes.

4. The respondents rated the subjective items on a Likert scale from 0 to 5 (0=most dissatisfied, 5=most satisfied). The results served to establish the reliability and validity of the full scale; validities were established by using Pearson r correlation between individual items and scale or domain total scores, and between domain and scale total scores (for details, please refer to Tables 15.4 and 15.5).

5. The ASI was further attested by a focus group study that consisted of service users from various elderly services for feedback—a bottom-up methodology advocated by UNESCAP—to cross-check and to validate the initial results of the ASI, i.e., if the item and total scores matched with what they would have rated.

6. Survey results and focus group checks showed a good match. Reliabilities and validities from the sample survey were acceptable, noting that the instrument is still to be refined with more varied and representative samples. A validated instrument consisting of 156 items collecting users' feedback is ready for use. The accumulated score as a percentage of the total score formed the second set indicator of the APIAS for Macao, i.e., the ASI (Fig. 15.1).

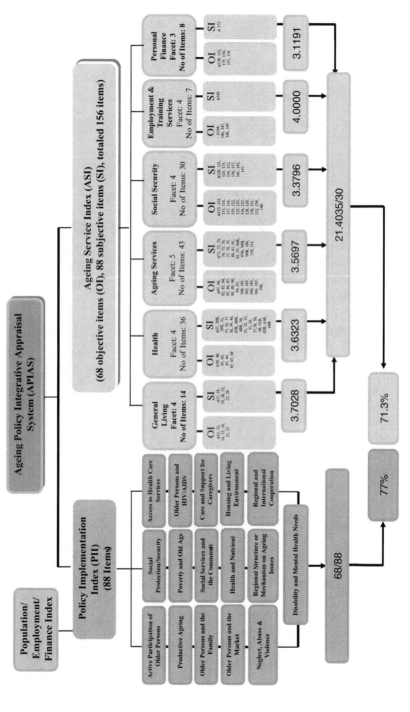

Fig. 15.1 Result of Ageing Policy Appraisal Index (APAI)

Results

The items of PII have been verified by policy bureaus and were piloted with the Social Services Bureau with almost 100% agreement of the consulting team's ratings (i.e., face validity). There were three items that required further clarification and were agreed upon after discussion. The ASI was found to have an overall reliability of 0.7, with the reliabilities for all domains ranging from 0.4 to 0.7 (please refer to Table 15.4 for details). The six domains were significantly correlated with the ASI composite score with the weakest being health at 0.4, which included the subjective appraisals on physical and mental health, habits and ability for daily living and social ability, and the strongest being personal finance at 0.7, which relates to the subjective appraisals of their income and expenditure (see Table 15.5). The results of the study confirm a validated instrument consisting of 156 items for 6 core service domains for collecting user feedback, which was in line with the bottom-up methodologies advocated by UNESCAP.

PII indicates how well a country or a region has followed the SIS directives for governments' consideration in preparing population ageing, ASI is working hand in hand with PII to reveal how well the policies, as a top-down process, among different priorities and through various outlets, have been implemented from the voices of the users. As in the case of Macao, a composite index of PII/ASI = 77/71.3 shows that Macao has done quite well both in policy implementation and in matching user feedback. The high–low of the score refers to the usual standardized score of 100 for easier grasp of performance, with 50 as the mid-point. The PII is recorded 77 means Macao has followed 77% of the key actions as advocated in the SIS, while the unfulfilled 23% may indicate areas for future policy development. Whilst the ASI is designed to reveal the people's satisfaction towards the services provided, as a result of policy be implemented, 71.3 as recorded for ASI implies the users are quite happy with the governments' provisions. The performance aligned with that of PII, i.e., what the government has thought and expected itself to do. The high PII/ high ASI matching is the most ideal outcome, with both policy-makers and users seeing things eye-to-eye. The government is doing things to meet the users' needs. However, it is worth noting that the first scores (i.e., PII and ASI) are merely indicative, given the fact it was not representative in samples nor they are comparable within the government or with other governments, but they have become the benchmark for the coming take in the fall of 2011.

Putting the two percentages of the PII and ASI side by side highlights the discrepancies, if any, between policy implementation and service performance evaluated by the users. In the case of Macao SAR, we see the most ideal outcome of high PII/high ASI matching. However, different combinations and interpretations may also be possible:

Higher PII/High ASI Discrepancy: This shows policy directives and services are on the right track but there are mismatching policy areas that could be improved for a better satisfaction of the users. In such a case, policy-makers may need to get into the items causing the discrepancy (i.e., item analysis) and take corresponding actions.

High PII/Low ASI Discrepancy: Policy-makers think they have achieved a lot but users rated low. Again, this is a case for detailed item analysis.

Low PII/Low (or lower) ASI Matching: This shows both policy-makers and users want improvements. Again, policy-makers need to get into those items causing the discrepancies and make necessary actions.

Low PII/high ASI Discrepancy: This happens when users are from a less demanding group or from a group expecting a low level of social provision from the government, as users are satisfied with what they have when compared to a lower living standard. In reality, new immigrants or older persons are likely to be these users as they do not expect or demand a lot from the government. However, policy-makers may also think that there are areas they could have covered but have not. For example, for universal design facilities for older persons in tourist sites (toilets, seats and etc.) users rated well but tourists thought there was a lot to be improved.

Conclusion

With the present versions of the indices attested by acceptable psychometric properties, the APIAS now has two core components: The PII and the ASI (using 68 of its subjectively appraised items). The discrepancy between the two scores (in percentages) easily should alert policy-makers to further probe into individual items for the source of the discrepancy and are thus made sensitive to policy actions needed for improvement. The indices are themselves benchmarks for service performance for internal (within country or district) or external (comparing with other governments in the region) uses. The application of the APIAS, as in the case of Macao, implies governments now have a policy appraisal tool to monitor its application, and eventually an effective means for good governance in the long run.

References

Statistics and Census Service of Macao SAR Government (2008). Macao Resident Population Project 2007–2031. Macao: Statistics and Census Service of Macao SAR Government.

Phillips, D. R., & Chan, C. M. A. (2002). National policies on ageing and long-term care in the Asia-Pacific: Issues and challenges. In D. R. Phillips & C. M. A. Chan (Eds.), *Ageing and long-term care: National policies in the Asia-Pacific*. Singapore: Asian Development Research Forum and Thailand Research Fund.

United Nations. (1982). *Vienna International Plan of Action on Ageing*. New York: United Nations.

United Nations. (1991). *United Nations principles for older persons*. New York: United Nations.

United Nations Economic and Social Commission for Asia and the Pacific (UNESCAP). (1999). *Macao Plan of Action on Ageing for Asia and the Pacific*. New York: United Nations.

United Nations. (1995b). Conceptual framework of a programme for the preparation and observance of the International Year of Older Persons in 1999: report of the Secretary-General.

United Nations. (2002). Madrid International Plan of Action on Ageing. In *Report of the second world assembly on ageing, Madrid, 8–12 April 2002*. New York: United Nations.

United Nations. (2002). *Report of the second world assembly on ageing, Madrid, 8–12 April 2002, A/CONF.197/9*.

United Nations. (2003). *Report of the Secretary-General "Follow-up to the second world assembly on ageing" (A/58/160)*. New York: United Nations.

United Nations Economic and Social Commission for Asia and the Pacific (UNESCAP) (2004). *Thematic background paper: Macao 2004 guidelines for the review and appraisal of the Shanghai Plan of Action on Ageing—Protocol adopted in regional seminar on follow-up to the Shanghai Implementation Strategy for the Madrid and Macao Plans of Action on Ageing on 18–21 October 2004, Macao, China*. ESCAP No. ESID/PSIS/AGEING/1, 18 October 2004/10/2.

United Nations. (2005). Older people in emergencies: The Darfur crisis. *Ageing in Africa, 24*, 5.

United Nations Population Division (UNPD). (2005). *World population prospects: The 2004 revision*. New York: United Nations.

United Nations Economic and Social Commission for Asia and the Pacific (UNESCAP). (2007). Report on the regional survey on ageing (2005). New York: United Nations.

World Health Organization. (2007). *Women, ageing and health: A framework for action, focus on gender*. Geneva: World Health Organization.

Index

A

Ageing Policy Integrative Appraisal System (APIAS)
  Asia-Pacific Region, 262–263
  bottom-up participatory approach
    APAI result, 271, 272
    ASI, 267, 269, 270
    correlations, 271
    different data collection, 267
    evaluation process, 266
    full scale and domain reliability, 271
    monitoring process, 266–267
    PII, 267–269
  higher PII/high ASI discrepancy, 273
  high PII/low ASI discrepancy, 274
  life expectancy, 261–262
  low PII/high ASI discrepancy, 274
  low PII/low/lower ASI matching, 274
  Macao SAR Government's response, 265–266
  reliability, 273
  United Nations' work
    16 action areas (AA), 265
    demographic change recognition, 263
    recommendations and principles, 263
    regional plan, 264
    SIS, 264, 265
    "society for all ages" framework, 264
Ageing Services Index (ASI), 267, 269, 270
Aging and housing stratification
  CNCA, 213–214
  community-based aging, 215–217
  constant property, 210
  cultural quality and social status, 211
  home-housed aging, 214–215, 219
  housing and encouraging communication, 212

  housing policies, 210
  institutionalized aging, 217–218
  living space, per capita, 210
  marketization impact, 213
  pay-as-you-go/own, 209
  polarization and segregation, 212
  poor communities, quality, 212
  price and taste, 211
  service system, built, 219
  social status, 211
  staff, professions, 219
  urban and suburban connection, 211–212
Aging Policy Appraisal Index (APAI), 9
  development, 265, 266
  result, 271, 272
  tool, 262
Agism process, 13
Ambivalence, aging
  bio-medical model, 15
  expert knowledge, 17
  gerontological "epistemes," 16
  gerontology, bio-medical, 15
  knowledge base, burdernsome population, 14
  medical problem, 16
  social construction, 14
  social regulation, 16
  sociology, 17
  Westernization, 14, 15
  Western science and rationality, 14
AOP. *See* Association for Old Persons
APAI. *See* Aging Policy Appraisal Index
APIAS. *See* Ageing Policy Integrative Appraisal System
Association for Old Persons (AOP), 67, 69–70

Printed by Publishers' Graphics LLC
MO20120501